Hydrology and Hydraulic Systems

Hydrology and Hydraulic Systems

Editor: Allison Sergeant

www.callistoreference.com

Callisto Reference,
118-35 Queens Blvd., Suite 400,
Forest Hills, NY 11375, USA

Visit us on the World Wide Web at:
www.callistoreference.com

ISBN: 978-1-64116-058-2 (Hardback)

Cataloging-in-Publication Data

Hydrology and hydraulic systems / edited by Allison Sergeant.
p. cm.
Includes bibliographical references and index.
ISBN 978-1-64116-058-2
1. Hydrology. 2. Hydraulics. 3. Hydraulic machinery. 4. Hydraulic engineering. I. Sergeant, Allison.
GB661.2 .H93 2019
551.48--dc23

Table of Contents

Preface

This book aims to highlight the current researches and provides a platform to further the scope of innovations in this area. This book is a product of the combined efforts of many researchers and scientists, after going through thorough studies and analysis from different parts of the world. The objective of this book is to provide the readers with the latest information of the field.

Hydrology as a scientific field is concerned with the study of existence, properties, quality and distribution of water on Earth. It branches out into subdomains of surface water, ground water and marine hydrology. There has been rapid progress in this field and its applications are finding their way across multiple industries. Some of the most prominent applications are power generation, irrigation, providing potable water for consumption, etc. This book is a compilation of chapters that discuss the most vital concepts and emerging trends in the field of hydrology. Some of the diverse topics covered in this book address the varied branches that fall under this category. It is an essential guide for both academicians and those who wish to pursue this discipline further.

I would like to express my sincere thanks to the authors for their dedicated efforts in the completion of this book. I acknowledge the efforts of the publisher for providing constant support. Lastly, I would like to thank my family for their support in all academic endeavors.

Editor

1

Land Use and Land Cover Changes under Climate Uncertainty: Modelling the Impacts on Hydropower Production in Western Africa

Salomon Obahoundje [1,*], Eric Antwi Ofosu [2], Komlavi Akpoti [1] and Amos T. Kabo-bah [2]

[1] Faculty of Science and Techniques, Master Research Program of the West African Science Service Center on Climate Change and Adapted Land Use (WASCAL), Climate Change and Energy, University abdou Moumouni of Niamey, P.O. Box 10662, Niamey 8000, Niger; akpotikomlavi@yahoo.fr

[2] Department of Energy and Environmental Engineering, University of Energy and Natural Resources, P.O. Box 214, Sunyani, Ghana; eric.antwi@uenr.edu.gh (E.O.A.); amos.kabobah@uenr.edu.gh (A.T.K.-b.)

* Correspondence: obahoundjes@yahoo.com

Abstract: The Bui hydropower plant plays a vital role in the socio-economic development of Ghana. This paper attempt to explore the combined effects of climate-land use land cover change on power production using the (WEAP) model: Water Evaluation and Planning system. The historical analysis of rainfall and stream flow variability showed that the annual coefficient of variation of rainfall and stream flow are, respectively, 8.6% and 60.85%. The stream flow varied greatly than the rainfall, due to land use land cover changes (LULC). In fact, the LULC analysis revealed important changes in vegetative areas and water bodies. The WEAP model evaluation showed that combined effects of LULC and climate change reduce water availability for all of demand sectors, including hydropower generation at the Bui hydropower plant. However, it was projected that Bui power production will increase by 40.7% and 24.93%, respectively, under wet and adaptation conditions, and decrease by 46% and 2.5%, respectively, under dry and current conditions. The wet condition is defined as an increase in rainfall by 14%, the dry condition as the decrease in rainfall by 15%; current account is business as usual, and the adaptation is as the efficient use of water for the period 2012–2040.

Keywords: rainfall variability; LULC; climate change; WEAP model; Bui dam; stream flow

---

1. Introduction

The world's environmental challenge today is climate change. Climate change is predicted to have major impacts on many aspects of human society, from agriculture and energy production to water supply [1]. Hydropower has a significant role to play in the era of Climate Change where Green Energy is the key to reducing global warming [2]. However, hydropower production may face challenges due to dependence on water availability and rainfall variability. According to [1,3], freshwater related-risks to climate change increase significantly with greenhouse gas emissions. Climate change is projected to reduce renewable surface water and groundwater resources significantly in most dry sub-tropical regions and will exacerbate competition for water among agriculture, ecosystems settlement, industry, energy, and food security. In addition to climate change, the stream flow regimes and runoff are key elements to be conserved in the planning and water management of the basin.

LULC can lead to changes in the infiltration capacity of the land, therefore, changing the dynamic of the runoff. The increase in population with increase in land use for agricultural, urbanization lead to direct impacts on runoff and stream flow [4]. Land cover plays an important role in the ecosystem, its change can lead to the modification of the micro-climate, and therefore, the hydrological cycle of that basin. (Andreini et al., 2002) [5] focused on the rainfall-runoff relationship over the Volta basin before and after the construction of the Akosombo dam. Their results showed that the coefficient of variation of runoff from the sub-watersheds is much more variable than rainfall, due to land use. According to the same study, it is apparent from the Volta water balance that land use and land cover changes in the uplands of the basin play a pivotal role in determining the future of the basin water resources. (Wei et al., 2013) [6] argued that the relative hydrological effects of forest changes and climatic variability are largely dependent on the change magnitudes and watershed characteristics. In some extents, impacts on the watershed of forest changes or land use changes can be as important as those from climatic variability. A key variable that can affect runoff generation could be land cover change [7]. Climate and land use change already have a large impact on the hydrological cycle in West African countries [8–10]. These changes can have direct or indirect impacts on the hydrological cycle of the watershed. According to [7], there is inter-annual variability in rainfall and runoff and the mean monthly potential evapotranspiration over the Black Volta.

The expansion of forest (afforestation) can result in reduction in the stream flow while the deforestation can lead to an increase in the stream flow at the small scale [11,12]. However, the afforestation-deforestation effects may be different in a larger-scale basin [4]. For instance, study in an upland watershed in Sri Lanka, showed that replacement of natural forests by other agricultural land uses led to decreased base flows and increased surface runoff generation, while the annual water yield remained more or less unchanged [4]. Change in land cover can affect the microclimate of the given area [13] and the infiltration rates can be reduced on cultivated land compared to natural land cover [10,14].

Hydropower production can be seriously reduced during extreme events. According to [15], in Kenya, a drought over the period 1998–2000 reduced hydropower production, while in Ghana in 1998 shortages in rainfall caused hydropower production to fall by up to 40%. Moreover, in both 2006 and 2011, Tanzania, experienced energy crises due to droughts. These examples give an insight on how climate variability and uncontrolled land use can impact hydropower potential. The objectives of the present work are: (i) to access the historical steam flow and rainfall variability; (ii) to access land use land cover change; and (iii) to evaluate the potential effect of the LULC and climate change on the Bui hydropower production via the WEAP model. The expected results of this study are to improve water management in the Black Volta basin.

2. Materials and Methods

2.1. Study Area Presentation

Black Volta is located in West Africa at latitude 7 N–15 N and longitude 5.5 W–1.5 W. It is a trans-boundary basin shared between Ghana, Ivory Coast, Burkina Faso, and Mali, and covers an area of 156,798 km^2 (see Figure 1). It is one of the sub-basins of the Volta basin. The basin is mainly located in the north western part of Ghana and the south western part of Burkina Faso. The basin includes northern and central parts of Ghana, Southern Burkina Faso, Southern Mali, and Northern Cote d'Ivoire.

The climate of the Black Volta, like the whole of West Africa's climate, is controlled by the movement of the Inter Tropical Convergence Zone (ITCZ). The southern part of the Black Volta presents a bimodal rainfall pattern (March–June and August–November). The Black Volta's annual rainfall varies between 1400 mm and 1000 mm and varies from south to north [16]. The pan evaporation is estimated at 2540 mm per year, and an average annual runoff coefficient of about 8.3% [17]. The mean monthly runoff from the sub-basin varies on average from about 623 m^3/s at the peak of the rainy season to about 2 m^3/s in the dry season [9]. During the rainy season, the rainfall is greater than evapotranspiration, and vice versa, during the dry seasons. The basin's mean, highest and lowest temperature are, respectively, 26 °C, 44 °C and 15 °C. The relative humidity varies between 20%–30% during the harmattan season and 70%–80% in the rainy season.

The Bui hydropower station is on the border of the Northern Region and the Brong-Ahafo Region in Ghana. The Bui hydropower plant comprises three Francis-type turbines with an installed capacity of $3\times$ 133 MW, producing about 400 MW, the second largest of Ghana's hydropower plant. The construction of this dam holds a total capacity of 12,570,000,000 m^3 and an active capacity of 7,720,000,000 m^3 is for multipurpose use: energy production and irrigation. The reservoir's maximum operating level is 185 m above sea level (a.s.l.) and the minimum is 167 m a.s.l.

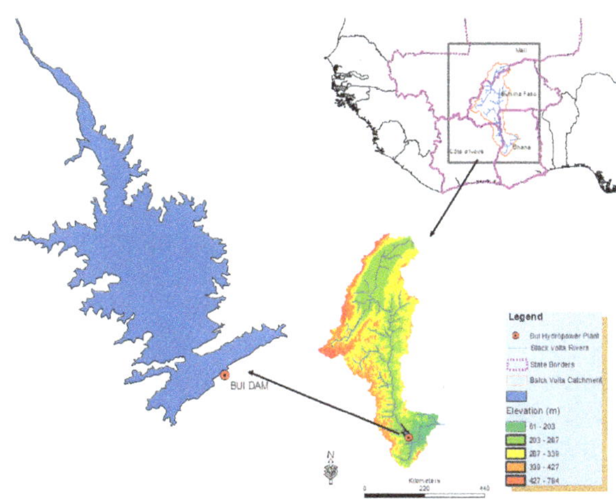

Figure 1. Study area.

2.2. *Land Use and Land Cover Change Analysis*

The canopy of vegetation may vary significantly between the south and the north parts of the Black Volta due to the differences in rainy season within the climatic zones in the basin. The months of January and February are the driest for all climatic zones and across the whole basin. These months also present less cloud cover, therefore, enhancing the quality of satellite images. Therefore, the Landsat 5, 7, and 8 data were obtained for the years 1986, 2000, and 2014, respectively, for the months of January and February. The land cover was classified in five groups, namely, water bodies, land use, shrubs, savanna and forest. The water bodies represent the stream line, small reservoirs, and dams and the land use referring to agricultural land, urban area, and bare land. The shrubs represent some woody plants, smaller than a tree, usually having multiple permanent stems branching from or near, the ground or woody plants of relatively low height, having several stems arising from the base and lacking a single trunk. The savanna represents grassland with scattered trees, grading into either open plain or woodland while the forest is referred to growth of trees and other plants covering a large area. In order to calibrate and validate the land cover classification the accuracy assessment was performed and the kappa coefficient was used as the statistical parameter.

2.3. *Rainfall and Stream Flow Analysis*

The rainfall data obtained from thirteen meteorological stations within Ghana (five stations) and Burkina Faso (eight stations) were considered in this study. The Thiessen method was applied to compute the mean rainfall. This mean rainfall amount was then converted into cubic meters per second (cms) in order to facilitate its comparison with the stream flow. Further analyses were performed in terms of annual and seasonal variation of the rainfall and stream flow. In addition, annual rainfall was aggregated in very dry, dry, normal, wet, and very wet years based on P-factor analysis.

The P-factor is the percentage at which the rainfall depth is lower or higher than the normal year. A year is called normal if its mean rainfall depth is equal, or closer, to the mean annual rainfall of the period of study and it has the value of 1. The P-Factor is given as:

$$\text{P-factor} = 1 \pm \frac{\text{value}}{100} \tag{1}$$

For example, if the dry years have 5% of mean rainfall depth less than the normal year, then it has the P-factor equal to 0.95 and if the very wet year has 9% of mean annual rainfall depth than the normal year, then it has the P-factor equal to 1.09. In fact, the value is defined as:

(Mean − standard deviation) × 100/Mean corresponding to very dry year,

(Mean − 0.5 × standard deviation) × 100/Mean corresponding to dry year,

(Mean + 0.5 × standard deviation) × 100/Mean corresponding to wet year and

(Mean + standard deviation) × 100/Mean corresponding to very wet year.

2.4. *Brief Description of the Black Volta WEAP Model Implementation*

The starting point of the model implementation was the watershed, river network and sub-catchment delineation. The water demand per sector and per sub-catchment were included considering the following water demand side sectors: domestic, livestock, irrigation, small Reservoir, and hydropower. The Bui reservoir was included in the Bui sub-catchment (see Figure 2). In WEAP, catchment processes such as evapotranspiration, runoff, infiltration, and irrigation demands, can be simulated based on four methods including: (a) the rainfall runoff method; (b) irrigation demand; (c) the soil moisture method; and (d) the MABIA method. The Soil Moisture Method was selected for the present work because of its specificities of including the characterization of Land Use and/soil types impacts on the catchment processes.

The land use classes were incorporated as a percentage share. Climate data including precipitation, temperature, relative humidity, cloud fraction, and wind speed have also been used.

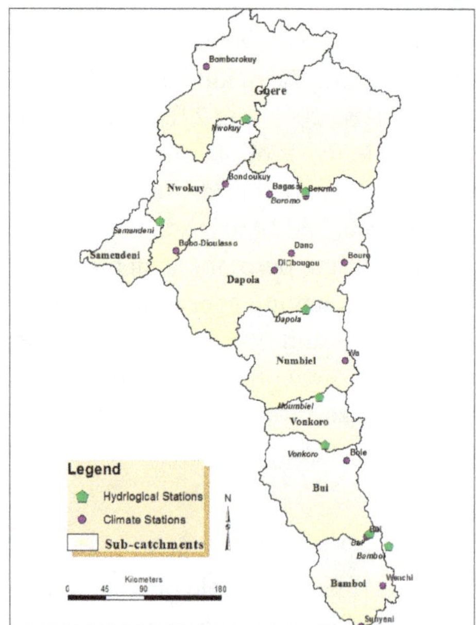

Figure 2. Different sub-catchments.

2.5. *The WEAP Model Evaluation*

The model calibration and validation was performed to determine values of a set of key parameters which represent the physical characteristics of the catchments. These parameters include soil water capacity, root zone conductivity and runoff resistance factor. To explore how well the model reproduced river flows as observed at any given gauging station, the Nash-Sutcliffe coefficient (NS) and the coefficient of determination (R^2) were computed. The model was calibrated and validated in two sub-catchments: Bui and Samandeni (Figure 2 shows the sub-catchment in Black Volta).

The Nash–Sutcliffe coefficient is defined by the Equation (2) and the coefficient of determination by Equation (3) as:

$$NS = 1 - \frac{\sum_{i=1}^{n} (Q_{sim,i} - Q_{obs,i})^2}{\sum_{i=1}^{n} (Q_{obs,i} - \overline{Q}_{obs})^2} \quad (2)$$

$$R^2 = \frac{\sum_{i=1}^{n} (Q_{obs,i} - \overline{Q}_{obs})(Q_{sim,i} - \overline{Q}_{sim})}{\left[\sum_{i=1}^{n} (Q_{obs,i} - \overline{Q}_{obs})^2\right]^{0.5} \left[\sum_{i=1}^{n} (Q_{sim,i} - \overline{Q}_{sim})^2\right]^{0.5}} \quad (3)$$

where: $Q_{obs,i}$ is the observed flow at time i (m^3/s), $\overline{Q}_{obs}$ is the mean of observed flow (m^3/s), $Q_{sim.i}$ is the simulated flow at time i (m^3/s), and $\overline{Q}_{sim}$ is the mean of simulated flow (m^3/s). The greater the values of NS and R^2, the better the model reproduced observations, with 1 being the ideal situation. The monthly data was used for the calibration and the validation. The periods 2000–2005 and 2000–2002 were used for calibration, while the periods 2006–2010 and 2003–2005 were used for validation, respectively, for the sub-catchment of Bui and Samandeni. The periods of calibration and validation were chosen according to data availability. Another reason for selecting these periods was the fact that our main focus was on the Bui dam, which was constructed between 2009–2012. This also explains our choice of the year 2000 as the starting year (current account/scenario year for the WEAP model).

2.6. Scenarios Development

2.6.1. Climate Change Scenarios

A compilation from the Intergovernmental Panel on Climate Change (IPCC) fourth report, 2007 (AR4) by the African Ministerial Conference on Environment (AMCEN) Secretariat suggests that average temperature will increase over the African continent by 1.5 °C to 3 °C by 2050 [18]. The IPCC global assessment of climate change, 2007 synthesis report noted that the West Africa region, within which lies the Black Volta basin, experienced a mean temperature rise which ranged from 0.2 °C to 1 °C between 1974 and 2004, increasing by at least 0.2 °C per decade [19]. Then, for this paper, we assumed that the rise in temperature is 1 °C for the projected period the same as in another study [20]. However, the simulation of large-scale patterns of precipitation has improved somewhat since the AR4, although models continue to perform less well for precipitation than for surface temperature [1]. Thus, the fifth assessment report of IPCC, 2014 (AR5) the observed temperature over West Africa was statistically significant and vary between 0.5 °C and 0.8 °C between 1970 and 2010 [21]. As for the precipitation over West Africa, the Sahel experienced a significant reduction overall, over the course of 20th century, with drought between 1970–1980 and with a recovery toward the last 20 years of the century [21].

This will be considered for all CC scenarios. Climate change scenarios are defined as:

Current condition or "as business as usual" scenario: in this scenario, there is no change in precipitation relative to historical trends.

Wetter scenario: This scenario assumes an increase in precipitation relative to historical trends. The average annual rainfall increases by 14.2%, and ranges from 7% to 20% during the 30-year simulation period from 2000 to 2040.

Drier scenario: This scenario assumes decrease in precipitation. The relative reduction in annual precipitation averaged 15.8%, ranging from 9% to nearly 21% over the simulation period.

Adaptation scenario: demand management programs will result in a reduction of the irrigation water usage by 20%, domestic loss reduces to 10% for rural and 2% for urban areas.

2.6.2. The WEAP Model Scenarios

The WEAP evaluation was based on five scenarios as follow:

Scenario 1. No LCC and No CC: no land use change and no climate change. This scenario inheriting from References business as usual.

Scenario 2. No LCC and CC: is based on the climate change condition.

Scenario 3. LCC and CC: it uses the projected LULC change and the climate change condition.

Scenario 4. LCC and No CC: uses the projected LULC change.

Scenario 5. Adaptation: under this scenario, water for irrigation is efficiently used and the rain water harvesting was put in place for domestic needs, while the demand management program will reduce the irrigation water by 20%, domestic water loss by 10% in rural areas and 2% for urban areas.

3. Results

3.1. Land Use Land Cover Change Analysis

Land classification can be subjected to error due to geometric errors, misclassifications, and undefined classes. To statistically quantify these errors, a random selection of pixels of the classified maps was performed to build a confusion matrix. The kappa coefficient K, a discrete multivariate technique used in accuracy assessments of thematic maps, is an efficient approach to derive information from an image via the confusion matrix [22–25]. $K > 0.80$ represents strong agreement and good accuracy, 0.40–0.80 is middle, and <0.40 is poor [26]. For this study, the kappa coefficient for the years 1986, 2000, and 2014 are respectively $K_{1986} = 81.5\%$, $K_{2000} = 89\%$, and $K_{2014} = 84.5\%$, showing strong agreement. In general, the LULCC were correctly classified with strong agreement.

The results from the LULC trend analysis (Table 1) shows that water bodies have increased rapidly in the basin. This increase can be explained by the increase in the construction of small reservoirs and the Bui dam. The land use occupied about 16%, 18%, and 30% of the study area, respectively, in 1986, 2000 and 2014. This represents an increase of 12.7% between 1986 and 2000, 59.9% between 2000 and 2014. This relatively high percentage change can be linked to a high rate of urbanization, increases in bare land, and an extension of agricultural land (note that land use class as defined in this paper refers to the ensemble of agricultural land, urban area, and bare land). The excessive use of trees for timber, for charcoal, wood fires, and the phenomenon of uncontrolled bushfire are the main factors contributing to land cover change. In fact, the fire strips the land of its vegetative cover by burning trees and grasses; thus, pastures for livestock are largely destroyed. The soil is then exposed to erosion as it lies bare for most of the season and then leads to land cover change. In addition, bushfires are cited among the causes of soil erosion in the Volta basin [27], while the charcoal and wood fuel production are considered as threats to Africa's forests [28].

The shrubs covered about 21.4%, 24.7%, and 22.6% in 1986, 2000, and 2014. There was an increase of about 15.6% between 1986 and 2000 and a decrease of about 8.9% between 2000 and 2014 in shrub cover. The period of the 1980s was considered as a drier period with less vegetation, while the year 2000 is considered as a normal year with high vegetation cover. The decrease in shrubs between the year 2000 and 2014 may be due to the expansion of agriculture (farmland and livestock breeding). The expansion of agriculture land also resulted in a decrease of savanna.

The forest occupied about 8.7%, 3.7%, and 3.8% of the study area, respectively, in 1986, 2000, and 2014. We noticed an important decrease of about 57% forest coverage between 1986 and 2000 many in the Ghana and Ivory Coast (see Figure 3a,b), and an increase about 4% between 2000 and 2014. The sharp decrease (1986–2000) can be explained by the excessive use of the forest for timber and agriculture. The small increase in forest cover during the 2000–2014 period may be due to some land conservation practices in the basin.

Table 1. LULC characteristics.

Land Cover Type	Area Coverage (km^2)			Area Coverage (%)			1986–2000		2000–2014		1986–2014	
	1986	2000	2014	1986	2000	2014	Change (%)	Change Rate (%/Year)	Change (%)	Change Rate (%/Year)	Change (%)	Change Rate (%/Year)
Water bodies	75.30	365.6	588.60	0.0049	0.2360	0.38	385.60	27.8	60.97	4.06	681.70	24.34
Land Use	25,393.20	28,628.4	45,785.60	16.30	18.50	29.50	12.70	0.85	59.90	4.00	80.30	4.02
Shrubs	33,243.30	38,416.1	34,992.10	21.40	24.70	22.50	15.60	1.04	−8.90	−0.59	5.26	0.19
Savannah	82,808.70	81,894.7	67,695.20	53.40	52.80	43.70	−1.10	−0.07	−17.30	−1.13	−18.25	−0.65
Forest	13,545.60	5761.3	6004.60	8.74	3.72	3.87	−57.47	3.80	4.20	0.28	−55.67	−1.90

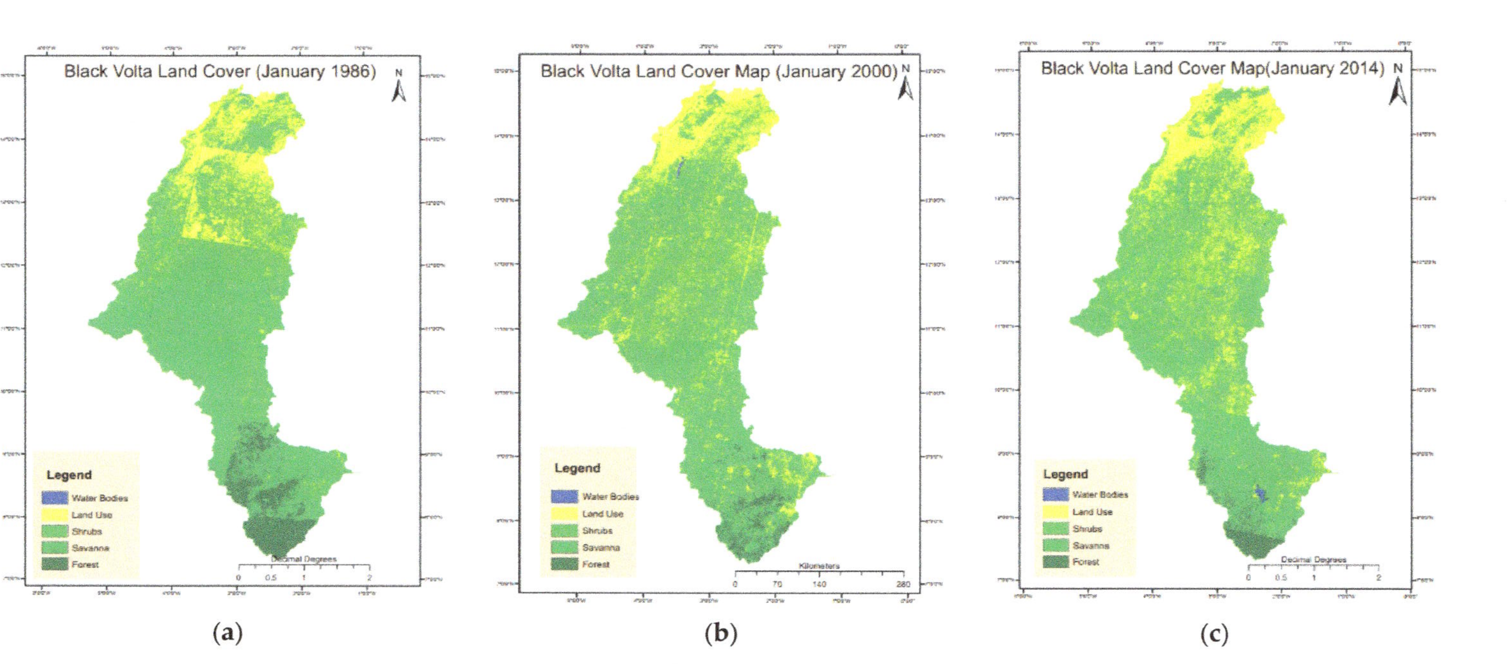

Figure 3. LULC maps. (**a**) LULC 1986; (**b**) LULC 2000; and (**c**) LULC 2014.

3.2. Rainfall-Stream Variability

3.2.1. Annual Variation

The annual variation of rainfall and stream flow show that there is an observed upward trend in rainfall pattern in general. However, there is a decrease in rainfall trend for the period 2000–2010. The stream flow trend shows a consistent trend and pattern as the rainfall (see Figure 4).

Table 2 shows the aggregation of the rainfall events according to the years. The results show that between 1982 and 2011, there were more dry years than very dry, normal, wet, and very wet years. While most of the years during the 1980s are classified as dry, recent years (2000s) are scattered through various rainfall events. These results suggest that the rainfall pattern of past decade and the recent years are erratic.

Although the relative rainfall trend decreases for the 2000–2010 period, we observed a sharp increase in the stream flow trend at Bui. This contrast for the last period may be due to not only land use land cover change but also the pre-Bui dam construction activities. The construction of the Bui dam, started early 2006 with the diversion of the Bui River resulting in an increase of stream flow more than the normal.

Basic statistics on rainfall and stream flow are reported in Table 3. The annual rainfall and stream flow have their coefficient of variation, respectively, 8.6% and 60.85%. The high variation in stream flow may be due to land use land cover change.

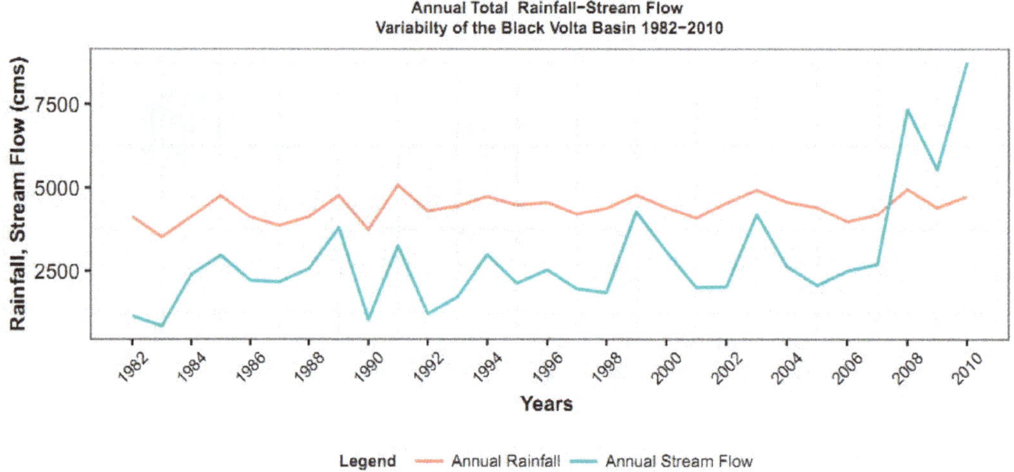

Figure 4. Annual Rainfall Stream flow variability.

Table 2. Rainfall characterization.

Events	Very Dry	Dry	Normal	Wet	Very Wet
Description	Below Mean − 1Std	[Mean − 1Std; Mean − 0.5Std]	Mean/Closer	[Mean + 0.5Std; Mean + 1Std]	Above Mean + 1Std
Years	1983 1987 1990 2011	1982 1984 1986 1988 1997 2001 2006 2007	1992 1993 1995 1998 2000 2005 2009	1985 1994 1996 2002 2004 2010	1989 1991 1999 2003 2008
P-Factor	0.9	0.95	1	1.045	1.09

Table 3. Basic statistics on Rainfall Stream flow.

	P (cms)				Q (cms)			
	Min	Max	Mean	CV (%)	Min	Max	Mean	CV (%)
January	0.00	97.62	7.62	240.46	1.50	82.70	25.21	86.77
February	0.00	89.53	23.22	107.60	0.40	63.30	14.06	103.49
March	3.67	241.31	102.22	59.19	0.10	53.50	10.78	120.99
April	71.82	391.4	243.73	35.09	0.40	42.30	12.92	86.77
May	248.03	963.8	415.81	32.88	2.10	138.50	40.67	79.51
June	343.40	835.82	581.77	20.67	25.03	343.70	105.88	76.60
July	406.01	1079.4	762.81	21.65	78.60	582.89	223.05	54.99
August	421.66	1234.92	941.03	20.42	196.40	1649	533.51	59.18
September	421.97	976.73	688.42	21.10	270.80	2905.32	1022.27	62.02
October	136.62	694.97	293.89	43.71	87.20	3147.55	689.76	88.42
November	1.49	196.71	36.45	111.89	18.30	500.00	164.25	72.28
December	0.00	51.12	8.92	145.82	3.80	200.00	59.4655	78.92
Annual	2054.67	6853.33	4105.89	8.60	684.63	9708.76	2901.83	60.85

3.2.2. Seasonal Variation

Dry Season (November–April 1982–2010)

The seasonal variation of both rainfall and stream flow shows an overall increasing trend during the dry season within the 1982–2010 period (Figure 5a). However, there was a decrease in rainfall for the 2006–2010 period compared to the overall trend. The coefficient of variation of rainfall and streamflow are, respectively, 25% and 63.6%.

Wet Season (May–October 1982–2010)

Both Rainfall and Stream flow increase during the wet season within the 1982–2010 period (Figure 5b). The same pattern is observed for both rainfall and streamflow except the period of 2006–2010 when there is an increase in stream flow, while there is a decrease in rainfall. The rate of increase in stream flow is higher than the rate at which the rainfall increases. During the wet season, the increase in the stream flow is generally due to high runoff with high rainfall events. The coefficient of variation of rainfall and stream flow are respectively, 14%, and 61.7%.

Driest Month (January)

Figure 5c shows an increase in both rainfall and stream flow but with high rate for the stream flow. Between 1982–1990 and 2008–2010, January shows nearly zero mm, rainfall while stream flow still increases. This may be explained by land use change and ground water contribution to stream flow. While the water contribution rate to the stream flow is clearly beyond the scope of this paper, we can speculate on its important contribution to stream flow. There was a slight increase in rainfall between 1991 and 1999, resulting in a sharp increase in stream flow. This may be explained by a change in land use in terms of the increase in the urbanization rate in the basin and soil degradation.

Wettest Month (August)

During the wettest month of the year for the 1982–2011 period, both rainfall and Stream flow increase, but the stream flow is at higher rate than the rainfall (see Figure 5d). From 1986 to 1999, the rainfall and stream flow have increased, respectively, from 916 cms to 1373.9 cms and 339.6 to 642.2 cms, with a coefficient of variation of 28.2% and 43.6% respectively. For the period of 2000–2010, the rainfall and stream flow have increased, respectively from 1042.4 cms to 1269.02 cms and 527 cms to 861.2 cms with a coefficient of variation 236% and 86.7%, respectively.

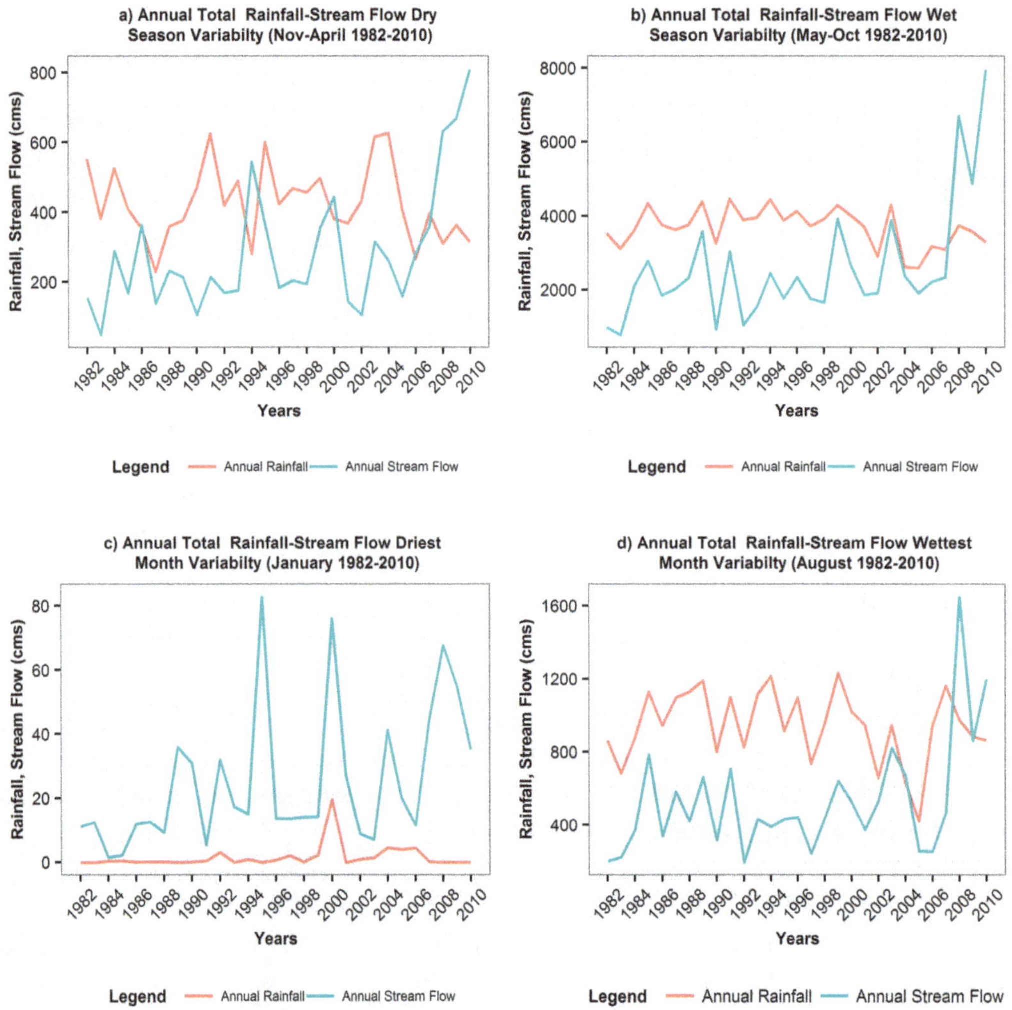

Figure 5. Seasonal variation of rainfall-stream flow.

3.3. *The WEAP Modeling*

3.3.1. The WEAP Model Performance Results

The calibration and validation of the WEAP model was performed at the sub-catchment of Bui (south of the Black Volta) and Samandeni (north of the basin). The model was calibrated at Bui for the 2000–2005 period and validation for the 2006–2010 period. The results show that during calibration at Bui, the model overestimates the stream flow between April and August and underestimates the stream flow between September and March (Figure 6a). The same pattern can be observed during validation where the model overestimates the stream flow between April and September, and underestimated the stream flow between October and May (Figure 6b).

The model was calibrated at Samandeni for the 2000–2002 period (Figure 6c) and validated for the 2003–2005 period (Figure 6d). The calibration results show overestimation of the stream flow between March and September and underestimate the stream flow between October and February. However, the validation shows an underestimation of stream flow from October to July and an overestimation of the stream flow from August to September. The results from the coefficient of determination and the Nash-Sutcliffe coefficient (Table 4) show very good agreement between observed and simulated flows.

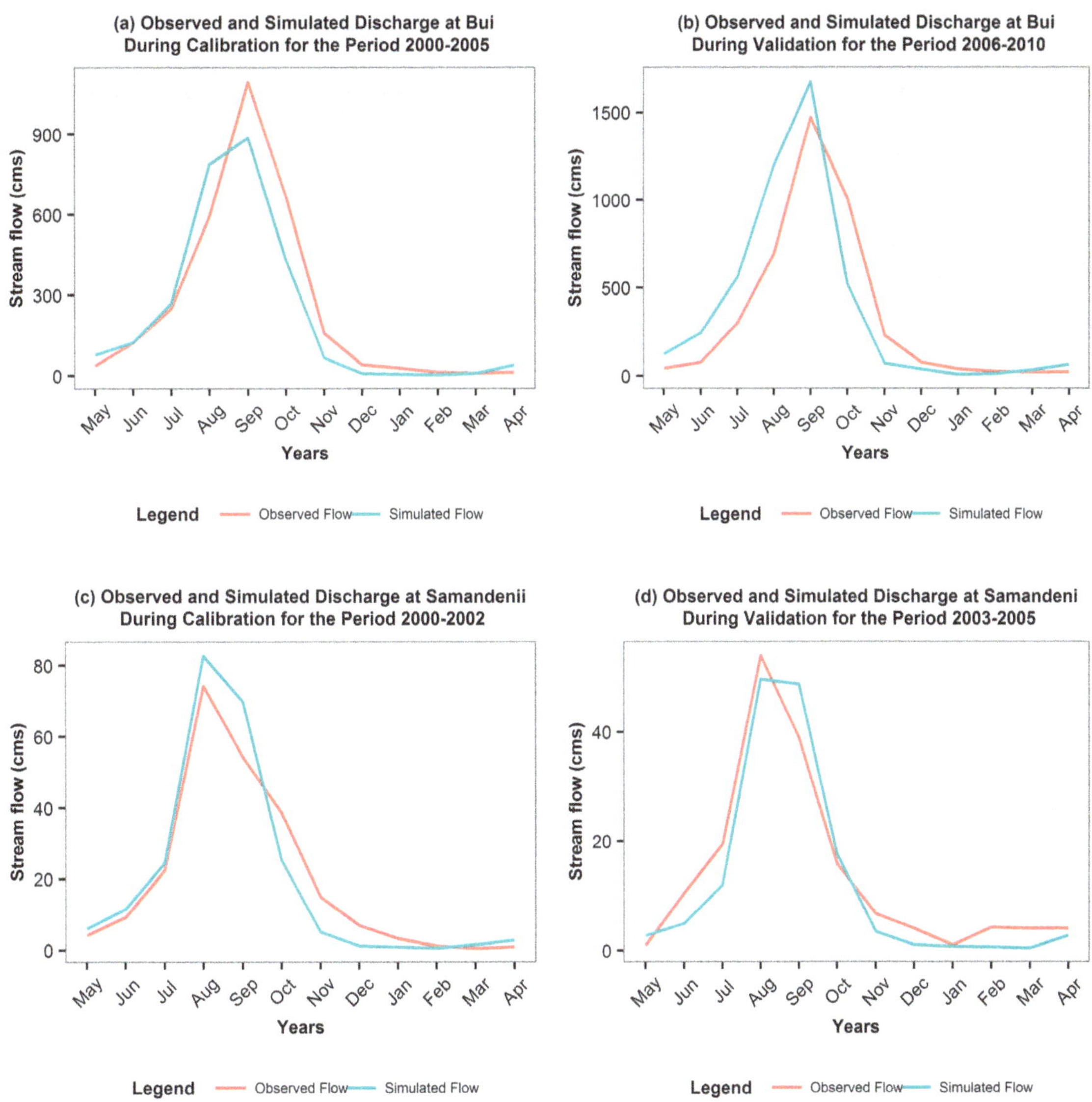

Figure 6. Calibration and Validation.

Table 4. WEAP model performance.

	Sub Catchment	R Square	Nash—Sutcliffe Coefficient (NS)
Calibration	Bui	0.89	0.889
	Samandeni	0.92	0.899
Validation	Bui	0.79	0.73
	Samandeni	0.94	0.97

3.3.2. Impact of LULCC on Projected Surface Runoff Coefficient and Evapotranspiration Coefficient (2012–2040)

The evapotranspiration coefficient is the percentage of rainfall water that is lost by evaporation and transpiration and the runoff coefficient is the percentage of rainfall water converted into runoff. The results are summarized in Table 5. Under current LCC and CC condition, the runoff coefficient (8.5%) is almost the runoff coefficient of Black Volta basin (8.3%) found in different studies [17,29].

Table 5. Evapotranspiration and runoff coefficient.

Scenarios	Evapotranspiration Coefficient (%)	Runoff Coefficient (%)
Reference	84.93	9.62
NoLCC&NoCC	84.93	9.62
NoLCC&CCWet	82.91	11.68
NoLCC&CCDry	90.06	5.75
NoLCC&CCcurrent	86.51	8.56
LCC&NoCC	83.70	10.70
LCC&CCWet	82.35	12.25
LCC&CCDry	89.32	6.48
LCC&CCcurrent	85.84	9.24
Adaptation CC&LCC	81.94	11.74

The results showed that wherever the land use effect is considered in the scenario, the runoff coefficient is higher. The highest runoff coefficient (12.25%) is found under the land use change and climate change wet condition, while the lowest (6.48%) is found under the land use change and climate change dry condition.

Evapotranspiration is also subjected to change under changing climate and changing land use. The highest value (90.06%) is found under the no land use change and dry climate change condition. In summary, the LULCC affects the surface runoff generation and the evapotranspiration coefficients in the basin. The adaptation scenario gives the lowest evapotranspiration and a moderate runoff coefficient.

3.3.3. The Potential Bui Hydropower Production under Different Scenarios

The Bui hydropower production was simulated in WEAP under different land use and climate change scenarios. The power output, as well as the changes in power production under various scenarios, are depicted in Figures 7 and 8. By comparing the land use and land cover change scenarios with no land use change scenarios all associated with climate change, it is clear that the land cover change is projected to favor the hydro power production at the Bui dam in the next 25 years under changing climatic conditions.

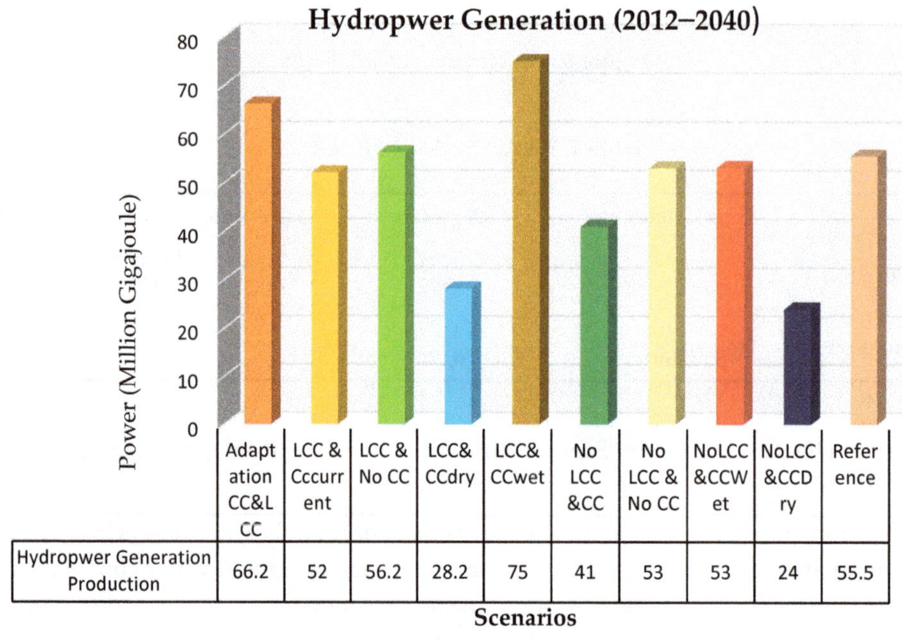

	Adaptation CC&LCC	LCC & Cccurrent	LCC & No CC	LCC& CCdry	LCC& CCwet	No LCC &CC	No LCC & No CC	NoLCC &CCWet	NoLCC &CCDry	Reference
Hydropwer Generation Production	66.2	52	56.2	28.2	75	41	53	53	24	55.5

Figure 7. Power generation.

However, under the climate change conditions, the LCC is projected to decrease the power production by 2.49%, while No LCC is projected to decrease the power production by 23.2%. Under the climate change dry condition, LCC is promised to decrease the power by 46%, while No LCC will decrease the power by 54%. In addition, under the climate change wet condition, the LCC is projected to increase the power production by 40.7%, while No LCC is promised to be neutral. Overall, under climate change wet condition, the land use land cover change is projected to increase the productivity of hydropower at Bui dam. However, it is projected to decrease the power production under dry and current conditions, but less than the effect of No LCC.

The low power production, for the No LCC under different climate change scenario, can be explained by less runoff generation due to the vegetative coverage, with an increase in evapotranspiration. Under any climatic condition, the more vegetative areas have greater evapotranspiration, while less vegetative areas have lower evapotranspiration. The negative effect of climate is lower with land use land cover change than with no land use land cover change. The highest power production under all of the developed scenarios will be observed during the wet scenario.

The land use land cover change under any changing climatic condition is promised to favor the Bui power production than the contrary. However, the main risk associated with dramatic land use change is the soil degradation, soil erosion, which will lead to silting of the river and, mainly, the reservoir. Another threat could be the water quality from sediment, as the climate change results in an increase in evapotranspiration, which could consequently reduce the amount of water available in the basin increasing the concentration of sediment. The increase in the concentration of sediment will affect the water quality for domestic use and livestock, and will also reduce the life time of power turbines.

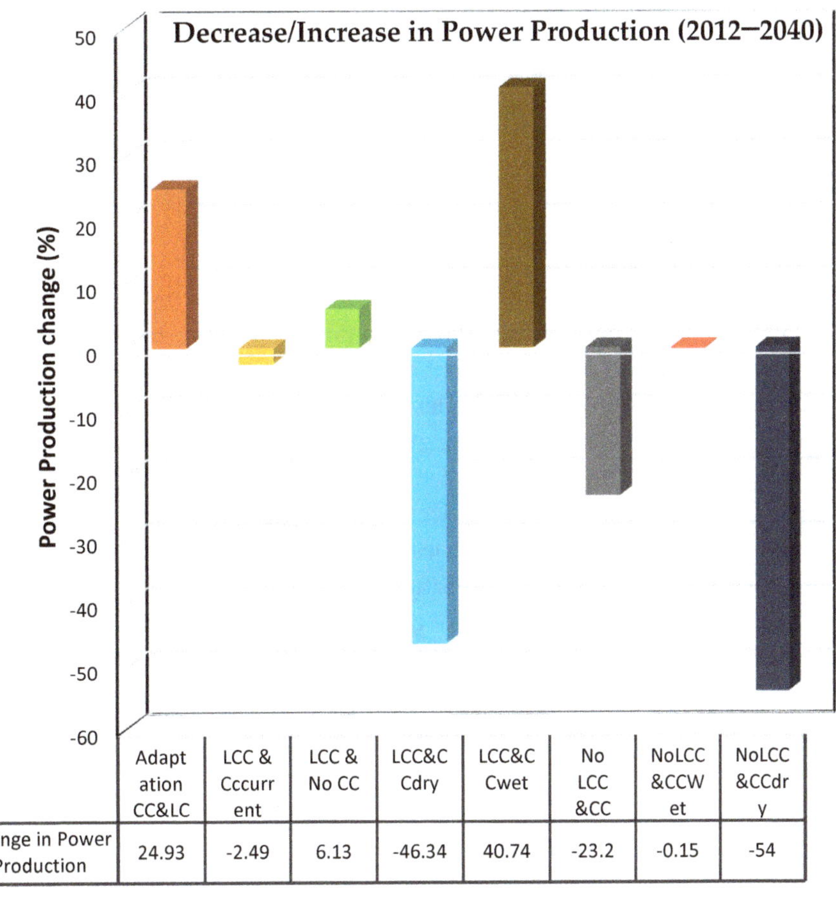

	Adaptation CC&LC	LCC & Cccurrent	LCC & No CC	LCC&CCdry	LCC&CCwet	No LCC &CC	NoLCC &CCWet	NoLCC &CCdry
Change in Power Production	24.93	-2.49	6.13	-46.34	40.74	-23.2	-0.15	-54

Figure 8. Change in power production.

4. Discussion

The analysis of annual rainfall data (1982–2010) reveals an overall increasing trend (but small or not statistically significant) in the Black Volta basin. This increasing trend is also confirmed by [30]. However, according to [31], the annual average rainfall is still as low as during the drought of the 1970s. The decrease of rainfall is higher in the Burkina Faso section and lower in the Ghana section. The increase of temperature over West Africa during the end of the 20th century induced an increase of potential evaporation, which might reduce the runoff. However, the joint effect of climate change and human activities on land cover over more than three decades is responsible for an increase of the runoff coefficients of the Black Volta basin, despite the low rainfall variability observed during the data analysis [31,32]. The runoff coefficients have increased and is projected to increase in Black Volta basin according to scenario of LCC and CC. The rapid change in land use and land cover may be attributed to population growth with the extension of agricultural lands, urbanization, and deforestation. The small increase in forest observed coverage between 2000 and 2014 may be due to the protection of some forest reserves [33]. The land use and land cover change trend found is in accordance with the work of [34] in the same study area and with research of [35] in Nakambe river (White Volta) of Burkina Faso.

Some research [4,14,34] showed that the increase in land use (Urbanization and farm land) is the main factor that contributes to the increase in runoff generation and runoff coefficient depends on the land cover types. For instance land use and water areas have higher runoff coefficient due to their low infiltration rate; while the grass land, shrubs and forest areas have low runoff coefficient [36]. The rainfall runoff experiments indicate that degraded and abandoned land generate surface runoff within a few minutes after the start of the rainfall event [35,37]. One of the paper cited by [35] showed that runoff coefficient of natural vegetation and fallows area, cultivated land and barren land are 13%, 20% and 50% respectively. Some studies cited by [35] found lower values for these types of land use in the Sahelian part of the basin.

In addition, it is found that evapotranspiration is increased due to the increase in air temperature observed, or the land use change of the last period and will be one factor that may affect water availability during the emerging climate change, with special concern regarding to the Bui hydropower production [16].

5. Conclusions and Recommendations

The increase in population of the Black Volta basin has caused the land cover and land use to change and that results in changes in the hydro climate of the basin. However, from a comparison analysis of rainfall and stream flow at different angles, it is clear that the increase in stream flow at the Bui catchment is mainly due to land use land cover change. The increase in stream flow does not mean more fresh water availability, as the growing population puts additional pressure on the water resources. Nevertheless, the rainfall and stream flow have the same pattern, except during the period 2006–2010 that the observed rainfall decreased, while the stream flow increased due to the Bui river diversion. Moreover, this study focuses only on the potential effects of land use land cover change on the dynamics of stream flow, and future research may look at the influence of land cover change on the local rainfall. The combination of climate change with land use land cover change may negatively affect the hydro power production in the future, except under wet condition. Nevertheless, the land use land cover on its own may favor power production more than no change in land cover under any changing climatic conditions. Therefore, there is a need for government to make a decision about the way the land is used in Black Volta basin. The limitation of this paper resides in the fact that the development conditions, were not included in the WEAP modelling. This could be a scope for further research.

Acknowledgments: We acknowledge with thanks the scholarship and financial support provided to me by BMBF, the Federal Ministry of Education and Research of Germany and the West African Science Service Centre on Climate Change and Adapted Land Use (**WASCAL**). We are also grateful to the Faculty of Science and Technology and the WASCAL MRP-CCE program of Niamey, Niger.

Author Contributions: Salomon Obahoundje conducted this research as part of his master thesis work. Eric Ofosu Antwi and Amos T. Kabo-bah were respectively first and second advisors. Komlavi Akpoti contributed in data analysis as well as writing of the paper.

Conflicts of Interest: The authors declare no conflict of interest. The founding sponsors had no role in the design of the study; in the collection, analyses, or interpretation of data; in the writing of the manuscript, and in the decision to publish the results.

References

1. Pachauri, R.K.; Allen, M.R.; Barros, V.R.; Broome, J.; Cramer, W.; Christ, R.; Dubash, N.K. *Climate Change 2014: Synthesis Report. Contribution of Working Groups I, II and III to the Fifth Assessment Report of the Intergovernmental Panel on Climate Change*; Pachauri, R., Meyer, L., Eds.; IPCC: Geneva, Switzerland, 2014; p. 151.
2. Bartle, A. Hydropower potential and development activities. *Energy Policy* **2002**, *30*, 1231–1239. [CrossRef]
3. Palmer, M.A.; Reidy Liermann, C.A.; Nilsson, C.; Flörke, M.; Alcamo, J.; Lake, P.S.; Bond, N. Climate change and the world's river basins: Anticipating management options. *Front. Ecol. Environ.* **2008**, *6*, 81–89. [CrossRef]
4. Bewket, W.; Sterk, G. Dynamics in land cover and its effect on stream flow in the Chemoga watershed, Blue Nile basin, Ethiopia. *Hydrol. Process.* **2005**, *19*, 445–458. [CrossRef]
5. Andreini, M.; Vlek, P.; Giesen, N.; van Lanen, H.A.J.; Demuth, S. Water sharing in the Volta basin. In *FRIEND 2002-Regional Hydrology: Bridging the Gap between Research and Practice. Proceedings of Fourth International Conference on FRIEND (Flow Regimes from International Network Data), Cape Town, South Africa, 18–22 March 2002*; IAHS Press: Wallingford, UK, 2002; pp. 329–335.
6. Wei, X.; Liu, W.; Zhou, P. Quantifying the relative contributions of forest change and climatic variability to hydrology in large watersheds: A critical review of research methods. *Water* **2013**, *5*, 728–746. [CrossRef]
7. Shaibu, S.; Odai, S.N.; Adjei, K.A.; Osei, E.M.; Annor, F.O. Simulation of runoff for the Black Volta Basin using satellite observation data. *Int. J. River Basin Manag.* **2012**, *10*, 245–254. [CrossRef]
8. Jung, G. Regional climate change and the impact on hydrology in the Volta Basin of West Africa. Doctoral Dissertation, University of Augsburg, Augsburg, Germany, 2006.
9. Kasei, R.A. Modelling Impacts of Climate Change on Water Resources in the Volta Basin, West Africa. Doctoral Dissertation, University of Bonn, Bonn, Germany, 2009.
10. Cornelissen, T.; Diekkrüger, B.; Giertz, S. A comparison of hydrological models for assessing the impact of land use and climate change on discharge in a tropical catchment. *J. Hydrol.* **2013**, *498*, 221–236. [CrossRef]
11. Nunes, C.; Auge, J.I. Land-Use and Land-Cover Change (LUCC): Implementation Strategy. Available online: http://digital.library.unt.edu/ark:/67531/metadc12005/m1/10/ (accessed on 20 June 2015).
12. Li, K.Y.; Coe, M.T.; Ramankutty, N.; De Jong, R. Modeling the hydrological impact of land-use change in West Africa. *J. Hydrol.* **2007**, *337*, 258–268. [CrossRef]
13. Calder, I.R. *Water-Resource and Land-Use Issues*; SWIM Paper 3; International Water Management Institute: Colombo, Sri Lanka, 1998.
14. Giertz, S.; Diekkrüger, B. Analysis of the hydrological processes in a small headwater catchment in Benin (West Africa). *Phys. Chem. Earth Parts A/B/C* **2003**, *28*, 1333–1341. [CrossRef]
15. Cole, M.A.; Elliott, R.J.; Strobl, E. Climate Change, Hydro-Dependency, and the African Dam Boom. *World Dev.* **2014**, *60*, 84–98. [CrossRef]
16. Kabo-bah, A.T.; Anornu, G.K.; Ofosu, E.; Andoh, R.; Lis, K.J. Spatial-temporal estimation of evapotranspiration over Black Volta of West Africa. *Int. J. Water Resour. Environ. Eng.* **2014**, *6*, 295–302.
17. *Diagnostic Study of the Black Volta Basin in Ghana*; Final Report; ALLWATER Consult Limited: Kumasi, Ghana, June 2012; Available online: http://www.gwiwestafrica.org/sites/default/files/6_gh8_hydrological_study_of_the_black_volta_basin.pdf (accessed on 20 June 2015).
18. Fact Sheet Climate Change in Africa—What Is at Stake, Compiled by AMCEN Secretariat. Available online: http://www.unep.org/roa/amcen/docs/AMCEN_Events/climate-change/2ndExtra_15Dec/FACT_SHEET_CC_Africa.pdf (accessed on 20 June 2015).
19. Climate Change 2007: Synthesis Report, an Assessment of the Intergovernmental Panel on Climate Change, p. 32. Available online: http://www.ipcc.ch/pdf/assessment-report/ar4/syr/ar4_syr.pdf (accessed on 20 June 2015).

20. Programme for Central and West Africa of International Union for Conservation of Nature (IUCN-PACO): Project for Improving Water Governance in the Volta Basin (PAGEV) Update of Water Audit of the Volta Basin, 2012.
21. Niang, I.; Ruppel, O.C.; Abdrabo, M.A.; Essel, A.; Lennard, C.; Padgham, J.; Urquhart, P. 2014: Africa. In *Climate Change 2014: Impacts, Adaptation, and Vulnerability. Part B: Regional Aspects. Contribution of Working Group II to the Fifth Assessment Report of the Intergovernmental Panel on Climate Change*; Barros, V.R., Field, C.B., Dokken, D.J., Mastrandrea, M.D., Mach, K.J., Bilir, T.E., Chatterjee, M., Ebi, K.L., Estrada, Y.O., Genova, R.C., et al., Eds.; Cambridge University Press: Cambridge, UK; New York, NY, USA; pp. 1199–1265.
22. Van Vliet, J.; Bregt, A.K.; Hagen-Zanker, A. Revisiting Kappa to account for change in the accuracy assessment of land-use change models. *Ecol. Modelli.* **2011**, *222*, 1367–1375. [CrossRef]
23. Stehman, S.V.; Czaplewski, R.L. Design and Analysis for Thematic Map Accuracy Assessment—An application of satellite imagery. *Remote Sens. Environ.* **1998**, *64*, 331–344. [CrossRef]
24. Foody, G.M. Status of land cover classification accuracy assessment. *Remote Sens. Environ.* **2002**, *80*, 185–201. [CrossRef]
25. Zavoianu, F.; Caramizoiub, A.; Badeaa, D. Study and Accuracy Assessment of Remote Sensing Data for Environmental Change Detection in Romanian Coastal Zone of the Black Sea. In Proceedings of the International Society for Photogrammetry and Remote Sensing, Istanbul, Turkey, 12–23 July 2004.
26. Carletta, J. Assessing agreement on classification tasks: The kappa statistic. *Comput. Linguist.* **1996**, *22*, 249–254.
27. UNEP-GEF Volta Project, 2013. *Volta Basin Transboundary Diagnostic Analysis*. UNEP/GEF/Volta: Accra, Ghana, 2013.
28. IEA (International Energy Association). *World Energy Outlook 2002*; International Energy Agency: Paris, France; p. 533. Available online: http://www.iea.org/textbase/nppdf/free/2000/weo2002.pdf (accessed on 21 June 2015).
29. Béné, C. *Diagnostic Study of the Volta Basin Fisheries. Part 1—Overview of the Fisheries Resources*; Volta Basin Focal Project Report No 6; WorldFish Center Regional Offices for Africa and West Asia, Cairo Egypt, and CPWF: Colombo, Sri Lanka, 2007; p. 31.
30. Akpoti, K.; Antwi, E.O.; Kabo-bah, A.T. Impacts of Rainfall Variability, Land Use and Land Cover Change on Stream Flow of the Black Volta Basin, West Africa. *Hydrology* **2016**, *3*, 26. [CrossRef]
31. Mahé, G.; Paturel, J.E. 1896–2006 Sahelian annual rainfall variability and runoff increase of Sahelian Rivers. *C. R. Geosci.* **2009**, *341*, 538–546. [CrossRef]
32. Roudier, P.; Ducharne, A.; Feyen, L. Climate change impacts on runoff in West Africa: A review. *Hydrol. Earth Syst. Sci.* **2014**, *18*, 2789–2801. [CrossRef]
33. *Comprehensive Assessment of Water Management in Agriculture: Comparative Study of River Basin Development and Management*; The Volta River Basin, 2005. Available online: http://www.iwmi.cgiar.org/assessment/files_new/research_projects/River_Basin_Development_and_Management/VoltaRiverBasin_Boubacar.pdf (accessed on 20 February 2015).
34. Giertz, S.; Junge, B.; Diekkrüger, B. Assessing the effects of land use change on soil physical properties and hydrological processes in the sub-humid tropical environment of West Africa. *Phys. Chem. Earth Parts A/B/C* **2005**, *30*, 485–496. [CrossRef]
35. Mahe, G.; Paturel, J.E.; Servat, E.; Conway, D.; Dezetter, A. The impact of land use change on soil water holding capacity and river flow modelling in the Nakambe River, Burkina-Faso. *J. Hydrol.* **2005**, *300*, 33–43. [CrossRef]
36. Shi, P.J.; Yuan, Y.; Zheng, J.; Wang, J.A.; Ge, Y.; Qiu, G.Y. The effect of land use/cover change on surface runoff in Shenzhen region. *China Catena* **2007**, *69*, 31–35. [CrossRef]
37. Molina, A.; Govers, G.; Vanacker, V.; Poesen, J.; Zeelmaekers, E.; Cisneros, F. Runoff generation in a degraded Andean ecosystem: interaction of vegetation cover and land use. *Catena* **2007**, *71*, 357–370. [CrossRef]

2

Revisiting Cent-Fonts Fluviokarst Hydrological Properties with Conservative Temperature Approximation

Philippe Machetel [1,*] and David A. Yuen [2,3]

1 Laboratoire Géosciences Montpellier, CNRS/UM2, 34095 Montpellier CEDEX 9, France
2 Minnesota Supercomputing Institute and Department of Earth Sciences, University of Minnesota, 310 Pillsbury Dr. SE, Minneapolis, MN 55455, USA; daveyuen@gmail.com
3 School of Environmental Studies, China University of Geosciences, 388 Lumo Road, Wuhan 430074, China
* Correspondence: philippe.machetel@laposte.net

Abstract: We assess the errors produced by considering temperature as a conservative tracer in fluviokarst studies. Heat transfer that occurs between karstic Conduit System (CS) and Porous Fractured Matrix (PFM) is the reason why one should be careful in making this assumption without caution. We consider the karstic aquifer as an Open Thermodynamic System (OTS), which boundaries are permeable to thermal energy and water. The first principle of thermodynamics allows considering the enthalpy balance between the input and output flows. Combined with a continuity equation this leads to a two-equation system involving flows and temperatures. Steady conditions are approached during the recession period or during particular phases of pumping test experiments. After a theoretical study of the error induced by the conservative assumption in karst, we have applied the method to revisit the data collected during a complete campaign of pumping test. The method, restricted to selected data allowed retrieving values of base flow, mixing of flow, intrusions of streams, and aquifer answer to drawdown. The applicability of the method has been assessed in terms of propagation of the temporal fluctuations trough the solving but also in terms of conservative assumption itself. Our results allow retrieving the main hydrological properties of the karst as observed on field (timed volumetric samplings, geochemical analyses, step pumping test and allogenic intrusion of streams). This consistency argues in favor of the applicability of the conservative temperature method to investigating fluviokarst systems under controlled conditions.

Keywords: groundwater hydrology; karstic hydrology; energy budget; water energy interactions; water resource; modeling; conservative tracer; temperature; mixing

---

1. Introduction and Presentation of Cent-Fonts Fluviokarst

This paper addresses the issue of using the water temperature as a conservative tracer for karstic functioning studies. Whereas costly investments are often necessary to evaluate the potential of karstic aquifers, temperature records may bring cheap complementary method for obtaining information. However underground flows undergo heat transfers with the embedding rocks by advection and by conduction. Despite of this difficulty several studies used temperature as a mixing tracer. Several works [1,2] study the water exchanges between underground flows and surface stream. Others [3] retrieved aquifer recharges solving heat transport equation constrained by measurements of vertical temperature. Slow and rapid equilibrations between ground water and aquifer rocks have been analyzed [4] to study the dynamics of exchanges in fractured carbonate systems. Genthon et al. [5] determined the deep preferential path of rainfall water in caves from annual temperature variations of spring. This author also used temperature to determine limestone drainage in lagoon by removing the tidal component and emphasizing its poor correlation with rain temperature [6]. This short list is far to be exhaustive.

As mentioned in the following theoretical part of this work, the formal applicability of the conservative temperature approximation depends on the existence steady conditions that are never perfectly verified in nature. The question consists of estimating the outcome of this assumption. Indeed, our work is not the first to address this problem. In spite of disturbances due to meteorological conditions, Karanjac and Altug [7] used temperature records to characterize the recharge area, transmissivity and hydraulic regime of karst. Following Stonestrom and Constantz [8], O'Driscoll and DeWalle [9] studied the stream-ground water temperature interactions from stream-air temperature fluctuations. They used weekly averaging and equilibrium temperature concepts [10–12]. All these studies rely on damping of diurnal or seasonal variations of temperature records. This is done either by time averaging [13] or by natural damping of temperature fluctuations by depth [14,15]. It is clear that the method presented in this work is not adapted to transient flow or rapidly varying water conditions. Conversely, many situations as recession in karst undergo naturally damping of the temporal fluctuations both for temperatures and flows.

These preliminary considerations justify the interest to study the karst systems as an Open Thermodynamic Systems (OTS), that is to say a "black box" that exchanges water and heat energy with its surrounding environment. However, let us try to describe the thread of this paper. The following part of the present Section 1 recalls the hydrological description of the Cent-Fonts site and its consistency with the White's fluviokarst model. Thus, we benefit of an exceptional set of high quality data recorded during the 2005 Cent-Font pumping tests and preliminary studies. This will provide a fair opportunity to test the method. Section 2 displays theory and resulting equations on a reduced form of the White's model considered as an OTS. Section 3 describes the available data that will be revisited and Section 4 illustrates the application of the method to these selected data. The results are discussed in the "Summary and discussion" Section 5.

The Cents-Fonts resurgence is the only free-flowing of an underground drainage basin that covers 40 km^2 in the median part of Hérault River (Figure 1). The surface of the watershed reaches 60 km^2 considering the Buèges stream. It stretches in outcrops of thick calcareous and dolomite massifs (Middle and Upper Jurassic). Since several decades, many geological, geochemical and hydrological studies focused on the potential spring water production of this site (i.e., [16–27]).

The Buèges stream and the Cévennes fault mark the boundaries of the watershed to the north and northwest, while the altitude of the Hérault River matches the karstic base level to the southeast (Figure 1). Morphologically, the area is a plateau, uplifted of 200 to 500 m during the late Quaternary and eroded during Oligocene. The watershed drains rainfalls through an epikarstic transition zone that supplies the Cent-Fonts resurgence base flow [22]. To the north, Buèges stream flows permanently on Triassic low permeability outcrops however it disappears à few kilometers after Saint-Jean de Buèges where the course crosses a Bathonian, calcareous dolomite swallow zone. After this area, the Buèges stream course follows a valley that remains dry most of the year except after major rainfalls. Then a surface course catches up with Hérault River at a confluence point north of Lamalou stream (Figure 1). Tracing experiments have confirmed the underground path of Buèges stream intrusion from swallow zone to the Cent-Fonts resurgence [17,20].

Bathonian, dolomitic layers of Middle Jurassic between 150 and 300 m thick embed the saturated zone of the Cent Fonts fluviokarst near the resurgence. The aquifer also possibly meets an underlying Aalenian-Bajocien layer. The near spring, Conduit System (CS) of the resurgence (Figure 2) has been explored thank to speleological diving that reached 107 m below the base level (Hérault River) [28]. The terminal part of the CS displays an outlet cave, roughly "Y" shaped, with two sub-horizontal branches joining above a deep sub-vertical chimney. The two ends of the "Y" branches open outside. These caves are seldom active except during peak flows. Most of the time, the Cent-Fonts resurgence flows into Hérault River through a shallow network of springs that gush out a few tens of centimeters above the base level. A spring also arises directly through the bottom of the Hérault River bed [20].

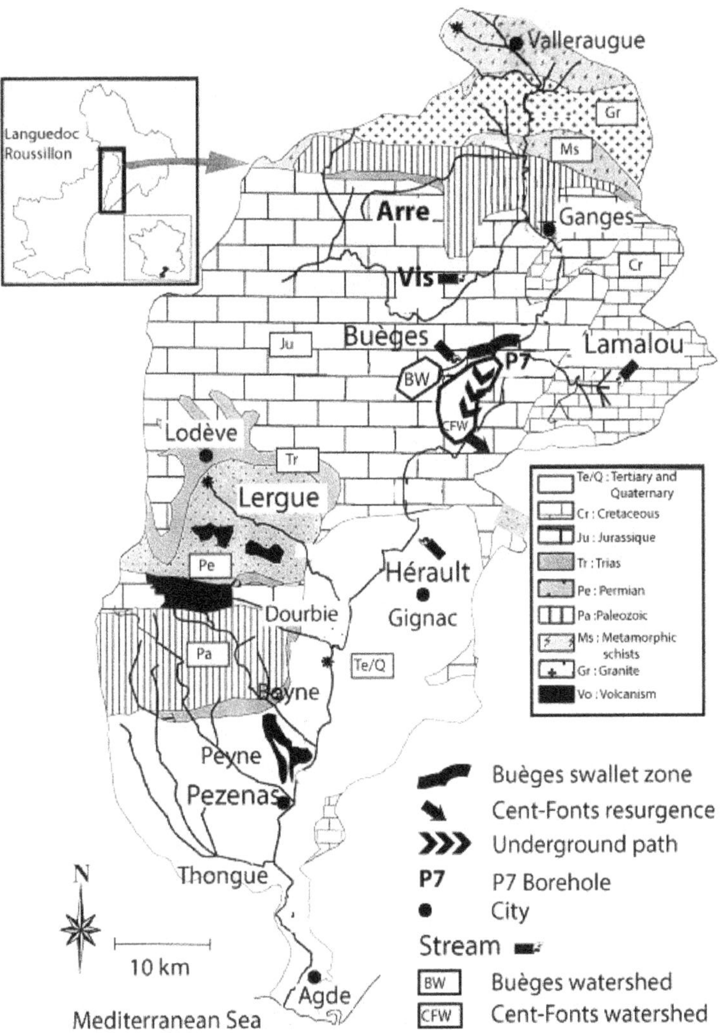

Figure 1. Simplified geological map of the Cent-Fonts karst, redrawn after Petelet et al. [16]. Note the location of the P7 borehole where far field temperature (T_{∞}) has been recorded.

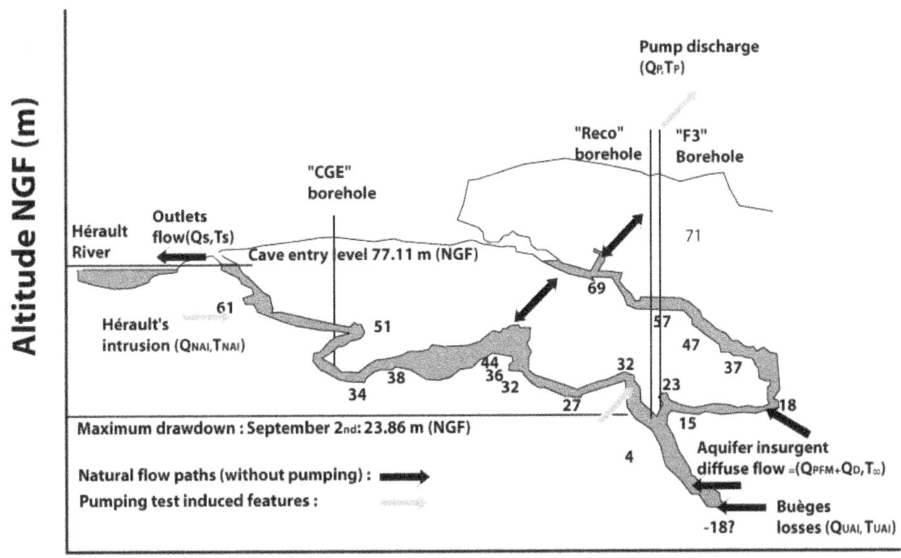

Figure 2. Unrolled 3-D speleological map of the Cent-Fonts CS near the resurgence. Altitudes are in m NGF (Nivellement Général de la France). Note the locations of the Cge, Reco and F3 boreholes.

The Cent-Fonts karst matches precisely the fluviokarst idea depicted by Smart [29]. Following his definition, fluviokarst consist: in "karst landscapes where the dominant landforms are valleys cut by surface rivers. Such original surface flow may relate either to low initial permeability before caves (and hence underground drains) had developed, or to reduced permeability due to ground freezing in a periglacial environment. In both cases the valleys become dry as karstification improves underground drainage" [30]. White [31,32] proposed a model where the hydrological behaviors are forced by the external boundary conditions. The various recharges in the CS are Allogenic Intrusions, internal Run-off and Porous Fractured Matrix flows. Intrusions of neighbor streams cause the first category of flows. The second results from sporadically floods that occur after heavy rainfalls. The third category gathers percolation flows through soils and epikarstic layer that reach the CS as a Diffuse Infiltration through the fractured and porous rocks of the aquitard (Figure 3a). In the local context of the Cent-Font watershed, the permanent course of Buèges stream matches the Upper Allogenic Stream. Later, at the swallow zone, the Buèges splits into a Surface Stream and an Upper Allogenic Intrusion. The Surface stream forms a non-perennial flow that joins the Hérault River north of Lamalou confluence. The Upper Allogenic Intrusion, Q_{UAI}, joins the CS after an underground journey. Table 1 recalls the notations and acronyms of this article. This model also considers the Diffuse Infiltration that percolates through the CS wall. The only output flow of the fluviokarst is the spring, Q_S, that falls into the Hérault River at the karstic base level.

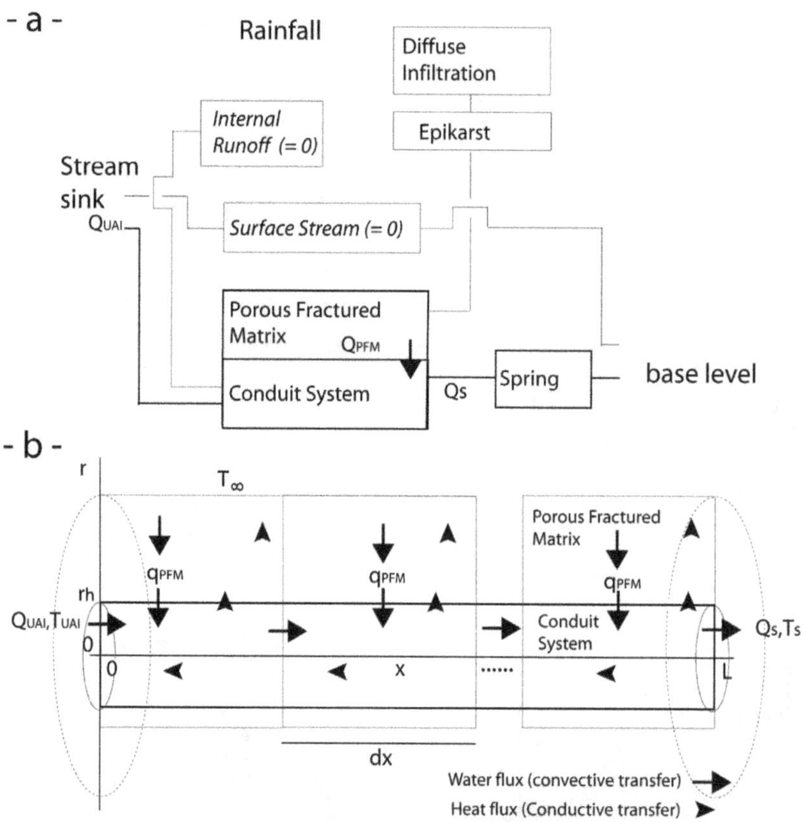

Figure 3. (**a**) Conceptual fluviokarst redrawn from White [31,32] but restricted to recession period (neither runoff nor surface flow). The karstic aquifer is embedded in a saturated PFM drained by a CS that gathers inflows. Outflow discharges at the base level of the neighbor stream through a spring (or pump). The hydrological system includes an upper allogenic stream, which flow joins the CS through a swallow zone. The CS also receives PFM, diffuse flow through an epikarstic layer; (**b**) An OTS is surrounded by pervious boundaries bounding the CV. Water and thermal energy inputs come from the allogenic streams and PFM. Outputs leave the CS through a spring (or pump). During recession no reverse flow occurs from CS to PFM.

Table 1. Chronology of pumping test phases and periods.

Periods	Pumping Operation	Beginning mm/dd (hh:mm)	Ending	Revisited Periods
a	Pre-pumping phase	1992	07/27 (07:25)	07/14 (00:00)–07/18 (23:55)
b	Step drawdown	07/27 (07:30)	07/30 (12:40)	-
c	Constant pumping	08/01 (07:10)	08/09 (07:20)	-
d	Recovering test	08/09 (07:20)	08/09 (13:00)	-
e	Constant pumping	08/09 (13:05)	09/02 (07:15)	08/13 (00:00)–09/31 (23:55)
f	Recovering	09/02 (07:20)	09/03 (07:40)	-
g	Equilibrium-pumping	09/03 (07:45)	09/06 (06:00)	09/04 (12:00)–09/05 (11:55)

As mentioned above, the Cent-Fonts resurgence has received much attention since decades (see Ladouche et al. [23] for a review). In September 1992, Cge and the city of Montpellier organized a series of pumping tests but heavy rainfalls and resulting runoff force to stop the experiments after only 16 days. This short time prevents reaching significant aquifer drawdowns despite the high rate of pumping (0.5 m^3/s). Thus, this experimentation failed to assess the hydrological properties of the karst and let believe in a high water production potential [33].

Many field observations, such as gauging data of the spring, of the Hérault River and of the Buèges stream have been recorded for several years before the summer 2005 pumping test experiments. New boreholes were drilled. New records of temperatures and discharges were collected in Buèges and Hérault. Temperatures and hydraulic heads were also recorded in the "P7" (Figure 1), "Reco", "Cge" and "F3" boreholes (Figure 2). The 2005 pumping test campaign started with a sequence of step-drawdown tests. Heavy pumping sequences followed with a high constant rate pumping and a drawdown constant pumping (see Table 1 for accurate chronology). The drawdown induced during these heavy pumping allowed speleological explorations of the Cent-Fonts chimney. It also opened the possibility of in situ timed volumetric gauging and geochemical sampling of Hérault River intrusions in the CS. This rich set of data has been extensively analyzed already [26,34]. The present work aims to revisit part of these data to assess the accuracy of conservative tracer assumption for water temperature. In the following, we will consider that two error sources mainly affect the applicability of the method: (1) The natural temporal instabilities that affect the data series; (2) the effect of conservative temperature assumption itself according to our previous work [35,36].

The next section of the paper recalls the theoretical context of Open Thermodynamic System for fluviokarst. Section 3 describes the data chosen from the 2005 pumping test campaign, Section 4 presents their analysis and accuracy assessments. The results are summarized and discussed in Section 5.

2. Theoretical Context

An Open Thermodynamic System (OTS) does not require a precise knowledge of the locations and shapes of its boundaries to study the balance of the fluxes that enter or quit its Control Volume (CV). In that sense, CS of fluviokarsts are similar to CV of OTS (Figure 3b). We will benefit of this analogy to study the balance of fluxes and energy in the Cent-Fonts fluviokarst CV despite the inaccurate knowledge of its boundaries [35].

In our previous work [36] we assessed the error resulting, in OTS, of conservative temperature assumption within the theoretical context of White's fluviokarst model. This analysis relied on two successive numerical solving of thermal behavior. The first considered the temperature as a conservative tracer, while the second, following works by Covington et al. [37,38] solved the complete set of energy equations including convective, diffusive and dispersive terms. The amplitude of differences between both results shows the first order of the error due to the conservative assumption. We drew abacus curves of this error versus thermal diffusivity ratio and the Peclet numbers, Prandtl numbers and Reynolds numbers of the CS. The error remains less than 1 percent in the dimensionless space and converges to zero for the most extremes values of karst features. This dimensionless error needs to be rescaled for each singular fluviokarst to calculate its physical amplitude.

In our study, we consider water as a Boussinesq fluid with constant thermal capacity, thermal expansion and density. As a result, water motion forms a zero divergence velocity field in the PFM and in the CS (Equation (1)).

$$div(\vec{v}) = 0 \tag{1}$$

It is possible to convert the volume integrals over fluviokarst CS in flux integrals over the PFM boundaries and hydraulic sections of conduits entering or leaving the CS. Then, the terms of Equation (2) are equal to the input flows into the CS (positive algebraic values, q_i) and to the flows escaping from the CS (negative algebraic values, q_o). Therefore, the continuity equation leads to the mass conservation equation that links the various flows.

$$\sum_{input} q_i + \sum_{output} q_o = 0 \tag{2}$$

The first law of thermodynamics stipulates that the internal energy change of the CV (Equation (3)) corresponds to the balance of the energy differences between all the incoming and outgoing flows (index j) and considering the work done. Thus, the internal energy change (δE) is equal to the summation of: the enthalpy by unit of mass (h_j), the potential energy ($e_{pot\,j}$), the kinetic energy ($e_{cin\,j}$), the external heat transfer ($\delta\Phi_j$) and the work exchanges (δW_j) with the surrounding [39,40].

$$\delta E = \sum_j (h_j + e_{pot,j} + e_{cin,j})dm_j + \sum_j \delta\Phi_j + \sum_j \delta W_j \tag{3}$$

In the following, we will consider no chemical contribution to enthalpy and flow transfers with negligible exchange between heat and work. We also will consider steady flux conditions in the CV (we discuss this point later). These hypotheses cancel the internal energy change but also the potential energy and kinetic energy changes. Then, Equation (3) becomes a balance between the specific enthalpies by unit of time of the flows entering (h_i) and escaping (h_o) the CS (Equation (4)).

$$\sum_i h_i \rho_i q_i + \sum_o h_o \rho_o q_o = 0 \tag{4}$$

Now, specific enthalpy depends only on thermal capacity (Cp) and temperature (referred to an arbitrary value T_a) (Equation (5)).

$$h_j = Cp_j(T_j - T_a) \tag{5}$$

Combination of Equations (4) and (5) leads to the enthalpy balance (Equation (6)).

$$\sum Cp_i(T - Ta)\rho_i q_i + \sum Cp_o(T_o - Ta)\rho_o q_o = 0 \tag{6}$$

Finally, with constant density and thermal capacity Equation (6) reduces to a classical mixing equation (Equation (7)) that links temperatures (°K or °C) and mass transfers in the CS.

$$\sum_{input} q_i T_i + \sum_{output} q_o T_o = 0 \tag{7}$$

Equations (2) and (7) form the basis of a linear system able to discover two unknown flows in the CS (so-called Q_k and Q_l in Equation (8)).

$$\begin{aligned} q_k T_k + q_l T_l &= -\sum_{\substack{o \neq k \\ o \neq l}} q_o T_o - \sum_{\substack{i \neq k \\ i \neq l}} q_i T_i \\ q_k + q_l &= -\sum_{\substack{o \neq k \\ o \neq l}} q_o - \sum_{\substack{i \neq k \\ i \neq l}} q_i \end{aligned} \tag{8}$$

Thus, these the two unknown flows are (Equation (9)).

$$q_k = \frac{1}{(T_k - T_l)} \begin{vmatrix} -\sum\limits_{\substack{o \neq k \\ o \neq l}} q_o T_o - \sum\limits_{\substack{i \neq k \\ i \neq l}} q_i T_i & T_l \\ -\sum\limits_{\substack{o \neq k \\ o \neq l}} q_o - \sum\limits_{\substack{i \neq k \\ i \neq l}} q_i & 1 \end{vmatrix} = \frac{-\sum\limits_{\substack{o \neq k \\ o \neq l}} q_o(T_o - T_l) - \sum\limits_{\substack{i \neq k \\ i \neq l}} q_i(T_i - T_l)}{(T_k - T_l)},$$
$$q_l = -\sum_{\substack{o \neq k \\ o \neq l}} q_o - \sum_{\substack{i \neq k \\ i \neq l}} q_i - q_k. \tag{9}$$

As mentioned above, we will use the theoretical results of Machetel and Yuen [36] to quantify the sensitivity of the model to the conservative enthalpy assumption. However since we assumed that no other sources of heat are present (neither chemical heat, nor work conversion to heat), enthalpy conservation comes down on to temperature conservation. We will also use the differential form (Equation (10)) of Equation (9) to assess the effects of temperature and discharge uncertainties solving Equation (9).

$$\delta q_k = \frac{1}{(T_k - T_l)} \left[\begin{matrix} -\sum\limits_{\substack{o \neq k \\ o \neq l}} \left[\delta q_o (T_o - T_l) - q_o \left((\delta T_o - \delta T_l) - \frac{(\delta T_k - \delta T_l)(T_o - T_l)}{(T_k - T_l)} \right) \right] \\ -\sum\limits_{\substack{i \neq k \\ i \neq l}} \left[\delta q_i (T_i - T_l) + q_i \left((\delta T_i - \delta T_l) - \frac{(\delta T_k - \delta T_l)(T_i - T_l)}{(T_k - T_l)} \right) \right] \end{matrix} \right]$$
$$\delta q_l = -\sum_{\substack{o \neq k \\ o \neq l}} \delta q_o - \sum_{\substack{i \neq k \\ i \neq l}} \delta q_i - \delta q_k \tag{10}$$

The applicability of the method developed above depends on a "reasonably" steady CS. This is never the case in nature where temperature and flow variations due to human activities or diurnal or meteorological cycles disrupt steadiness. This is a recurring problem for hydrological studies. It can be significantly with careful choice of working periods and a 24-h moving averaging of data impacted by diurnal effects [10–12]. The thermal inertia of soil also damps the meteorological or seasonal effects with deepening [14,15].

However, we also have to trust on the common sense of operators and analysts to "instinctively" avoid the most unstable periods for data collection. Thus, the conservative temperature assumption is unsuitable for runoff flow studies or all other kinds of events that imply transient and unstable thermal or water fluxes in the CS. This is why, the present work use a restriction of the White's model to the recession period, with no run-off flow, and with a complete loss of the Upper Allogenic Stream (no surface stream). The 2005 Cent-Fonts pumping tests take place during the summer season when rainfalls are scarce on the watershed. However, even during that time, we focused the method on "steadiest" periods for water temperatures and flows. This is also why, despite available data until November, we stopped our analyses on 6 September (16:55) when a runoff due to a heavy thunderstorm flooded the boreholes (Figure 4).

3. Presentation of the Cent-Fonts Pumping Test Data

3.1. *Period (a): Data Collected Prior the Beginning of the Pumping Test*

Several years of flows and water temperatures have been recorded at the karst spring and in Buèges since 1997 to study the recession of the base flows during dry periods. This knowledge is essential to understand the answers of the aquifer during the pumping tests. The base flow (Q_S) gathers the upper allogenic intrusion of Buèges stream (Q_{UAI}) and the diffuse infiltration (Q_{PFMB}). The study of the karst recession conducted from 1997 to 2001 used a modified Mangin method to distinguish the parts played by Q_{PFMB} and Q_{UAI} to the base flow [23]. Q_S is described thank to three terms calculated from a Maillet homographic function [41]. Hence, two recession coefficients appear that characterizes the respective recession evolutions of Q_{PFMB} or Q_{UAI}. Ladouche et al. [23] calculated recession coefficients of 0.0080 (1998); 0.0088 (2000) and 0.0088 (day^{-1}) (2001) for the Q_{PFMB} contribution to the base flow [23] (p. 64).

Complementary flows and water temperatures have been recorded in the weeks preceding the step-drawdown sequence on 27 July (see Table 1 for an accurate chronology of the pumping tests). Figure 4 presents the far field (T_∞), the Upper Allogenic Intrusion (T_{UAI}, Buèges), the Neighbor Allogenic Intrusion (T_{NAI}, Hérault) and the CS (T_{CS}, in Cge borehole) temperatures recorded from 1 July to 6 September. During the pumping tests, Cge's devices provided CS hydraulic heads and temperatures until their disconnection by drawdown. T_{CS} temperature was also recorded in the F3 borehole at the output of the pump (see Figures 2 and 4).

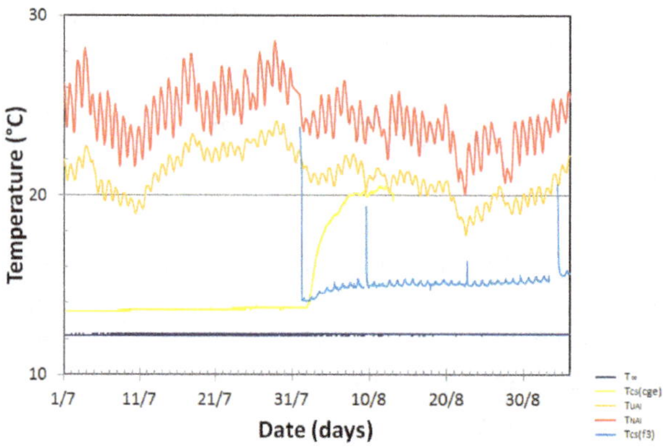

Figure 4. The temperature series used in the study are T_{UAI} (recorded at the Buèges losses), T_{NAI} (Hérault), T_{CS} (recorded in CGE borehole until disconnection on 13 August 12h15 and at pump output), and T_∞ (recorded in the P7 borehole).

3.2. Period (b): Data Collected during the Step-Drawdown Sequence (27 to 30 July)

Step-drawdown sequences are one of the most often performed pumping test to find out the behavior of wells and aquifer features. For the Cent-Fonts pumping test campaign, the pump has been placed directly inside a large CS conduit. The hydraulic heads recorded in F3, Reco and Cge will display such close curves they will be undistinguishable at the Figure 5 scale despite the bottleneck between Cge and F3 boreholes (Figures 2 and 5). This superposition of curves reveals the hydraulic connectivity in this final part of the CS. Thus, the step-drawdown sequence will efficiently find out the resurgence yield by overestimating or underestimating the rate of pumping drying the spring.

Table 2 recalls the step-drawdown chronology, the pumping rates and the drying effects on griffons. Q_P of 0.2 and 0.3 m^3/s did not achieve the completed drying of griffons while it was reached for 0.4 and 0.5 m^3/s [26] (pp. 55–59). These results allow inferring that between 27 July and 30 July the base flow of the resurgence spring (that is Q_{PFMB} + Q_{UAI}) was ranging between 0.3 and 0.4 m^3/s.

Four "coma-shaped" events (due to sudden deepening followed by recovering) occurred on the hydraulic head for each pumping of the step-drawdown sequence (Figure 5, Period (b)). During that time, the CS temperature remains constant to a few tens of degree while, on the opposite, T_{UAI} (Buèges) and T_{NAI} (Hérault) display diurnal temperature oscillations that reach 1 to 3 degrees (Figure 4). These diurnal temperature variations oscillate over 4 degrees, of meteorological trend that affect both T_{UAI} and T_{NAI} between 11 July and 30 July. The meteorological trend has the same amplitude on T_{UAI} and T_{NAI}. The amplitude of T_{UAI} diurnal oscillations remains lower than those of T_{NAI} because of shortness Buèges course and its low emergence temperature (12.5 °C). The stability of T_{CS} despite these oscillations results from the damping effects of soils and 10 days underground transfer from the losses area to the Cent-Font resurgence [20].

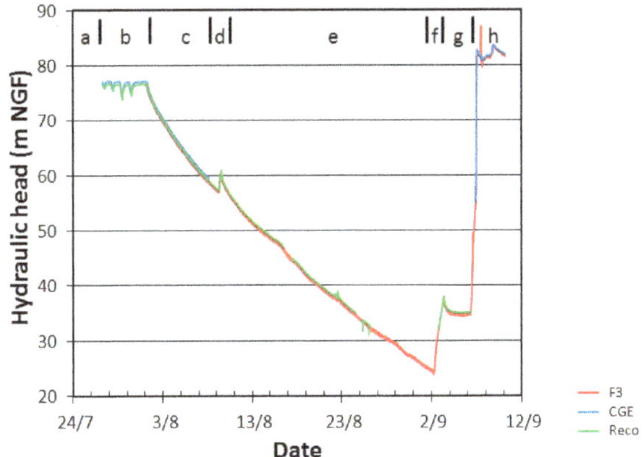

Figure 5. Hydraulic head recorded in the CGE, F3 and Reco boreholes. Letters (a to g) refer to periods of Table 1. Note the four "coma-shaped" events (due to sudden deepening followed by recovering) that occurred on the hydraulic head for each of the pumping of the step-drawdown sequence (period b); the almost constant rate of hydraulic head increase during the constant high rate pumping (periods c and e) and the rapid stabilization of the hydraulic head during the equilibrium pumping (period g).

Table 2. Step drawdown pumping, Q_{PFMB} from Equation (13), Q_{UAI} recorded at Buèges, total recovering.

Step-Drawdown Phase	Mean Q_P (m³/s)	Outlets Drying [26]	Mean Q_{PFMB} (m³/s)	Mean Q_{UAI} (m³/s)	Q_{PFMB} + Q_{UAI} (m³/s)
07/27 (07:30–13:20)	0.203	Partial drying of outlet n°4 No drying of other outlets	0.305	0.042	0.347
07/28 (06:35–12:00)	0.301	Drying of outlets except outlet n°4 (partial drying)	0.303	0.034	0.337
07/29 (06:20–12:20)	0.500	Drying of all outlets	0.300	0.031	0.331
07/30 (06:30–12:35)	0.402	Drying of all outlets	0.297	0.033	0.330

3.3. *Period (c): Constant High Rate Pumping (1 to 9 August)*

Long high rate pumping is used to assess the answer of the aquifer to drawdown. From 1 August to 9 August, the hydraulic head in the CS increases linearly with time (note that to avoid concave curves in Figure 5, the data are not plotted versus the log of time as usual). However, the drawdown induces a reversal of the hydraulic head that triggers the intrusion of Hérault in the CS and a new contribution that adds to the base flow Q_{PFMB} coming from the PFM. Two new flows, Q_{NAI} and Q_{PFMD} are added to Q_{PFMB} and Q_{UAI} while spring drying let Q_P be the only discharge of the CS. During that time, T_{UAI} and T_{NAI} display diurnal and meteorological variations while T_∞, recorded 25 m below the surface in the P7 borehole, remains remarkably constant (Figure 4, Period (c)).

T_{CS} is recorded both in the Cge and F3 boreholes. The first is measured close to the arrival of the neighbor allogenic intrusion of Hérault (Figure 2). The second T_{CS} is recorded deeper, close to the arrivals of Q_{UAI} and Q_{PFMB}. These two flows carry temperatures T_{UAI} and T_∞ lower than this of T_{NAI}. Thus, the values of the two T_{CS} series diverge rapidly as soon as the drawdown triggers arriving of hot intrusive Hérault water in the CS branch near the Cge borehole. T_{CS} (Cge recorded) increases rapidly a few hours after the starting of the high rate constant pumping from its 13.7 °C constant value since 2 August, 12h05. After a few days of transient evolutions, the temperatures in Cge and F3 boreholes stabilize respectively around 20 and 15 °C. Their temporal variations are correlated with the meteorological and diurnal trends observed for T_{NAI} and T_{UAI} (Figure 4, Period (c)).

The second series of T_{CS} records (F3) stabilizes rapidly around 15 °C. It displays diurnal oscillating changes that are clearly due to the mixing of hot Q_{NAI} with the cold Q_{PFMB}, Q_{PFMD} and Q_{UAI}

in the vicinity of the pump. This stabilization of the T_{CS} increase indicates that Q_{NAI} acts like an almost constant vadose flow despite the increase of the hydraulic head between the water table and Hérault [26] (p. 191).

According to Maréchal et al. [34] we will considerer in the following the dewatering of the conduit network as a supplementary outgoing flow from the CS, Q_{CS}.

3.4. Period (d): Recovering Test (9 August)

The 6 h pump stop of 9 August allowed a recovering of 3.43 m (Figure 5, Period (d), and [26] (p. 65). During the interruption, Q_{NAI} brought warm water of Hérault that accumulated in the chimney above the pump. After pump re-starting, the temperature, T_P displayed a short peak 9 August, 13h05 consecutive to the rapid extraction of this warm water. Similar phenomena occurred at the restarting 22 August, 13h30 and 3 September, 07h45 (Figure 4, Periods e and g).

3.5. Period (e): Constant High Rate Pumping (9 August to 2 September)

After 6 h of recovering, high rate pumping has been restarted from 9 August until 2 September. During Period (e), the hydraulic head decreased almost linearly with a slope similar to the one of Period (c). On 13 August at 12h15, the drawdown reached the level of the temperature probe in Cge borehole (51.6 m NGF), causing its disconnection and the loss of the T_{CS} signal recorded there. At the end of Period (e), the drawdown approached the level of the pump in F3 that caused its stopping (Table 1).

As mentioned above for Period (c), two diurnal and meteorological trends are noticeable in the T_{NAI} and T_{UAI} temperature records. These variations are also present in the last part of T_{CS} (Cge recorded) and in T_{CS} (F3 recorded) (Figure 4, Period (e)).

From the beginning of August, the drawdown allowed speleological explorations of the resurgence branches (Figure 6). Surprisingly, while significant Q_{PFMD} infiltrations were expected in CS, no water was apparently percolating through the chimney wall (Figure 6a–c). However, several cascading vadose flows were observed between marks 657–658 of the lifeline (Figure 6d) and mark 670 (Figure 6e). Partial catchments of these Hérault intrusions at point 657 collect 0.045 to 0.050 m^3/s [26] (p. 40). This value gives a rough lower bound value of Q_{NAI} since other hidden or deeper entries were probably active. These speleological explorations also opened the opportunity to collect geochemical samples to quantify the mixing of Hérault intrusion in the CS.

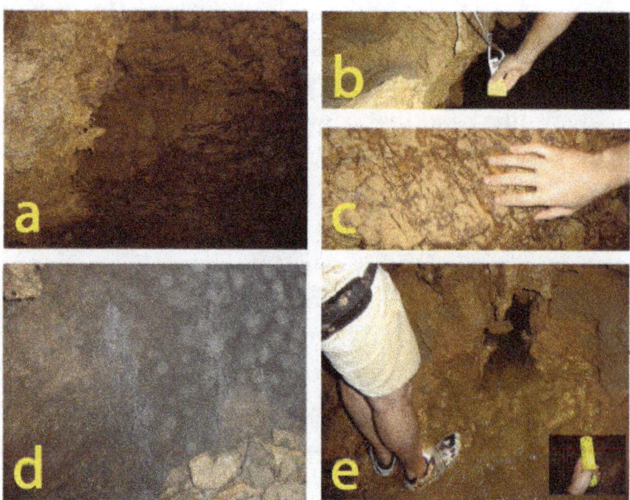

Figure 6. Speleological views of CS chimney. (**a**) CS wall at lifeline mark n° 658 (8 August); (**b**) lifeline mark 658; (**c**) "dry" CS wall at mark n° 658; (**d**) Cascading Hérault intrusion at mark 658 (CS1 in Table 3); (**e**) Second Hérault intrusion at lifeline mark 670 (1 September) (CS2 in Table 3).

Water samples were collected 1 September at the above intrusion points. We used PP® bottles previously washed with chlorydric acid, then bromydric acids, to prevent contamination [42]. Bottles and corks have been rinsed four times on site. The solutions have been filtered, acidified and prepared in two dilutions for analysis. Complementary samples were collected the same day at the Buèges spring, in Hérault and at the pump.

Table 3. CS1 and CS2: Hérault intrusions in CS, HR: Hérault, BS: Buèges, PO: Pump output.

ppB	CS1	CS2	HR	PO	BS	CS1	CS2	HR	PO	BS
Rb	1.054	1.131	1.582	0.460	0.193	1.157	1.166	1.366	0.443	0.188
Sr	76.854	82.443	91.054	58.334	52.596	83.390	81.500	77.440	58.510	53.551
Ba	52.231	52.577	94.786	12.723	3.736	55.773	52.676	80.564	12.935	3.791

The samples were analyzed for the Rb, Sr, and Ba on the VG Plasmaquad II turbo ICPMS of Montpellier 2 University (Table 3 and Figure 7). Sr, Rb, and Ba have been chosen as field tracers for water circulation and mixing [4] Petelet et al. 2003). Figure 7 shows the alignment of the samples in a (Ba/Sr) vs. (Rb/Sr) graph. The alignment of point denotes a mixing between two poles [43]. According to the regression curves and the rate of pumping $Q_P = 0.4$ m^3/s, Q_{NAI} ranges from 0.039 to 0.048 m^3/s for Ba, from 0.060 to 0.083 m^3/s for Sr, and from 0.077 to 0.087 m^3/s for Rb. However, the lower values of the Ba/Sr ratio may reflect a sorption effect onto mineral-water interfaces [44]. The averaged value of these six measures gives $Q_{NAI} = 0.066$ m^3/s.

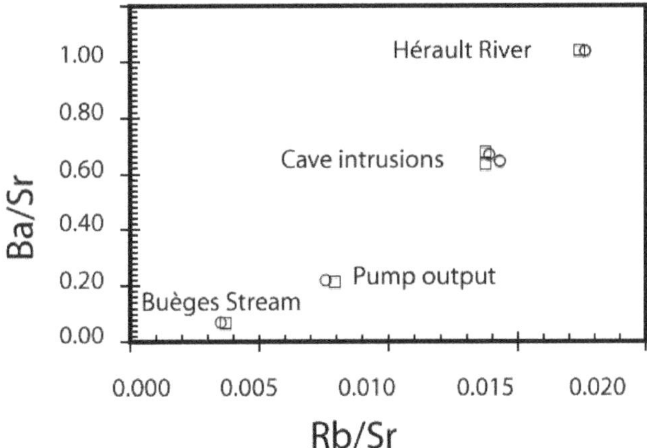

Figure 7. Plot of Ba/Sr vs. Rb/Sr (see Table 3). The alignments of points are characteristic of a two-pole mixing between Buèges and water Hérault. Two dilutions have been applied on the samples before ICPMS analyses (open squares: dilution by a factor two, open circles: no dilution). Assuming that Buèges water composition is characteristic of the far field water composition, we can write, after Vidal [43]: $Q_{NAI} = Q_P\,(c_P - c_{UAI})/(c_{NAI} - c_{UAI})$, where c_P, c_{UAI} and c_{NAI} respectively stand for mass concentrations of Rb, Ba and Sr in Q_P, Q_{UAI} and Q_{NAI} (Table 3).

3.6. Period (f): Recovering Test (2 September to 3 September)

A second stop of the pump lasts 24 hours and 25 minutes, from 2 September to 3 September, that induced a 13.49 m recovering in the chimney (Figure 5).

3.7. Period (g): Drawdown Constant Pumping (3 September to 6 September)

The equilibrium-pumping (Period (g)) followed one month of constant pumping at constant rate that resulted in an important drawdown. After the recovering of Period (f), the equilibrium-pumping aimed to assess the dynamics of aquifer answer to hydraulic head. It started on 3 September at 7h45 with Q_P = 0.324 m^3/s. The pumping rate has been dropped to Q_P = 0.304–0.305 m^3/s a few hours later to stabilize the hydraulic head at 35.0 ± 0.1 m NGF (Figure 5).

Since, this value is around 40 m deeper than the karst base level the Hérault intrusion remained active (the first intrusion of Hérault occurred for 76.7 m NGF on 1 August at 07h10).

4. Revisiting the 2005 Pumping Test Data

The challenge of studying temperature as a conservative tracer also faces the natural complexity of karsts, which CS mixes water coming from low resistance conduits and low permeability PFM. A few decades ago, studies were often considering medium where CS was not disrupting the aquifer but continuous models were unsatisfactorily. Since a few years, improved analytical models take better into account for the different behavior of the two types of reservoirs [34] or even three reservoirs [45]. However, despite the importance of calibration for the models, temperature has not been used probably because of the non-conservative character of this signal [26,34]. In the following we will take benefit of particular periods of the pumping test to get new constraints for the models.

Our process consists, firstly, to revisit the data of Period (a) to recalculate the base flow Q_{PFMB} that forms the background over which the assessment of Q_{PFMD} is possible. Secondly, we benefit of the base flow knowledge to reevaluate the intrusion of Hérault, Q_{NAI}, on Period (g). Thirdly, relying on Q_{PFMB} and Q_{NAI}, we re-assess Q_{PFMD}, the answer of the aquifer to drawdown, during the constant pumping (Periods (c) and (e)).

4.1. Revisiting Q_{PFMB} from Period (a)

4.1.1. Revisiting the Data

Unlike Maréchal et al. [34], but following the results of [26] (p. 64), we consider that the recession of the Cent-Fonts base flow is better described separating the Buèges contribution Q_{UAI}. Then, the recession of Q_{PFMB} is calculated using Equation (11) in which we need to set the amplitude coefficient $Q_{PFMB}(t_o)$.

$$Q_{PFMB}(t) = Q_{PFMB}(t_0)e^{-0.0088(t-t_0)} \tag{11}$$

Without pumping, the only flow leaving the CS is the one of the spring. In July 2005, as the surface course of Buèges fully disappears, and since no pumping affects the resurgence, no drawdown occurs in the CS. Under these natural conditions Q_S gathers Q_{UAI} and Q_{PFMB} and the water table in CS sits on top of the base level by a few centimeters. This weak hydraulic head is enough to prevent Hérault to intrude.

Then, replacing $q_k = Q_{PFMB}$, $T_k = T_\infty$, $q_l = -Q_S$, $T_l = T_{CS}$, $q_i = Q_{UAI}$ and $T_i = T_{UAI}$ in Equation (8) leads to the Equation (12) below where the spring discharge Q_S and the base flow Q_{PFMB} are unknowns.

$$\begin{aligned} Q_{PFMB} &= \frac{Q_{UAI}(T_{UAI}-T_{CS})}{(T_{CS}-T_\infty)} \\ Q_S &= Q_{PFMB} + Q_{UAI} \end{aligned} \tag{12}$$

It is, therefore, possible to fix $Q_{PFMB}(t_o)$ in the recession curve with the values Q_{UAI}, T_{UAI}, T_{CS} and T_∞ recorded for several weeks. Figure 8 displays an enlargement of these records from 14 July (00:00) to 19 July (23:55). T_{UAI} displays diurnal oscillations of 0.2 to 0.4 °C around its 24-h moving averaging and a small meteorological increasing trend of a few degrees (Figure 8, top). In bottom panel of Figure 8, Q_{UAI} also displays diurnal oscillations due to the water catchments upstream of the swallow zone. Concurrently, T_∞ and T_{CS} remain remarkably stable (Figure 8, top). The Q_{PFMB} curve in the bottom panel of Figure 8 displays the results of Equation (12) solving. We calculated its mean values (0.336 m^3/s) and affected it at the median time to = 16 July (12:00) to extrapolate the PFM base flow for the remainder of our study (Equation (13)).

$$Q_{PFMB}(t) = 0.336e^{-0.0088(t-t_0)} \tag{13}$$

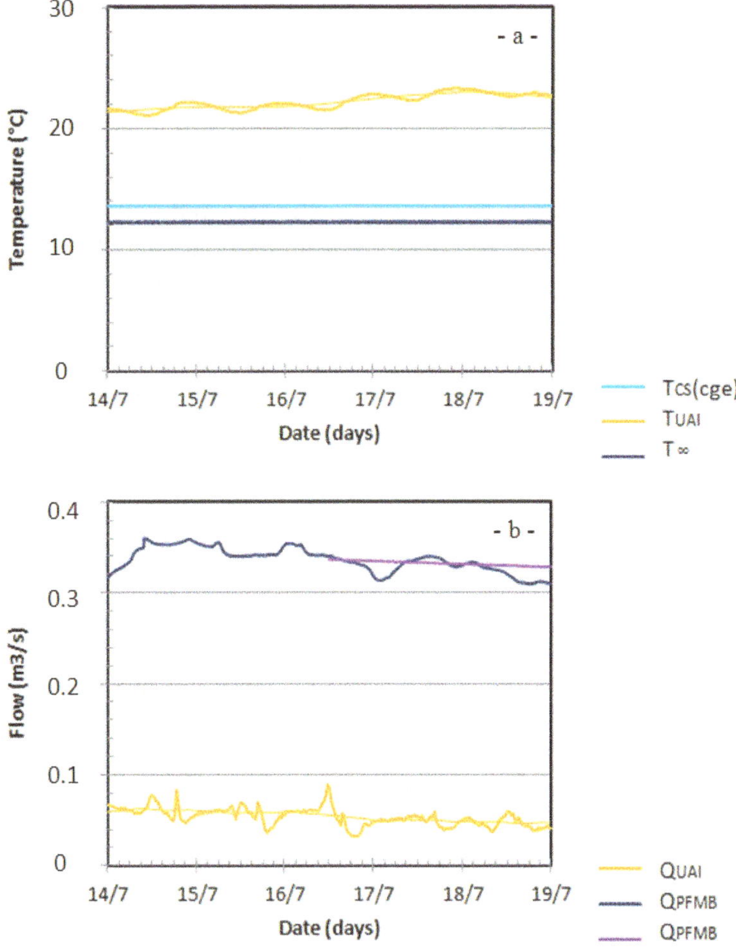

Figure 8. (**a**) Temperatures and (**b**) flows recorded or calculated during the pre-pumping period. The conduit system temperature T_{CS} recorded in the CGS borehole and the matrix-conduit flow temperature T_∞ are almost constant over this period while the upstream allogenic intrusion temperature T_{UAI} displays both diurnal and meteorological trends that are smoothed by the 24-h averaging. The upstream allogenic intrusion, Q_{UAI}, also displays a diurnal behavior due to water catchments upstream of the swallow zone. Q_{PFM}, is solved with Equation (14).

4.1.2. Assessment of Error Due to Data Variability

Two kinds of inaccuracy may affect the measures and, therefore, the use of Q_{UAI}, T_{UAI}, T_{CS} and T_∞ for the method described in this article. The first are the precisions of the measures while the second relate to the difference between the steady state and the physical conditions in the CV. In the following, we will consider that modern thermometers result in negligible errors (less than 0.1 °C) in front of those resulting of unsteady behaviors. For Buèges gauging, the level error δQ_{UAI} at the swallow zone is not explicitly mentioned [26]. However, it is reduced by the total swallowing that limits it to the one of the zone entry. It is also reduced by the low flow context that allows more accurate gauging [46]. Therefore, without better assessment of this error, it seems reasonable to consider that it remains lower than a few liters by second. This level matches the one of standard errors induced by the diurnal variations (Table 4).

Equation (10) provides a powerful tool to assess how errors spread through the solving. According our preceding comments, and considering the stability of T_{CS} and T_∞, we have reported $\delta T_\infty = 0$ and $\delta T_{CS} = 0$ in Equation (10). The differential form of Equation (9) becomes Equation (14), for which the numerical coefficients have been calculated by using the mean values and the Standard Errors of temperatures and flows.

$$\delta Q_{PFMB} = \frac{1}{(T_{CS}-T_{\infty})}[(T_{UAI} - T_{\infty})\delta Q_{UAI} + Q_{UAI}\delta T_{UAI}]$$

$$\delta Q_S = \delta Q_{PFMB} + \delta Q_{UAI} \tag{14}$$

$$\delta Q_{PFMB} = 6.21\,\delta Q_{UAI} + 0.038\,\delta T_{UAI}$$

More realistic physical conditions have been searched for in the CS by operating 24-h moving averaging on the diurnal variations of raw data (Table 4, columns 6 and 8). Such numerical process accounts for the natural thermal and kinetic inertia acting along underground water flows and allows damping the error that could arise from their neglecting. Thus, over a few days, a meteorological trend increases the temperature recorded at the Buèges swallow zone by one to two °C (Figure 8). However, the final standard error remain limited to 0.64 °C on raw data (Table 4, column 6) and to 0.55 °C after 24-h averaging of the data (Table 4, column 8). The natural damping of temperature fluctuations that affects Q_{UAI} during its underground travel to the CS prompts to consider the standard error as an upper bound.

Equation (14) shows that 1 °C of error on T_{UAI} induces 38 L/s (liters by second) of error on Q_{PFMB}; and that one L/s of error on Q_{UAI} induces 7 L/s on the final result. We can therefore expect that Q_{PFMB} is obtained to a few tens of L/s (around 20%) (Table 4, columns 7 and 9). This is consistent with the step drawdown tests that revealed a spring flow Q_{PFMB} + Q_{UAI} comprised between 0.3 and 0.4 m^3/s.

Table 4. Error assessments: Parameter, Status (M)easured, (E)xtrapolated or (C)alculated; Figure or Equation; Mean value; Standard Error; Max Error; Standard Error 24 h averaging, Min Error.

Periods	Q, T	Status	Figure	Mean	SE	Error	24mSE	24mError
(a)	T_{CS}	M	4, 8	13.6	0	-	0	-
	Q_{UAI}	M	8	0.053	0.010	-	0.005	-
	T_{UAI}	M	8	22.3	0.64	-	0.55	-
	T_{∞}	M	4, 8	12.2	0	-	0	-
	Q_{PFMB}	C	8	0.336	-	0.086%–25.6%	-	0.052%–15.5%
	Q_S	C	8	0.391	-	0.096%–24.7%	-	0.057%–14.7%,
(g)	Q_P	M	9	0.304	0	-	0	-
	T_P	M	9	15.73	0	-	0	-
	T_{∞}	M	4, 9	12.2	0	-	0	-
	Q_{UAI}	M	9	0.015	-	-	-	-
	T_{UAI}	M	9	22.0	-	-	-	-
	T_{NAI}	M	9	25.1	0.52	-	-	-
	Q_{NAI}	C	9	0.070	-	0.004%–5.7%		
	Q_{PFM}	C	9	0.219	-	0.003%–1.4%		-
(e)	Q_P	M	10	0.396	0.008	-	-	-
	T_P	M	10	15.16	0.015	-	-	-
	T_{∞}	M	4, 10	12.2	0	-	-	-
	Q_{UAI}	M	10	0.0184	0.006	-	0.002	-
	T_{UAI}	M	10	20.1	0.89	-	0.84	-
	Q_{NAI}	E	9, 10	0.070	-	-	-	-
	T_{NAI}	M	10	23.2	1.21	-	0.97	-
	T_{CS}	E	4	T_P	-	-	-	-
	Q_{PFMB}	E	Equation (13)	var.	-	-	-	-
	Q_{PFM}	C	10	0.248	-	0.016%–5.7%		0.014%–5%
	Q_{CS}	C	10	0.278	-	0.029%–97%		0.021%–70%

4.1.3. Assessment of Error due the Conservative Temperature Assumption

Another way to explore the effects of non-stationarity on the solutions consists in confronting karst numerical models that consider (or not) temperature as a conservative tracer. This has been done in a previous study where several numerical models have been calculated over very broad ranges of karst morphological and hydrological parameters. According to Equation (15) of [36] a first order of the error induced by the conservative temperature assumption is reached by the following Equation (15):

$$\delta T = \varepsilon' \left[(T_{CS} + \frac{T_\infty}{(T_{UAI} - T_\infty)}) \right] \tag{15}$$

The error, ε', is calculated versus the hydrological and morphological properties as thermal diffusivity ratio (9.93), Conduit Peclet number (1.5×10^8), Prandtl number (6.99) and Conduit Reynolds number (4.29×10^4). Tables 1 and 5 recall the form and the physical values compatible with the Cent-Fonts karst. Then, the abacus curves of Figure 4 of [36] indicate $\varepsilon' = 0.00613$ at the exit of the CS. With $\varepsilon' = 0.00613$, $T_{CS} = 286.75$ K, $T_{UAI} = 295.45$ K and $T_\infty = 285.35$ K (Table 4, Period (a)) the first order of the error $\delta T = 1.93$ °C. Coming back to Equation (14), we can see that it may induce 0.073 m^3/s of error for Q_{PFMB}. This result is higher but remains consistent with the previous estimate and the results of the step drawdown tests.

Table 5. Assessment of error due the conservative temperature assumption. Period, characteristic length, hydraulic radius, characteristic velocity, Dimensionless thermal diffusivity, Peclet number, Prandtl number, Conduit Reynolds number, dimensionless error, rescaled thermal error.

Periods	L (m)	r (m)	V(m/s)	D	Pe	Pr	Red	ε	δT
(a)	5×10^3	5	4.29×10^{-3}	9.93	1.5×10^8	6.99	4.29×10^4	0.00613	1.93
(c, e, g)	2×10^2	5	2.78×10^{-3}	9.93	3.9×10^6	6.99	2.78×10^4	0.002	0.62

4.2. Revisiting Q_{NAI} from Périod (g)

4.2.1. Revisiting the Data

As recalled above, the equilibrium-pumping that started on 3 September (7h45) and ended on 6 September (6h00) followed one month of constant pumping and a one day recovering. After a few hours, the initial rate $Q_P = 0.324$ m^3/s has been lowered to $Q_P = 0.304$–0.305 m^3/s that stabilized the hydraulic head at 35.0 ± 0.1 m NGF (Figure 5). This hydraulic head is around 40 m deeper than the base level Hérault. Consequently, the Hérault intrusion Q_{NAI} remained fully active during all the equilibrium-pumping.

During Period (g), the incoming flows in the CS are Q_{UAI}, Q_{NAI} and both basic and drawdown induced PFM contribution $Q_{PFM} = Q_{PFMB} + Q_{PFMD}$. These input flows equilibrate the output flow Q_P while the stabilization of the hydraulic implies that the dewatering of the CS stops ($Q_{CS} = 0$). Hence, Equation (9) can be rewritten as Equation (16) to calculate Q_{NAI}, and Q_{PFM}.

$$\begin{aligned} Q_{NAI} &= \frac{Q_P(T_P - T_\infty) - Q_{UAI}(T_{UAI} - T_\infty)}{(T_{NAI} - T_\infty)} \\ Q_{PFM} &= \frac{Q_P(T_{NAI} - T_P) - Q_{UAI}(T_{NAI} - T_{UAI})}{(T_{NAI} - T_\infty)} \end{aligned} \tag{16}$$

We focus on a remarkably stable 24-h data range from 4 September (12:00) to 5 September (12:00) (Figure 9). Indeed, the sinusoidal shapes of T_{UAI} and T_{NAI} are fully damped by the 24-h moving averaging, while T_∞, T_P, and Q_{UAI} remain rather constant.

The bottom panel of Figure 9 displays the results obtained for Q_{NAI} and Q_{PFM}. They lead to mean values of $Q_{NAI} = 0.070$ m^3/s and $Q_{PFM} = 0.219$ m^3/s. The first agrees well with the geochemical results presented in Section 3 while the second is only a few L/s higher than the base flow $Q_{PFMB} = 0.216$ m^3/s obtained from Equation (11) for t = 5 September (00:00). This seems indicate that, a few hours after stabilization of the hydraulic head, the PFM contribution due to drawdown, Q_{PFMD}, is very low.

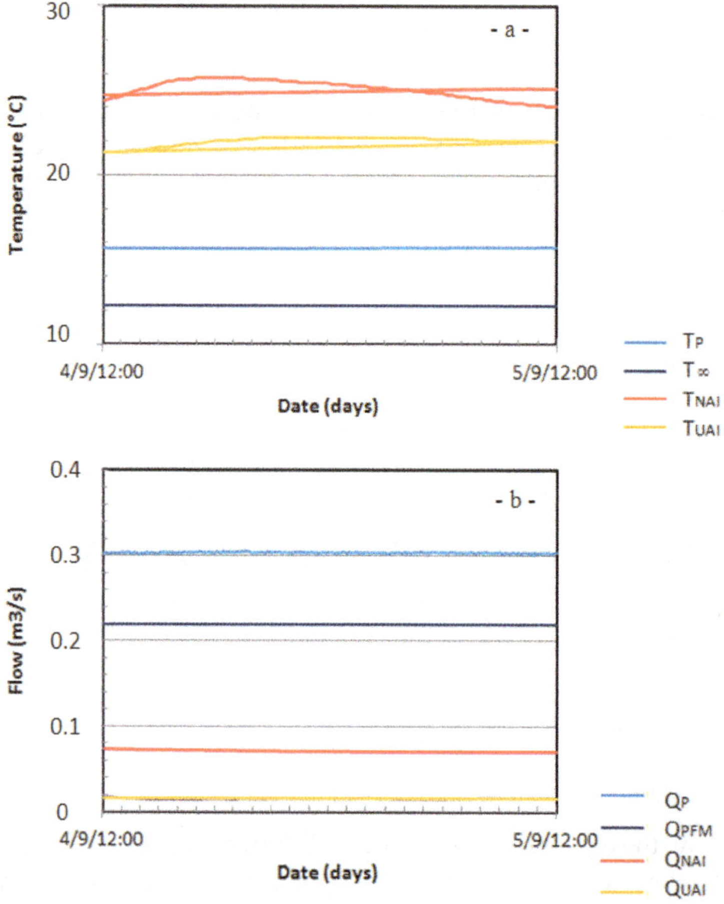

Figure 9. (**a**) Temperatures and (**b**) flows recorded or calculated during the equilibrium-pumping (see Table 1). The conduit system temperature (T_{CS}) and the matrix-conduit flow temperature (T_P) are stable over this period while the Upstream Allogenic Intrusion temperature (T_{UAI}) and the Neighbor Allogenic Intrusion temperature (T_{NAI}) display sinusoidal diurnal trends that are smoothed by the 24-h moving averaging. Bottom: The Upstream Allogenic Intrusion recharge (Q_{UAI}) and the pump discharge (Q_P) remain almost constant during Period (g). Solving of Equation (16) allows calculating the Neighbor Allogenic Intrusion (Q_{NAI}) and the recharge of the Matrix-Conduit flow (Q_{PFM}).

4.2.2. Assessment of Error Due to Data Variability

A procedure similar to the one described in Period (a) has been applied to Period (g) assessing the error due to data variability. The mean values of the parameters obtained on this interval have been introduced in the reduced form of Equation (10) to calculate the propagation of these errors through the resolution process (Equation (17)).

$$\begin{aligned} \delta Q_{NAI} &= \frac{1}{(T_{NAI}-T_{\infty})^2}\begin{bmatrix} (Q_{UAI}(T_{UAI}-T_{\infty}) - Q_P(T_P-T_{\infty}))\,\delta T_{NAI} \\ -\delta Q_{UAI}(T_{UAI}-T_{\infty})(T_{NAI}-T_{\infty}) \\ -Q_{UAI}(T_{NAI}-T_{\infty})\delta T_{UAI} \end{bmatrix} \\ \delta Q_{NAI} &= -0.0056\,\delta T_{NAI} - 0.760\,\delta Q_{UAI} - 0.0012\,\delta T_{UAI} \\ \delta Q_{PFM} &= 0.0056\,\delta T_{NAI} - 0.240\,\delta Q_{UAI} + 0.0012\,\delta T_{UAI} \end{aligned} \quad (17)$$

In this analysis, we will consider that pump gauging error δQ_P is negligible in front of δQ_{UAI}. Equation (17) shows that 1 °C of error on T_{NAI} induces 5.6 L/s of error on Q_{NAI} or Q_{PFM}; that of 1 L/s on Q_{UAI} results in 0.760 L/s; and that 1 °C on T_{UAI} induces 1.2 L/s. When the standard error of Table 4 are introduced in Equation (17), the error falls to a few L/s (a few %) (Table 4, Period (g)).

4.2.3. Assessment of Error Due the Conservative Temperature Assumption

During the equilibrium pumping, but also during constant pumping, most of the CS mixing occurs near the pump, in the top part of the chimney just beside Hérault (Figure 2). This proximity causes the diurnal oscillation of TP that occur with a delay of 20 h (Figure 4). The drawdown changes significantly the configuration of the CV with a mixing zone close to Hérault. The distance between the river and the chimney that contains the pump is less than 200 m. We have to take these consequences of the drawdown to assess the error due to the conservative temperature approximation during heavy pumping. Therefore, using these delay and distance to estimate the properties of the CV, the CS Peclet number and the Conduit Reynolds number fall respectively to $Pe = 3.9 \times 10^6$ and $Re = 2.78 \times 10^4$ (Table 5). The ratio of (PFM to Water) thermal diffusivities and the Prandtl number remain unchanged. We will use this new dimensionless configuration revisiting the data on Periods (e) and (g). (With these new values the abacus curves of Figure 4 of [36] tell that ε' reaches around 0.002 at the exit of the CS.

$$\delta T = \varepsilon' \left[(T_P + \frac{T_\infty}{(T_{NAI} - T_\infty)}) \right] \quad (18)$$

With $\varepsilon' = 0.002$, $T_P = 288.88$ K, $T_{NAI} = 298.25$ K and $T_\infty = 285.35$ K (Table 4, Period (g)) δT reaches 0.62 °C. Through Equation (17), it induces a 0.003 m^3/s error for Q_{PFM}. Similarly to the comparison of error on Period (a), both methods of error assessment lead to consistent results.

4.3. Revisiting Q_{PFMD} from Period (e)

4.3.1. Revisiting the Data

The knowledge of Q_{PFMB} and Q_{NAI} makes possible a backward analysis of the data recorded during Periods (c) and (e). Indeed, Q_{PFMB} forms the background over which it is possible to assess Q_{PFMD}. On the other hand, the vadose character of the Hérault intrusion results in a constant amplitude despite an increasing drawdown [26] (p. 191). These two previous results bring two "corner stones" situations separated by the constant pumping sequence. This allows us calculating two unknown flows: Q_{PFM} that includes the drawdown induced contribution Q_{PFMD} and the dewatering of the CS (Q_{CS}). The linear increase of the hydraulic head with time over Periods (c) and (e) (Figure 5) suggests a constant dewatering rate, bolstering us to assume a near steady CV situation. The input flows are Q_{UAI}, Q_{NAI}, Q_{PFMB}, Q_{PFMD} and the output flows are Q_{CS} and Q_P.

In order to maintain these "constant" conditions at best, our data processing skipped the data 24 h before and after the stopping. This allowed avoiding the transient phenomena observed at the restarting of the pump for temperature (Figure 4) and discharges. Consequently, we focused our analysis from 10 August (00:00) to 1 September (19:10). Within the context, Equation (9) can be rewritten as Equation (17).

$$\begin{aligned} Q_{PFM} &= \frac{Q_P(T_P - T_{CS}) - Q_{UAI}(T_{UAI} - T_{CS}) - Q_{NAI}(T_{NAI} - T_{CS})}{(T_\infty - T_{CS})} \\ Q_{CS} &= \frac{Q_P(T_\infty - T_P) - Q_{UAI}(T_\infty - T_{UAI}) - Q_{NAI}(T_\infty - T_{NAI})}{(T_\infty - T_{CS})} \end{aligned} \quad (19)$$

The brutal increase of the temperature records in the Cge borehole shows that, as soon as Hérault intrudes through the horizontal shallow branch of the CS, T_{CS}(Cge) is less representative of the dewatering temperature T_{CS}. Therefore, we will alternatively consider that the temperature recorded at the pump may represents another assessment T_{CS}(f3). This assumption seems reasonable since T_{CS}(f3) is only a few degrees higher than the T_{CS} temperature obtained on Period (a) before the mixing with the hot Hérault water.

The results of calculations for Q_{PFM} and Q_{CS} are presented on the lower diagram of Figure 10. From Q_{PFM} and Q_{PFMB} (Equation (13)), it is easy to calculate Q_{PFMD}. For both T_{CS}(Cge) and T_{CS}(f3) assumptions, Q_{PFMD}(Cge) and Q_{PFMD}(f3) curves display diurnal oscillations but do not display clear increasing trends despite of drawdown deepening. Figure 10 shows that Q_{PFMD}(Cge), Q_{PFMD}(f3), Q_{CS}(Cge) and Q_{CS}(f3) are clearly affected by the diurnal and meteorological oscillations on Q_{NAI} and Q_{UAI}. Considering the most advantageous situation T_{CS}(Cge), Q_{PFMD} ranges between 0.030 and 0.050 m^3/s. However, this drawdown induced contribution is probably overestimated because of a too hot dewatering temperature assumption T_{CS}. Indeed, the calculation forces the system to equilibrate on the pump temperature. This will increase the low temperature contribution Q_{PFM} at the expense of the hot dewatering Q_{PFM}. On the other hand Q_{PFMD}(f3), computed taking the pump temperature as T_{CS} seems really too low since it would induces a zero or even negative contribution of Q_{PFM} to drawdown. In any case, these results seem consistent with short transient flows coming with the increase of the hydraulic head. These conclusions are consistent with the ones of Maréchal et al. [34], and consistent with the shortness of the flows observed at the beginning of the equilibrium-pumping phase.

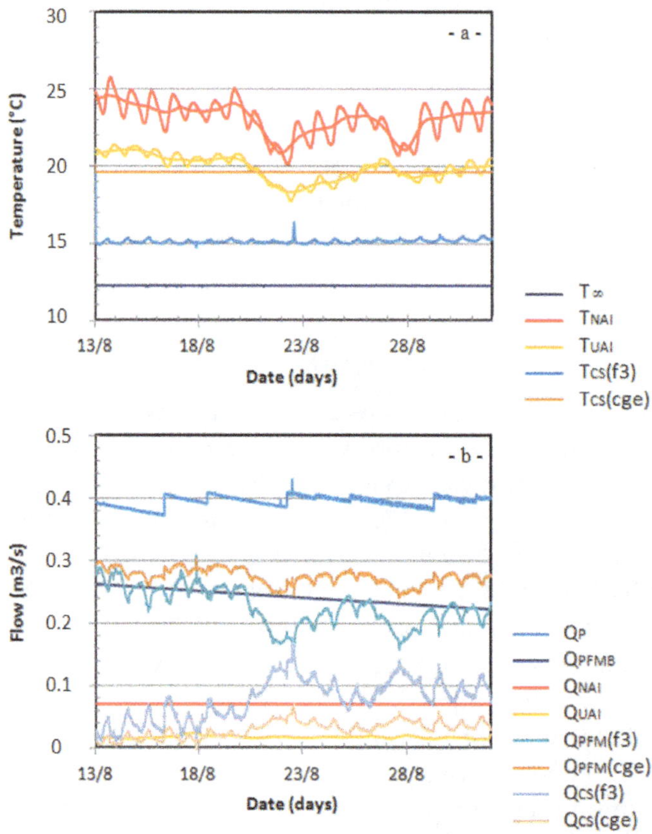

Figure 10. (**a**) Temperatures and (**b**) flows recorded or calculated during the constant pumping (see Table 1). Top: The records of the conduit system temperature (T_{CS}) in CGE borehole have been interrupted after its drawdown disconnection 13 August (12:15). Two extreme hypotheses have been considered. The first extrapolating the last value measured T_{CS}(CGE) = 19.67 °C; the second assuming T_{CS}(F3) = T_P. The temperature recorded at the pump output displays a low amplitude sinusoidal oscillation due to the direct intrusion of the Hérault intrusion. Concurrently, the matrix-conduit flow temperature remains constant. Bottom: Assessment of the matrix-conduit flow Q_{PFM} and of Q_{CS} corresponding to the dewatering of the CS. Q_{NAI} is considered as constant. The matrix-conduit flow Q_{PFM} gathers the base flow of the resurgence (Q_{PFMB}) and the supplementary contribution of matrix-conduit flow induced by drawdown Q_{PFMD}.

4.3.2. Assessment of Error Due to Data Variability

Following the same procedure, the mean values of the parameter over this interval (Table 4—Period (e)) have been introduced into the reduced form of Equation (10) to establish the formal and numerical forms of Equation (20).

$$\delta Q_{PFM} = \frac{1}{(T_{CS}-T_\infty)^2}\begin{bmatrix} \delta T_{CS}(Q_P(T_P - T_\infty) - Q_{UAI}(T_{UAI} - T_\infty) - Q_{NAI}(T_{NAI} - T_\infty)) \\ -\delta Q_{UAI}(T_{UAI} - T_{CS})(T_\infty - T_{CS}) - \delta T_{UAI}Q_{UAI}(T_\infty - T_{CS}) \\ -\delta Q_{NAI}(T_{NAI} - T_{CS})(T_\infty - T_{CS}) - \delta T_{NAI}Q_{NAI}(T_\infty - T_{CS}) \end{bmatrix} \tag{20}$$

$$\delta Q_{PFM} = 0.0044\,\delta T_\infty + 0.039\,\delta Q_{UAI} + 0.0024\,\delta T_{UAI} + 0.447\,\delta Q_{NAI} + 0.0092\,\delta T_{NAI}$$

$$\delta Q_{CS} = -0.0044\,\delta T_\infty - 1.289\,\delta Q_{UAI} - 0.0024\,\delta T_{UAI} - 1.697\,\delta Q_{NAI} - 0.0092\,\delta T_{NAI}$$

One °C of error on T_{CS} value induces 4.4 L/s of error for Q_{PFM}. Finally, 1 °C of error on T_{UAI} induces 2.4 L/s of error and one °C on T_{NAI} results in 9.2 L/s of error. The introduction of the Standard error (24SE—Table 4) in Equation (20) leads to 14 to 21 L/s of error on Q_{PFM} and on Q_{CS}. Thus, small amplitude of Q_{PFMD} makes the error is of the same order than Q_{MCD} itself.

4.3.3. Assessment of Error due to the Conservative Temperature Assumption

As mentioned above, we consider that the hydrological configuration induced by drawdown is the same the one of Period (g). Then, the CS Peclet number and the Conduit Reynolds number remain the same and, ε′ keeps the same value around 0.002 at the exit of the CS.

$$\delta T = \varepsilon'\left[(T_P + \frac{T_\infty}{(T_{NAI} - T_\infty)})\right] \tag{21}$$

Therefore, the error on temperature balance due to the conservative approximation δT remains of the order 0.62 °C. Such temperature error would induce several tens liters by second of errors.

5. Summary and Discussion

The heat and matter exchanges occurring through the karstic boundaries impose a consideration of the Open Thermodynamic System (OTS). Within this framework, the first principle of thermodynamics leads to an enthalpy balance between flows entering and leaving the CV. Combining this property with mass conservation leads to systems of two equations involving the flows and temperatures. If formal physical conditions (steady states) are never achieved in nature, they are approached during particular periods as recession or certain phases of the pumping tests. These periods have been used these to calculate "corner stones" descriptions of the hydraulic regime between which we extend the results. The restriction of our model to the quietest part of the recession period cuts off the risk of disruption an approximate steady state by flooding and the possibility of reverse flow from CS to PFM. That way the periods of data revisiting were chosen outside of the recovering and the analysis has been stopped as soon as the flood event of 6 September occurred.

Revisiting the data recorded during these three periods with our theoretical analyses leads to a recalculation of consistent hydrological behavior of the karstic system. Thus, the speleological, hydrological or geochemical observations validated by previous studies [26,34] are not refuted despite of a less optimistic yield of the resurgence. This is particularly the case for spring drying during the step drawdown sequence, the results of geochemical analyses or the base flow recession of the resurgence.

Even though it is never perfectly reached in nature, the steady state approximation is necessary to use the method. While such state is assumed, temperatures equilibrate at the PFM/Conduit interface. However, this situation does not mean cancelling of embedding rocks/water heat transfer at the interface. Indeed, as shown in Machetel and Yuen 2015, PFM temperature gradient is maintained by the advection of cold, far field, water that counteracts the heat diffusion from CS to PFM through the wall. During the recession period, mixing of intrusive flows hotter than far field temperature, results in CS temperature hotter than in PFM. However, when a steady state is reached (or approached), all these local conductive effects (inside PFM and CS but also between PFM and CS) are taken into account by the final enthalpy balance. This is the most interesting property of the OTS approach that refers to the comparison of integrated incoming and leaving external heat sources and not on the local thermal properties inside of the "black box". It is clear that what we called "*conservative temperature assumption*" may better be called as "*conservative enthalpy approximation*". However since we assumed that no other sources of heat are present (neither chemical heat, nor work conversion to heat); enthalpy conservation comes down on to temperature conservation.

Thus, the application of the method to the first period (before pumping) of the Cent-Fonts pumping test experiments allowed assessing the basic recession flow of the resurgence. Then, the equilibrium-pumping allowed assessing the Hérault intrusion and the supplementary contribution induced by drawdown. The analysis shows that errors induced by unsteadiness reach a few tens of liters by second. They are of the same order of magnitude than the errors induced by conservative approximation itself.

In conclusion, we have confirmed the validity for using the thermometric method by the field observations that never contradict the results. It would be advantageous; insofar data exists, to check the method on other sites. It could also be interesting and of relatively low additional costs to develop recording and processing of temperature profiting of next pumping experiments. Indeed, as it does not require sophisticated equipment or procedures, the additional costs should remain low compared to drilling and pumping operations.

Our works may open the opportunity of using the steadiest part of the pumping test sequences to calibrate the global operating mode of complex resurgence system. Combining energy equation and mass conservation equations, temperature measurements in surface waters and boreholes may allow assessing the flow properties in borehole or mixing in karst CS. It seems therefore constitute an efficient tool to separate and calculate the karstic properties and could be a promising tool worthwhile to apply on other sites.

Acknowledgments: The authors thank "le Conseil Général de l'Hérault" and the PREVHE association for providing the complete set of digital data of the 2005 Cent-Fonts pumping test campaign and financing ICPMS analyses. We also thank Jean-Marc Luck and Olivier Bruguier for their active participation to the analysis of water samples on the ICPMS "Turbo PQ2 Quadruple Plasma VG Plus" of Montpellier University and their interpretation. We also thank the two anonymous referees, which comments contributed to improve this paper. This work benefited of a grant from China University of Geosciences.

Author Contributions: Philippe Machetel and David Yuen are both involved in numerical solving of non-linear differential equations, thermodynamics and physics of fluids.

Conflicts of Interest: The authors declare no conflict of interest.

Appendix A

A.1. Acronyms

CS	Conduit System of fluviokarstic system
CV	Control volume of the OTS
OTS	Open Thermodynamic System
PFM	Porous Fractured Matrix

A.2. Cent-Fonts Fluviokarst Flows, White's and Pumping Test Notations

A.2.1. CS Inflows (m^3/s)

Q_{UAI}	Upper allogenic stream intrusion (Buèges intrusion at swallow zone—M)
Q_{NAI}	Neighbor allogenic stream intrusion (Hérault intrusion at base level—C)
Q_{PFM}	Porous Fractured Matrix to CS flow (C)
Q_{PFMB}	Porous Fractured Matrix to CS base flow (C)
Q_{PFMD}	Porous Fractured Matrix to CS drawdown induced flow (C)
Q_D	Drawdown induced Porous Fractured Matrix to CS flow (C)

A.2.2. CS outflows (m^3/s)

Q_S	Spring discharge (Resurgence discharge in Hérault M or C)
Q_P	Pump discharge (Pump discharge in Hérault during pumping tests—M)
Q_{CS}	CS Dewatering corresponding to the lowering of water table (C)

A.2.3. Temperatures (°C)

T_{UAI}	Upper allogenic stream temperature (M)	-Buèges temperature (M)
T_{NAI}	Neighbor allogenic stream temperature (M)	-Hérault temperature (M)
T_P	Pump output temperature (M)	
T_{CS}	CS water temperature (C or M)	
T_∞	Far field temperature-P7 borehole temperature (M)	

A.3. Equations Notations (units)

v	Fluid velocity vector	(m/s)
q_i	CS algebraic value (positive for incoming flow)	(m^3/s)
q_o	CS algebraic value (negative for outgoing flow)	(m^3/s)
δE	CS Internal Energy	(J)
j	CS incoming or outgoing flows	(-)
h_j	Specific enthalpy of CS flow j	(J/kg)
$e_{pot,j}$	Potential energy of CS flow j	(J/kg)
$e_{cin,j}$	Kinetic energy of CS flow j	(J/kg)
$\delta\Phi_j$	CS-PFM Thermal exchanges	(J)
δ_{Wj}	CS-PFM Work exchanges	(J)
ρ_j	Water density in flow j	(kg/m^3)
C_{pj}	Specific thermal capacity in flow	(J/kg K)
T_i	Temperature of CS incoming flow	(°K or °C)
T_o	Temperature of CS incoming flow	(°K or °C)

A.4. Error Assessment Notations, (Units), Values

D_m	PFM thermal diffusivity	(m^2/s)	1.42×10^{-6}
D_w	Water thermal diffusivity	(m^2/s)	1.43×10^{-7}
v	Water kinematic viscosity	(m^2/s)	10^{-6}
L	CS Length	(m)	5×10^3
r_h	Half hydraulic radius	(m)	5
V	Velocity scale	(m/s)	
D	Thermal diffusivity ratio (*Dm/Dw*)	(-)	
Pe	Conduit Peclet number (*LV/Dw*)	(-)	
Pr	Prandtl number (*v/Dw*)	(-)	
Red	CS Reynolds number (*2 V·rh)/v*	(-)	
δT	Conservative hypothesis error	(C or K)	

References

1. Constantz, J. Interaction between stream temperature, streamflow, and groundwater exchanges in Alpine streams. *Water Resour. Res.* **1998**, *34*, 1609–1615. [CrossRef]
2. Constantz, J. Heat as a tracer to determine streambed water exchanges. *Water Resour. Res.* **2008**, *44*, 2008. [CrossRef]
3. Tabbagh, A.; Bendjoudi, H.; Benderitter, Y. Determination of recharge in unsaturated soils using temperature monitoring. *Water Resour. Res.* **1999**, *35*, 2439–2446. [CrossRef]
4. Benderitter, Y.; Roy, B.; Tabbagh, A. Flow characterization through heat transfer evidence in a carbonate fractured medium: First approach. *Water Resour. Res.* **1993**, *29*, 3741–3747. [CrossRef]

5. Genthon, P.; Bataille, A.; Fromant, A.; D'Hulst, D.; Bourges, F. Temperature as a marker for karstic waters hydrodynamics. Inferences from 1 year recording at La Peyrere cave (Ariège, France). *J. Hydrol.* **2005**, *311*, 157–171. [CrossRef]
6. Genthon, P.; Wirrmann, D.; Hoibian, T.; Allenbach, M. Steady water level and temperature in a karstic system: The case of the coral Lifou Island (SW Pacific). *C. R. Geosci.* **2008**, *340*, 513–522. [CrossRef]
7. Karanjac, J.; Altug, A. Karstic spring recession hydrograph and water temperature analysis—Oymapinar Dam Project, Turkey. *J. Hydrol.* **1980**, *45*, 203–217. [CrossRef]
8. Stonestrom, D.A.; Constantz, J. Using temperature to study stream-ground water exchanges. *US Geol. Surv. Fact Sheet* **2004**, *3010*, 4.
9. O'Driscoll, M.A.; DeWalle, D.R. Stream-air temperature relations to classify stream-ground water interactions. *J. Hydrol.* **2006**, *329*, 140–153. [CrossRef]
10. Mosheni, O.; Stefan, H.G. Stream temperature/air temperature relationship: A physical interpretation. *J. Hydrol.* **1999**, *218*, 128–141. [CrossRef]
11. Bogan, T.; Mosheni, O.; Stefan, H.G. Stream temperature-equilibrium temperature relationship. *Water Resour. Res.* **2003**, *39*, 1245. [CrossRef]
12. Bogan, T.; Stefan, H.G.; Mosheni, O. Imprints of secondary heat sources on the stream temperature equilibrium temperature relationship. *Water Resour. Res.* **2004**, *40*, W12510. [CrossRef]
13. Sinokrot, B.A.; Stefan, H.G. Stream temperature dynamics measurements and modeling. *Water Resour. Res.* **1993**, *29*, 2299–2312. [CrossRef]
14. Luetscher, M.; Jeannin, P.Y. Temperature distribution in karst systems: The role of air and water fluxes. *Terra Nova* **2004**, *16*, 344–350. [CrossRef]
15. Dogwiller, T.; Wicks, C. Thermal variations in the hyporheïc zone of a karst stream. *Int. J. Speleol.* **2006**, *35*, 59–66. [CrossRef]
16. Petelet, E.; Luck, J.M.; Ben Ohtman, D.; Negrel, P.; Aquilina, L. Geochemistry and water dynamics of a medium-sized watershed: The Hérault, southern France 1. Organization of the different water reservoirs as constrained by Sr isotopes, major, and trace elements. *Chem. Geol.* **1998**, *150*, 63–83. [CrossRef]
17. Dubois, P. Étude des réseaux souterrains des rivières Buèges et Virenque (le Languedoc Bas). In Proceedings of the 2e Congrès International de Spéléologie, Salerne, Italy, 2–12 October 1958.
18. Paloc, H. *Carte Hydrogéologique de la France, Région Karstique Nord-Montpelliéraine, Notice Explicative*; Bureau de Recherches Géologiques et Minières: Orléans, France, 1967.
19. Camus, H. Formation des réseaux karstiques et creusements des vallées: L'exemple du Larzac méridional, Hérault, France. *Karstologia* **1997**, *29*, 23–42.
20. Schoen, R.; Bakalowicz, M.; Ladouche, B.; Aquilina, L. Caractérisation du Fonctionnement des Systèmes Karstiques Nord-Montpelliérains. Rap. BRGM R40939RP. 1999, Volume III, p. 91. Available online: http://infoterre.brgm.fr/rapports/RR-40939-FR.pdf (accessed on 27 November 2016).
21. Aquilina, L.; Ladouche, B.; Bakalowicz, M.; Schoen, R.; Petelet, E. Caractérisation du Fonctionnement des Systèmes Karstiques Nord-Montpelliérains. Synthèse Générale. Rap. BRGM R40746. 1999, p. 50. Available online: http://infoterre.brgm.fr/rapports/RR-40746-FR.pdf (accessed on 27 November 2016).
22. Petelet-Giraud, E.; Dörfliger, N.; Crochet, P. RISK: Méthode d'évaluation multicritère de la vulnérabilité des aquifères karstiques. Application aux systèmes des Fontanilles et Cent-fonts (Hérault, sud de la France). *Hydrogéologie* **2000**, *4*, 71–88.
23. Ladouche, B.; Dörfliger, N.; Pouget, R.; Petit, V.; Thiery, D.; Golaz, C. Caractérisation du Fonctionnement des Systèmes Karstiques Nord-Montpelliérains, Rapport du Programme 1999–2001. Buèges, Rap. BRGM 51584 Fr RP. 2002. Available online: http://infoterre.brgm.fr/rapports/RP-51584-FR.pdf (accessed on 27 November 2016).
24. Petelet-Giraud, E. Dynamic scheme of water circulation in karstic aquifers as constrained by Sr and Pb isotopes. Application to the Hérault watershed, Southern France. *Hydrogeol. J.* **2003**, *11*, 560–573. [CrossRef]
25. Aquilina, L.; Ladouche, B.; Dörfliger, N. Recharge processes in karstic systems investigated through the correlation of chemical and isotopic composition of rain and spring-waters. *Appl. Geochem.* **2005**, *20*, 2189–2206. [CrossRef]
26. Ladouche, B.; Maréchal, J.C.; Dörfliger, N.; Lachassagne, P.; Lanini, S.; Le Strat, P. Pompage D'essai sur le Système Karstique des Cent-Fonts (Commune de Causse de la Selle, Hérault), Présentation et Interprétation

des Données Recueillies. Rap. BRGM RP54426-FR. 2005. Available online: http://infoterre.brgm.fr/rapports/RP-54426-FR.pdf (accessed on 27 November 2016).

27. Aquilina, L.; Ladouche, B.; Dörfliger, N. Water storage and transfer in the epikarst of karstic systems during high flow periods. *J. Hydrol.* **2006**, *327*, 472–485. [CrossRef]
28. Elguero, E. *Les Grandes Cavités Héraultaises*; A.V.L. Diffusion: Montpellier, France, 2004; p. 44.
29. Smart, C.C. A deductive model of karst evolution based on hydrological probability. *Earth Surf. Process. Landf.* **1988**, *13*, 271–288. [CrossRef]
30. Malcom, A. *Lexicon of Cave and Karst Terminology with Special Reference to Environmental Karst Hydrology. Supercedes EPA/600/R-99/006, 1/'99*; National Center for Environmental Assessment, Office of Research and Development, U.S. Environmental Protection Agency: Washington, DC, USA, 2002; p. 214.
31. White, W.B. Karst hydrology: Recent developments and open questions. *Eng. Geol.* **2002**, *65*, 85–105. [CrossRef]
32. White, W.B. Conceptual models for karstic aquifers. *Speleog. Evol. Karst Aquifers* **2003**, *1*, 6.
33. Jambac, F. *Essais Complémentaire sur le Site des Cent-Fonts*; Technical repport; Compagnie Générale de Eaux: Montpellier, France, 1994.
34. Maréchal, J.C.; Ladouche, B.; Dörfliger, N.; Lachassagne, P. Interpretation of pumping tests in a mixed flow karst system. *Water Resour. Res.* **2008**, *44*, W05401. [CrossRef]
35. Machetel, P.; Yuen, D.A. Open thermodynamic system concept for fluviokarst underground temperature and discharge flow assessments. In Proceedings of the H11F-1257 2012 Fall Meeting AGU, San Francisco, CA, USA, 3–7 December 2012.
36. Machetel, P.; Yuen, D.A. Evaluation of first order error induced by conservative-tracer temperature approximation for mixing in karstic flow. In *Sinkholes and the Engineering and Environmental Impacts, Proceedings of the Fourteenth Multidisciplinary Conference, Rochester, MN, USA, 5–9 October 2015*; Doctor, D.H., Land, L., Stephenson, J.B., Eds.; National Cave and Karst Research Institute: Carlsbald, NM, USA, 2015; pp. 537–548.
37. Covington, M.D.; Luhmann, A.; Gabrovsek, F.; Saar, M.O.; Wicks, C.M. Mechanisms of heat exchange between water and rock in karst conduit. *Water Resour. Res.* **2011**, *47*, W10514. [CrossRef]
38. Covington, M.D.; Luhmann, A.J.; Wicks, C.M.; Saar, M.O. Process length scales and longitudinal damping in kart conduits, Mechanisms of heat exchange between water and rock in karst conduit. *J. Geophys. Res.* **2012**, *117*, P01025. [CrossRef]
39. Van Wylen, G.J.; Sonntag, R.E. *Fundamental of Classical Thermodynamics*; John Wiley and Sons: New York, NY, USA, 2013.
40. Vidal, J. *Thermodynamique, Application au Génie Chimique et à L'industrie Pétrolière, Technipp*; Institut Français du Pétrole: Paris, France, 1997; p. 500.
41. Samani, N.; Ebrahimi, B. Analysis of spring hydrographs for hydrogeological evaluation of a karst aquifer system. *Theor. Appl. Karstol.* **1996**, *9*, 97–112.
42. Patterson, C. Lead in sea water. *Sciences* **1974**, *183*, 553–558. [CrossRef] [PubMed]
43. Vidal, P. *Géochimie*; Dunod: Paris, France, 1998; p. 190.
44. Tunusoglu, O.; Shahwan, T.; Eroglu, A.E. Retention of aqueous Ba(2+) ions by calcite and aragonite over a wide range of concentrations: Characterization of the uptake capacity, and kinetics of sorption and precipitate formation. *Geochem. J.* **2007**, *41*, 379–389. [CrossRef]
45. Lu, C.; Shu, L.; Wen, Z.; Chen, X. Interpretation of a short-duration pumping test in the mixed flow karst system using a three-reservoir model. *Carbonates Evaporites* **2013**, *28*, 149–158. [CrossRef]
46. Opsahl, S.P.; Chapal, S.E.; Hicks, D.W.; Wheeler, C.K. Evaluation of ground-water and surface-water exchanges using streamflow difference analyses. *J. Am. Water Resour. Assoc.* **2007**, *43*, 1132–1141. [CrossRef]

3

Spatial and Temporal Variability of Potential Evaporation across North American Forests

Robbie A. Hember [1,2,*], Nicholas C. Coops [1] and David L. Spittlehouse [3]

[1] Faculty of Forestry, University of British Columbia, 2424 Main Mall, Vancouver, BC V6T 1Z4, Canada; nicholas.coops@ubc.ca

[2] Pacific Forestry Centre, Canadian Forest Service, Natural Resources Canada, 506 West Burnside Road, Victoria, BC V8Z 1M5, Canada

[3] Competitiveness and Innovation Branch, Ministry of Forests, Lands and Natural Resources Operations, Victoria, BC V8W 9C2, Canada; dave.spittlehouse@gov.bc.ca

* Correspondence: robert.hember@canada.ca

Abstract: Given the widespread ecological implications that would accompany any significant change in evaporative demand of the atmosphere, this study investigated spatial and temporal variation in several accepted expressions of potential evaporation (PE). The study focussed on forest regions of North America, with 1 km-resolution spatial coverage and a monthly time step, from 1951–2014. We considered Penman's model (E_{Pen}), the Priestley–Taylor model (E_{PT}), 'reference' rates based on the Penman–Monteith model for grasslands (E_{RG}), and reference rates for forests that are moderately coupled (E_{RFu}) and well coupled (E_{RFc}) to the atmosphere. To give context to the models, we also considered a statistical fit (E_{PanFit}) to measurements of pan evaporation (E_{Pan}). We documented how each model compared with E_{Pan}, differences in attribution of variance in PE to specific driving factors, mean spatial patterns, and time trends from 1951–2014. The models did not agree strongly on the sensitivity to underlying drivers, zonal variation of PE, or on the magnitude of trends from 1951–2014. Sensitivity to vapour pressure deficit (D_{a}) differed among models, being absent from E_{PT} and strongest in E_{RFc}. Time trends in reference rates derived from the Penman–Monteith equation were highly sensitive to how aerodynamic conductance was set. To the extent that E_{PanFit} accurately reflects the sensitivity of PE to D_{a} over land surfaces, future trends in PE based on the Priestley–Taylor model may underestimate increasing evaporative demand, while reference rates for forests, that assume strong canopy-atmosphere coupling in the Penman–Monteith model, may overestimate increasing evaporative demand. The resulting historical database, covering the spectrum of different models of PE applied in modern studies, can serve to further investigate biosphere-hydroclimate relationships across North America.

Keywords: potential evaporation; evaporative demand; pan evaporation; climate trends

---

1. Introduction

The evaporating power of the atmosphere, or simply evaporative demand, exerts a primary control on the distribution, abundance, and diversity of life on land [1–3]. On one hand, the atmospheric demand for water allows plants to draw water above the surface without exerting vast amounts of energy. On the other, fluctuations in evaporative demand simultaneously pose a risk to plants if soil water supply and transport efficiency cannot meet peak evaporative demand [4–6]. These compensating factors appear to put forest biomes at risk under anticipated climate change [7–11]. While declining precipitation must impose major constraints on productivity of trees across the forest-grassland ecotone, there is a growing appreciation that evaporative demand also strongly constrains productivity of humid forest biomes [12–14]. For these reasons, evaporative demand is an important flux in the surface water balance and a primary determinant of land climates [1,3,15,16].

Potential evaporation (PE) is a common measure of evaporative demand and is defined as the rate of evaporation that would occur from a surface with unlimited water supply and no resistance to transfer of water. It is, therefore, determined by the energy available to change the phase of water from liquid to gas [17–20]. Although PE has a conceptual definition, it is not easy to measure directly above land surfaces [21]. Inferences can be made from measurements of evaporation from lakes, lysimeters, evaporation pans, or models. For example, Class A pan evaporation was measured continuously at many meteorological stations across Canada and the U.S. and has been used to assess climatological variability of PE [22,23].

Despite extensive investigation of evaporation from pans, lakes, and lysimeters, empirical studies have not resulted in a universal approach to predict PE. Some of the earliest models were based on temperature and were developed with the limited availability of radiation, humidity, and wind measurements in mind [16,24,25]. Other models were simultaneously developed to predict PE based on a broader number of driving variables, including net radiation (R_n), temperature (T_a), vapour pressure deficit (D_a), and wind speed (U) [26,27]. Later, Penman's model was modified to produce the Penman–Monteith equation, which accounts for varying surface conductance to water transfer [28]. While the Penman–Monteith model incorporates an aerodynamic component, other models were developed that omit an aerodynamic component. For example, it was later argued that at regional scales PE could be approximated by multiplying the equilibrium rate of evaporation by a constant value, $\alpha = 1.26$ [29]. Although empirical studies suggest that the overall magnitude of PE using this Priestley–Taylor coefficient of $\alpha = 1.26$ will bring estimates within closer proximity to models that include an aerodynamic component, spatial and temporal variability in estimates will only reflect variation in T_a and R_n. In forests, α must be modified [30,31], yet it is still a useful concept if it is recognized that there is no universal value of α. Lastly, other studies express PE based solely on a function of R_n [32], or solely based on D_a [33].

The above modelling approaches all continue to be used broadly in hydrological and ecological research. For example, temperature-based models are applied in relatively basic surface water balance models [34]. Use of Penman's model has also been applied in ecological studies [2]. The Penman–Monteith model is widely applied [35–38]. Other studies represent potential evapotranspiraiton based on the Priestley–Taylor model [30,39], or close equivalents based on R_n and T_a [25,40], or R_n alone [32], or D_a alone [41,42]. Despite its importance in modelling the surface water balance and subsequent application in ecological research, differences in the variation of evaporative demand resulting from these approaches, including long-term trends at the regional scale, are not well documented.

Historical and projected changes in PE may arise from natural climate variability, and changing atmospheric greenhouse gas and aerosol concentrations. However, the response of PE to regional climate change involves scale- and surface-dependent relationships with multiple meteorological drivers that are not always directly measured. As such, continuous high temporal and spatial resolution maps of PE and projections of future change required to study many ecological processes remain relatively scarce. For regions with long-term continuous measurements of pan evaporation, some studies report a decrease in pan evaporation [43–45]. Other studies resort to use of models to understand trends. For example, Penman's model suggested a decrease in historical PE across some regions of India [43], but also projected long-term future increases of PE. Across Australia, trends in PE inferred from a temperature-based model and Class-A pan evaporation did not agree in direction [46]. Similarly, future projections of PE based on several models, including the Priestley–Taylor model and the Penman–Monteith model, concluded that differences among models partially led to differences in the direction of future trends in aridity index [47]. Trends in PE based on the Penman–Monteith equation generally indicate increasing trends in evaporative demand [7,48–50]. Positive future trends in PE are largely attributed to increasing saturation vapour pressure that accompanies global warming [48]. Comparison of historical trends in drought using Penman's model with previous studies that applied Thornthwaite's temperature-based model of PE emphasized that the latter might exaggerate increasing trends in aridity [51]. Conversely, time trends in global land aridity of Coupled Model Intercomparison Project Phase 5 (CMIP5) model simulations based on estimates of PE inferred from just net radiation were commonly not significant [32].

There is considerable demand for a spatial and temporal index of change in evaporative demand to understand how forests may have responded over time to changing atmospheric conditions [52,53]. Consequently, we investigated the spatial and temporal variability of PE across North America using six different model-based estimates of PE. To put the magnitude and underlying mechanics of model-based estimates of PE into context, we compared predictions against measurements of pan evaporation (E_{Pan}) for 28 sites across Canada and the conterminous U.S.. One of the six estimates of PE was defined by a statistical model fit to E_{Pan} measurements, while the other five estimates reflected approaches that are widely used in applied research, but that require no calibration. First, we evaluated the agreement between models and E_{Pan}. Second, we evaluated the variance in PE that is attributed to each meteorological driver in each of the models. Lastly, to understand spatial and temporal variation in the model estimates, we developed spatially-continuous monthly mean values of the driving variables covering North America from 1951–2014. Maps were then generated to document long-term averages ('climatologies') of PE, and historical time trends in PE from 1951–2014.

2. Data and Methods

2.1. Pan Evaporation Measurements

Daily total pan evaporation (E_{Pan}) measurements from Class A pans were collected across North America. Data collected at stations in Canada were requested and received from Environment Canada during March 2014. Data collected at stations in the U.S. were downloaded from the National Oceanic and Atmospheric Administration during March 2016. Any stations with available daily measurements between 1950 and 2014 were considered.

In Canada, large periods of the pan evaporation measurements were accompanied by simultaneous measurements of pan water temperature, air temperature (T_a), and wind speed (U). Separate databases of daily mean downward global shortwave radiation (R_{sgd}) and dew-point temperature were also available. Daily mean vapour pressure deficit of the air (D_a) was calculated from air temperature and dew-point temperature. From the initial databases, ten stations in Canada had multiple years with overlapping measurements of E_{Pan} and meteorological variables (all of R_{sgd}, T_a, D_a, and U). For all U.S. stations, daily mean values of R_{sgd} were downloaded from two National Renewable Energy Laboratory (NREL) databases [54]. All other meteorological variables were downloaded from the Global Summary of the Day (GSOD) database [55]. Relatively few U.S. stations simultaneously measured pan evaporation and meteorological drivers. By manually cross-referencing between databases, we identified eighteen stations that collected pan evaporation measurements and were also within close proximity to stations with the necessary meteorological measurements (a distance of less than 50 km in most cases). Daily values of R_{sgd}, T_a, D_a, and U for each station with pan evaporation measurements were then estimated from inverse-distance weighting interpolation of measurements from the ten nearest stations. As a quality control measure, multivariate regression was used to relate pan evaporation against meteorological variables at each candidate station. Despite reliance on off-site meteorological variables in the U.S., we did not observe a significant difference in the skill or behaviour of the models between Canada and the U.S. (not shown). Combining the stations from Canada and U.S. produced a total sample of 28 stations, each consisting of multiple years of measurements for analysis (Table 1). Stations were relatively evenly distributed across North America, with the exception of Mexico (Figure 1). Although the emphasis was on documenting variability of PE for forest biomes of North America, we chose to analyze all 28 available stations.

Table 1. Station information: Identifiers (ID) from Environment Canada or NOAA, geographic coordinates, mean annual temperature (MAT) and mean annual precipitation (MAP) for the 1971–2000 base period, number of days with measurements of pan evaporation (E_{Pan}) used in the analysis (n), and time period of containing E_{Pan} measurements.

Station ID	Station Name	Lat. (°N)	Long. (°W)	Elev. (m a.s.l.)	MAT (°C)	MAP (mm·year^{-1})	N (d)	Period
2101300	Whitehorse A	60.71	135.07	706	−0.6	274	2787	1974–1996
2202800	Norman Wells	65.28	126.80	73	−4.6	294	2573	1964–1984
2403500	Resolute Cars	74.72	94.99	66	−15.2	127	2231	1962–2004
301222F	Edmonton Stony Plain	53.55	114.11	766	2.7	530	2654	1988–2004
5023222	Winnipeg Richarson Int'l A	49.92	97.23	239	3.2	545	5422	1962–1996
5060600	Churchill A	58.74	94.06	29	−6.9	385	3672	1963–2000
7025250	Pierre Elliott Trudau Intl A	45.47	73.75	36	6.3	1059	1173	1987–1994
7093GJ3	La Grande IV A	53.76	73.68	306	−5	732	955	1987–1995
7095480	Nitchequon	53.20	70.90	536	−7.6	845	2261	1966–1985
7117825	Schefferville A	54.80	66.82	522	−7.7	772	818	1962–1969
040232	Antioch Pump Plant 3	37.98	121.73	18	15.7	346	7966	1955–1978
050834	Bonny Dam 2 NE	39.63	102.18	1113	10.8	466	7601	1951–2006
090181	Allatoona Dam 2	34.17	84.73	299	16	1347	10,578	1952–2006
100448	Arrowrock Dam	43.62	115.92	988	8.7	569	6630	1951–1998
151080	Buckhorn Lake	37.35	83.38	287	13	1199	6049	1961–1994
200089	Alberta Ford For Cen	46.65	88.48	400	3.2	867	246	1969–1971
241044	Bozeman Montana SU	45.67	111.05	1482	5.9	487	3401	1951–1969
260800	Beowawe 49 S U of N Ranch	39.90	116.58	1751	8.4	263	2208	1981–1998
290417	Animas 3 ESE	31.95	108.82	1348	15.3	305	3427	1967–1979
310375	Aurora 6 N	35.38	76.78	6	16.3	1273	10,845	1973–2014
340184	Altus Dam	34.88	99.30	467	14.8	735	7927	1979–2014
391076	Brookings 2 NE	44.32	96.77	503	6	634	9470	1953–2014
410174	Alpine	30.35	103.67	1360	16.5	438	3550	1971–1982
410613	Beaumont Resrch Ctr	30.07	94.28	9	19.3	1471	12,181	1979–2014
410639	Beeville 5 NE	28.45	97.70	76	20.6	832	9254	1979–2014
420336	Arches National Park HQS	38.62	109.62	1259	14	250	3666	1980–2004
450587	Bellingham 3 SSW	48.72	122.52	9	10.3	964	4947	1985–2014
481000	Boysen Dam	43.42	108.20	1467	7.6	245	3772	1951–1976

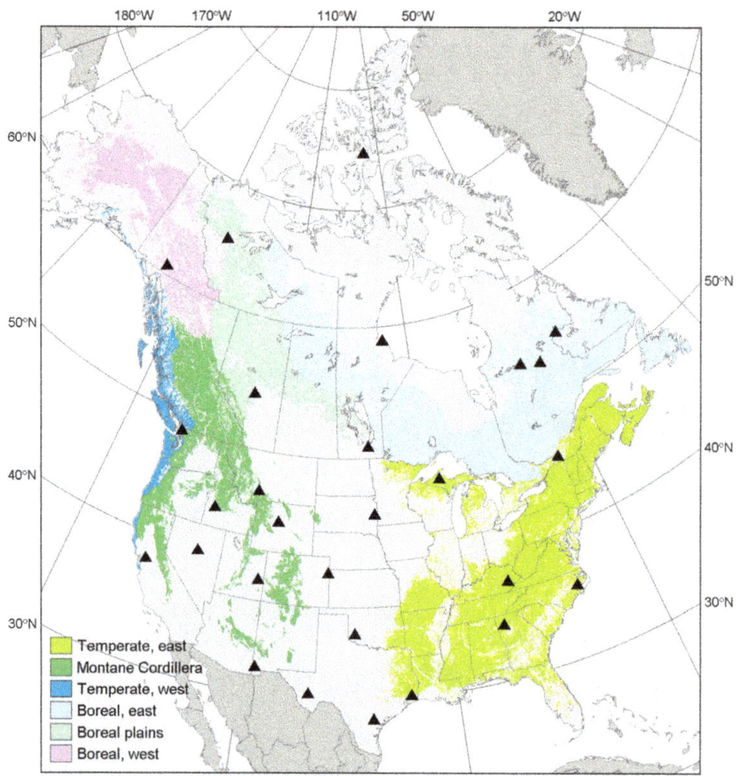

Figure 1. Location of 28 stations with pan evaporation measurements where it was possible to reconstruct meteorological variables (either from measurements at the local climate station, or from climate stations within 50 km of the pan evaporation measurements). The map also identifies the six forest ecozones for which mean trend estimates of potential evaporation from model predictions were reported.

A pan coefficient, k_{pan}, was applied to measurements of evaporation from Class A pans to account for heat transfer through the sides and bottom of the pans. Early studies in the U.S. suggested a coefficient of k_{pan} = 0.70 [23] and this was also adopted in estimates of open-water evaporation [22]. Another study for pan measurements in the conterminous U.S. reports a value of k_{pan} = 0.77 [45], while irrigation manuals report a value of k_{pan} = 0.80 for moderate wind and moderate humidity [56]. In this study, we applied the mid-range value, k_{pan} = 0.77.

2.2. Models of Potential Evaporation

Although temperature-based models may have practical value, they have limitations that have been addressed in detail elsewhere [21,25,57]; thus we restrict focus on methods of estimation that explicitly consider a radiative component. Penman's model [27] combines radiative and aerodynamic drivers of evaporation:

$$E_{\mathrm{Pen}} = \frac{s(R_{\mathrm{n}} - G) + \gamma f(U)(e_{\mathrm{s}} - e_{\mathrm{a}})}{s + \gamma} \frac{1}{\lambda_{\mathrm{L}}} \quad (1)$$

where s is the slope of the saturation vapour pressure-over-temperature relationship for air (kPa·K^{-1}), R_n is net radiation (W·m^{-2}), G is the ground heat flux (W·m^{-2}), γ is the psychrometric constant (kPa·K^{-1}), e_s is saturation vapour pressure (kPa), e_a is actual vapour pressure (kPa), $f(U)$ is the wind function, here given by $f(U) = 1.38\,U + 1.31$ [17], and λ_L is the latent heat of vapourization (J·kg^{-1}).

The equilibrium rate of evaporation was calculated according to [1,17,19,20]:

$$E_{\mathrm{EQ}} = \frac{s(R_{\mathrm{n}} - G)}{s + \gamma} \frac{1}{\lambda_{\mathrm{L}}} \quad (2)$$

where G, s, γ, and λ_L are as defined above. The Priestley–Taylor model of potential evaporation was then calculated from [29]:

$$E_{PT} = \alpha E_{EQ}. \quad (3)$$

Above wet surfaces, they argued that α = 1.26 yields a strong predictor of potential evaporation at the regional scale.

A 'reference' rate of evaporation (E_{Ref}) is commonly defined for a specific type of surface vegetation. It is generally defined as an intermediate variable between PE and actual evapotranspiration. In essence, it indicates the maximum rate of actual evapotranspiration [19], or the expected rate of actual evapotranspiration at maximum hydraulic conductivity of plants and soil. For example, the reference evapotranspiration for a hypothetical surface of vegetation with a characteristic height and surface conductance can be calculated from the Penman–Monteith equation as [35]:

$$E_{Ref} = \frac{s(R_n - G) + \rho_a c_p g_a (e_s - e_a)}{s + \gamma\left(1 + \frac{g_a}{g_s}\right)} \frac{1}{\lambda_L} \quad (4)$$

where ρ_a is the density of air (kg·m^{-3}), and c_p is the specific heat of the air (J·kg^{-1}·K^{-1}), g_a is aerodynamic conductance (m·s^{-1}), and g_s is surface conductance (m·s^{-1}). In reality, surface conductance is highly variable across forest types and environments, yet here we demonstrate the behaviour of predictions from the Penman–Monteith equation when g_a and g_s are set as constants. For example, for a reference grassland, conductance values are commonly set to g_a = 0.010 m·s^{-1} and g_s = 0. 014 m·s^{-1}. In so doing, the reference evapotranspiration rate specifically for grasslands (E_{RG}) represents the expected rate of actual evaporation in the absence of water stress.

Values of aerodynamic and surface conductance for the reference grassland are not representative of forest canopies [58,59]. We, therefore, considered application of the Penman–Monteith model to simulate reference values for forest canopies. Variation in forest canopy structure affects the conversion of sensible heat from the surrounding air into latent heat [60–63]. As surface roughness increases, so does the potential extraction of sensible heat and the surface is said to be increasingly 'coupled' to the atmosphere. Conversely, as surface roughness decreases, less sensible heat can be extracted from the atmosphere and the surface is said to be increasingly 'uncoupled' from the atmosphere [28,60,64]. We considered two parameterizations, including forest canopies that are moderately coupled to the atmosphere (E_{RFu}) and those that are well coupled to the atmosphere (E_{RFc}). For canopies that are moderately coupled to the atmosphere, we set aerodynamic conductance as g_a = 0.058 m·s^{-1}, consistent with average conditions recently reported for coniferous, deciduous, and mixed-wood forests using eddy-covariance systems [31]. For canopies that are well coupled to the atmosphere, we set aerodynamic conductance as g_a = 0.150 m·s^{-1}, more consistent with the value commonly assumed in models [59,65]. In choosing to focus on these two levels of aerodynamic conductance for forest canopies, it was not intended to advocate either one, but rather to understand the implications of each level. In both estimates for reference forests, we applied an equivalent value of surface conductance, g_s = 0. 010 m·s^{-1}, reflecting moderately high use of water under optimal conditions [58].

As a compliment to the PE models, we additionally developed a statistical model that was fit against daily measurements of E_{Pan}. This estimate was developed by applying a linear mixed-effects (LME) model with a combination fixed and random effects at the level of each station. We assumed that daily values of E_{Pan} would be related to four variables, including downward global shortwave radiation, air temperature, vapour pressure deficit, and wind speed, yielding:

$$E_{PanFit} = (\beta_0 + b_{0,\,i}) + (\beta_1 + b_{1,i})R_{sgd} + (\beta_2 + b_{2,i})T_a + (\beta_3 + b_{3,i})D_a + (\beta_4 + b_{4,\,i})U \quad (5)$$

where $\beta_0 ... \beta_4$ are fixed effects and $b_{0,i} ... b_{4,i}$, are random effects applied to i = 1...n stations. To the extent that the assumed drivers and the assumption of additive linear effects can explain PE, this estimate also reflects the upper limit of how well models should perform. In all cases, the response variable,

E_{Pan}, and the predictor variables, R_{sgd}, T_a, D_a, and U, were standardized (i.e., 'z-scored') by subtracting the sample mean and dividing by the sample standard deviation. The resulting coefficients express the magnitude of change in the response variable (in terms of number of standard deviations) that is imposed by an increase of one standard deviation in each predictor variable. All models were fitted in R (R Development Core Team, 2015) using the *lmer* function from lme4 package [66]. As interest in E_{PanFit} was to describe the behavior of the calibrated model rather than as a predictive tool, we evaluated the performance of Equation (5) based simply on the marginal coefficient of determination (i.e., the variance explained by the fixed effects alone) and the standard error of the coefficients. For all models, values of PE were converted to report flux densities in units of mm·day^{-1}. Unless otherwise specified, we focused on reporting mean daily PE over a consistent period during the warm season when most forest regions of North America are above freezing, defined as the months from May to September.

2.3. Comparison between Models and Pan Evaporation

Comparison between models and pan evaporation was first performed on a daily time scale by regressing model predictions of daily total PE with daily total E_{Pan}. For each comparison, we produced scatterplots and reported the coefficient of determination (R^2), the model efficiency coefficient (MEC), the root mean squared error (RMSE), and the mean error (ME) to describe overall model performance.

Model comparisons were then performed on a monthly time scale, as the ability to predict monthly mean PE has broader practical value in applied research. Monthly mean values of daily total pan evaporation were calculated from the daily values, only considering months with no missing daily measurements. Monthly mean values of the predictor variables were calculated at each station. To understand the general integrity of the measurements, monthly measurements of pan evaporation were plotted against individual monthly mean values for each of the driving variables. Scatterplots and performance statistics were reported for predictions of monthly mean potential evaporation based on E_{Pen}, E_{PT}, E_{RG}, E_{RFu}, E_{RFc}, and E_{PanFit}.

2.4. Analysis of Variance Attribution

The relative importance of the drivers was inferred from estimates of sensitivity assessed first on a daily time scale. For the fitted model, the sensitivity of drivers was simply inferred from the fixed effects coefficients of the LME model. For example, the sensitivity of daily E_{Pan} to daily R_{sgd} was inferred from β_1 in Equation (5). To produce comparable sensitivities from the models of PE, we fitted LME models (identical to Equation (5)) to daily values of E_{Pen} and E_{PT}, E_{RG}, E_{RFu}, and E_{RFc}. By comparing coefficients, we were able to infer whether attribution in the models was consistent with that of statistical fits to pan evaporation measurements. We additionally performed the same analysis, only for monthly mean values. That is, we re-fit LME models to monthly mean E_{Pan} to produce a second (monthly) version of E_{PanFit}, as well as sets of coefficients for monthly mean E_{Pen}, E_{PT}, E_{RG}, E_{RFu}, and E_{RFc}. The coefficients were compared to understand any differences in attribution between methods. As the coefficients were developed at daily and monthly time scales, the analysis was able to assess whether attribution was scale-dependent.

2.5. Model Application

To apply the models across North America, monthly climate data were assembled for a 1 km grid covering North America between 1951 and 2014. Where possible, the database was developed by compiling existing datasets. Other variables, developed from raw data sources, are described in more detail below. For each variable, processing and storage of grids was performed separately for long-term monthly normals and time series of monthly anomalies for 1951–2014. The former reflect twelve grids indicating the mean monthly values for January to December for the base period 1971–2000 for each variable. The latter (anomalies) indicate the monthly deviations from the 1971–2000 normal for all months during 1951–2014.

Transient monthly variation in R_{sgd} was taken from the National Centers for Environmental Predition/National Center for Atmosphere Research (NCEP/NCAR) global reanalysis [67]. Similar to the methodology we applied previously [68], a first approximation of monthly normal R_{sgd} was developed by combining the synoptic and zonal variation captured by values from the North American Regional Reanalysis [69] with topographically-driven variation captured by values from the ArcGIS Solar Radiation Tool [70]. In the first step, estimates of R_{sgd} from North American Regional Reanalysis (NARR) were downloaded and resampled from the original resolution of 32 km down to a resolution of 1 km using cubic interpolation. In the second step, potential R_{sgd} was calculated using the ArcGIS Solar Radiation Tool with default settings (eight zenith divisions, eight azimuth divisions, uniform sky diffuse model type, a diffuse proportion of 0.3 and a transmittivity of 0.5) for each month at 1 km resolution. The tool was run with a digital elevation model developed by the U.S. Geological Survey the HYDRO1k project (https://lta.cr.usgs.gov/HYDRO1K). The ArcGIS Solar Radiation Tool does not take latitude as an input argument so the function was executed for 0.5°-wide zonal bands with 0.15° overlapping buffers on the north and south sides so as to avoid edge effects. In the third step, the two datasets were combined to produce a first approximation of R_{sgd} by correcting the ArcGIS Solar Radiation Tool estimates so that they were equivalent to the NARR estimates at the original scale of the NARR dataset (i.e., 32 km). This was achieved by applying a two-dimensional 32-km average filter ("fspecial" and "filter2" functions, The Mathworks Matlab 7.9.0) to the difference between NARR and ArcGIS estimates and adding it to the estimates of potential radiation from ArcGIS. The intended outcome was an unbiased estimate of the initial NARR dataset with added subgrid-scale variation caused by topographic effects. That is, the ArcGIS Solar Radiation Tool accounts for expected variation in irradiance caused by earth–sun geometry and shading by neighbourhood terrain, while NARR estimates express the synoptic-scale variation.

To expand on previous accuracy assessment [68], we compared the first approximation with measurements of monthly R_{sgd} across North America provided by Environment Canada and from the National Solar Radiation Data Base for the United States produced by the National Renewable Energy Laboratory [54]. The comparison indicated that initial NARR estimates systematically overestimated R_{sgd} by 15%–30% depending on month of the year. We made the assumption that the estimates from station networks were closer to the true values and developed a second approximation of R_{sgd} that corrected the first approximation to better match the station values. This was achieved in two steps: First, the initial monthly biases between NARR and the station observations (an average offset of 1.73 $MJ \cdot day^{-1}$) was subtracted from the first approximation. Second, the remaining residual errors at each station were interpolated to the grid using a polynomial surface fit with spatial variation defined by the errorgram and solved using singular value decomposition [71,72]. The resulting continuous grid of differences suggested that NARR dataset overestimated R_{sgd} in the southwest U.S. and large regions of central Canada and underestimated it in the northeast conterminous U.S. The second (and final) approximation of R_{sgd} was then developed by adding the residual error field to the first approximation.

To validate the second approximation of R_{sgd}, nearest grid cell estimates were compared with 25 independent Fluxnet tower observations. Tower observations spanned different years between 1998 and 2010 so a proper validation of the base period was not possible. Deviations from the long-term mean from the NARR dataset, corresponding with the period of comparison with each tower were, therefore, added to the second approximation grid to ensure the comparison was for estimates representative of the same period. Removing regional biases between the first approximation and Environment Canada/NREL station networks increased the level of explained variance from 75% to 85% and reduced the root-mean squared error (RMSE) from 3.96 to 0.97 $MJ \cdot m^{-2} \cdot day^{-1}$. Mean relative error between the second approximation and the tower observations was consistently below 10% for each month of year.

A first approximation of monthly normal T_a was compiled from the ClimateNA dataset [73,74]. Transient variability in monthly temperature was derived from three sources, including Environment Canada [75], the U.S. Historical Climate Station Network (USHCN) [76] for the conterminous U.S.

and from the Global Summary of the Day (GSOD) product [55] for Alaska. For Canada, daily climate records were extracted for 697 stations that included all stations listed in Environment Canada's reference climate station network, all stations in the Adjusted and Homogenized Canadian Climate Data archive [77], and some additional remote stations with long records. Canadian station time series were scanned for inhomogeneities by eye. Non-climatic steps (that were obviously not natural) were identified and the shorter of the resulting time series segments were adjusted to eliminate the difference. Seven stations were removed in the process due to poor record quality. Gaps in the original records were filled using regression analysis to relate each station record with that of four nearest neighbours. Monthly mean air temperature was then calculated from the gap-filled records of daily minimum and maximum air temperatures.

Transient monthly variation of D_a was then derived from resampling of the Climate Research Unit estimates of vapour pressure [78] and saturation vapour pressure (e_a) derived from T_a. A first approximation of long-term mean monthly D_a was developed based on a function of minimum and maximum temperature [79,80], with the temperature data coming from ClimateNA [74]. As with radiation, the first approximation was compared with measurements compiled from the Global Summary of the Day dataset [55]. Environment Canada measurements were also added after noticing that not all Environment Canada data appears to have been entered into the GSOD database. The difference between nearest grid cells of the first approximation and stations was then calculated and several outliers were censored from the station record. The differences were then interpolated to produce continuous maps of the error, showing considerable differences over the southern conterminous U.S. Adding the difference field to the first approximation to produce the second approximation led to improved agreement with independent observations of vapour pressure deficit at Fluxnet towers. The level of explained variance in warm-season mean vapour pressure deficit increased from 54% to 76% and the RMSE decreased from 1.79 to 1.40 hPa.

Values of s, γ, and λ_L were calculated as functions of air temperature. The above preparation of climate input grids included R_{sgd}, yet models of PE required input grids of R_n. Net radiation was derived from incident shortwave radiation in two steps. First, we converted incident shortwave radiation to net shortwave radiation to account for differences in mean albedo of various forest surface types. This was achieved by producing a lookup table linking surface albedo values for land cover classes. Land cover classes were derived from the 2002 version of the North American Land Cover Characteristics map produced by the National Center for Earth Resources Observation and Science, United States Geological Survey [81]. Average surface albedo values for each land cover class were drawn from a previous synthesis of values [82]. Net shortwave radiation (R_{sgn}) was then calculated by multiplying R_{sgd} by (1-albedo). In a second step, we developed linear mixed effects models that related monthly mean observations of net shortwave radiation ($W \cdot m^{-2}$) to observations of monthly mean net radiation ($W \cdot m^{-2}$). In the model, we included a random effect for the intercept and slope. Data were compiled from 14 forested Fluxnet station sites across North America from the Oak Ridge National Laboratory Fluxnet Database [83] (Sites codes: CA-Ca1, CA-Ca2, CA-Ca3, US-Ha1, US-MRf, US-Me2, US-NR1, CA-Qfo, CA-Oas, CA-Obs, CA-Ojp, US-MMS, US-Wrc). This gave a sample size of 1054 monthly observations that were representative of deciduous and evergreen forest stands from boreal and temperate forest biomes. The global model yielded a relationship, $R_n = 0.830R_{sgn} - 31.364$. The slope and intercept had standard errors of 0.006 and 2.551, respectively, and were both significant ($p < 0.001$). The coefficient of determination, excluding random effects, was $R^2 = 0.95$, and the root mean squared error was 13.5 $W \cdot m^{-2}$. We additionally tested whether differences existed between deciduous and evergreen stands, or between boreal and temperate forest biomes. We found that parameters were not statistically different between boreal and temperate stands ($p = 0.23$), but we did find that parameters differed between deciduous and evergreen stands ($p = 0.009$). The model for deciduous stands yielded a relationship, $R_n = 0.882R_{sgn} - 39.500$. The slope and intercept had standard errors of 0.011 and 2.300, respectively, and were both significant ($p < 0.001$). The model for evergreen stands yielded a relationship, $R_n = 0.817R_{sgn} - 29.512$. The slope and intercept had standard errors of 0.031 and 5.894, respectively, and were both significant ($p < 0.001$). Parameters for deciduous and evergreen forest were linked to those corresponding land cover classes in the USGS land cover classification map, while parameters from the global model were applied in all other land cover classes, including mixed forest.

Although Class A evaporation pans and natural land surfaces may exhibit diurnal and seasonal cycles in heat capacity, for simplicity, we assumed that the ground heat flux was zero at a decadal time scale. Evaluation of the long-term mean ground heat flux from 12 of the aforementioned Fluxnet sites located in forest ecosystems suggested that this assumption may lead to an overestimate in net radiation of 0.11 percent.

We summarized long-term warm-season mean values of PE from each method by forest region according to ecozones identified in the unified Level 1 ecozones of Canada and the U.S. produced by the Commission for Environmental Cooperation [84]. We additionally summarized the zonal variation in warm-season mean values over the range of latitude in North America. We summarize time trends by forest region. There was one station in the western temperate forest ecozone, two stations in the Montane Cordillera ecozone, one station in the western boreal ecozone, two stations in the Boreal Plain ecozone, four stations in the eastern boreal ecozone, and six stations in the eastern temperate forest ecozone. There is a high degree of uncertainty in values of g_a and g_s at this scale. We therefore investigated the sensitivity of trends to the parameterization of the Penman–Monteith model by calculating time trends in the resulting values of PE for a broad range in values of g_a and g_s and then applied multivariate regression analysis to derive a general relationship between time trends in PE and values of g_a and g_s used in the Penman–Monteith model.

3. Results

3.1. Daily Model Comparison

All candidate predictor variables in the LME model fit to daily E_{Pan} were significant ($p < 0.001$) and fixed effects coefficients ranged from 0.402 for R_{sgd} to 0.126 for U (Table 2). Variance inflation factors for the predictor variables were all well below 10.0 so we assumed multicollinearity did not confound interpretation of the coefficients. Large systematic differences, perhaps reflecting measurement error and instrumentation, was implied by substantial variation in the random effect on the intercept, ranging from −0.632 to 0.429 (Table 2). Relationships between models of PE and E_{Pan} exhibited low precision (Figure 2). Coefficients of determination were similar across models. E_{Pen} exhibited a model efficiency coefficient of 0.27, E_{PT} exhibited a model efficiency coefficient of 0.33, E_{RFu} and E_{RFc} both exhibited model efficiency coefficients close to zero (i.e., no better than applying the mean as a predictor), while E_{PanFit} exhibited a model efficiency coefficient of 0.40. In terms of overall bias, E_{Pen} exhibited a mean error of 0.5 mm·day^{-1} (10%) (i.e., E_{Pen} overestimated E_{Pan} by 10 percent), E_{RG}, E_{RFu}, and E_{RFc} exhibited larger biases (i.e., mean errors of 0.3, −0.1, and 0.1 mm·day^{-1} (7%, −3% and 2%), respectively, while E_{PT} and E_{PanFit} both exhibited a mean error of 0.0 mm·day^{-1} (−1%).

Table 2. Fixed and random effects for parameters in the linear mixed effects model of daily pan evaporation measurements. The marginal coefficient of determination (i.e., without application of the random effects) was $R^2 = 0.41$). The mean (observation minus prediction) error was -0.04 ± 0.01 mm·day^{-1} (-0.96 ± 0.31%). The root mean squared error was 1.88 mm·day^{-1}.

	Intercept	R_{sgd}	T_a	D_a	U
Fixed effects					
Estimate	0.033	0.402	0.209	0.156	0.126
Standard error	0.060	0.032	0.023	0.026	0.031

Table 2. *Cont.*

	Intercept	R_{sgd}	T_a	D_a	U
Random effects					
Whitehorse A	−0.030	−0.008	−0.179	0.149	−0.020
Norman Wells	0.295	0.062	−0.046	0.038	0.263
Resolute Cars	0.230	−0.064	0.013	0.208	0.029
Edmonton Stony Plain	0.406	0.091	−0.059	0.001	0.434
Winnipeg Richarson Int'l A	0.299	0.079	0.039	−0.016	0.211
Churchill A	0.101	−0.093	0.027	0.156	−0.020
Pierre Elliott Trudau Intl A	−0.020	0.052	0.003	−0.057	−0.075
La Grande IV A	0.379	0.057	0.072	0.039	0.297
Nitchequon	−0.187	−0.036	0.012	−0.032	−0.075
Schefferville A	0.136	0.034	0.048	0.055	0.237
Antioch Pump Plant 3	0.082	0.275	−0.040	−0.164	−0.144
Bonny Dam 2 NE	0.394	0.021	0.096	−0.102	−0.006
Allatoona Dam 2	−0.632	−0.262	−0.034	−0.033	−0.031
Arrowrock Dam	−0.052	0.105	0.028	−0.208	−0.181
Buckhorn Lake	−0.565	−0.211	−0.069	−0.086	−0.035
Alberta Ford For Cen	−0.611	−0.166	−0.028	−0.007	−0.111
Bozeman Montana SU	−0.208	0.043	−0.099	−0.029	−0.228
Beowawe 49 S U of N Ranch	0.049	0.010	0.161	−0.096	−0.011
Animas 3 ESE	0.456	0.376	−0.058	−0.124	−0.073
Aurora 6 N	−0.077	−0.162	0.051	0.148	0.003
Altus Dam	−0.264	−0.268	0.083	0.219	0.070
Brookings 2 NE	0.136	−0.120	0.106	−0.104	−0.110
Alpine	−0.137	−0.029	−0.307	0.177	−0.009
Beaumont Resrch Ctr	0.037	0.121	−0.087	−0.126	−0.121
Beeville 5 NE	−0.299	−0.136	0.062	0.073	0.042
Arches National Park HQS	0.429	0.378	−0.078	−0.066	−0.189
Bellingham 3 SSW	−0.352	−0.155	0.056	0.181	0.038
Boysen Dam	0.006	0.008	0.225	−0.195	−0.183

3.2. *Monthly Model Comparison*

Monthly variability of E_{Pan} ranged from 1.0 to 11.1 mm·day^{-1}, with no major outliers (Figure 3). Of the individual meteorological drivers, the relationship was strongest between E_{Pan} and R_{sgd} and this relationship showed some indication of curvature (Figure 3a). Relationships with T_a and D_a were weaker (Figure 3b,c), while there was no relationship between monthly E_{Pan} and U (Figure 3d).

Consistent with fits previously reported at the daily scale, all candidate predictor variables in the LME model fit to monthly E_{Pan} were significant ($p < 0.001$) (monthly LME model coefficients not shown). The marginal coefficient of determination for the monthly fit to pan evaporation was $R^2 = 0.68$, with a mean (observation minus prediction) error of -0.16 ± 0.06 mm·day^{-1}, and a root mean squared error of 0.93 mm·day^{-1}.

Models listed in decreasing order of performance (i.e., agreement with monthly E_{Pan} measurements based on model efficiency coefficients), were E_{PanFit} (MEC = 0.67), E_{PT} (MEC = 0.64), E_{Pen} (MEC = 0.53), E_{RG} (MEC = 0.31), E_{RFu} (MEC = 0.10), and E_{RFc} (MEC = −0.20) (Figure 4). The rate of evaporation for a reference forests with canopies that are moderately coupled to the atmosphere substantially underestimated E_{Pan} (Figure 4d). By design, these values were not expected to be unbiased expressions of PE, but we plotted them here to compare precision and explained variance for context. E_{PanFit} slightly outperformed E_{PT}, explaining 68% of the variance, with a mean error of −3% (Figure 4f). It is interesting to note that, despite slightly greater overall performance of E_{PanFit} and E_{PT}, these models still had limited capacity to reproduce the highest monthly values of E_{Pan} above 7.0 mm·day^{-1} (Figure 4b,f).

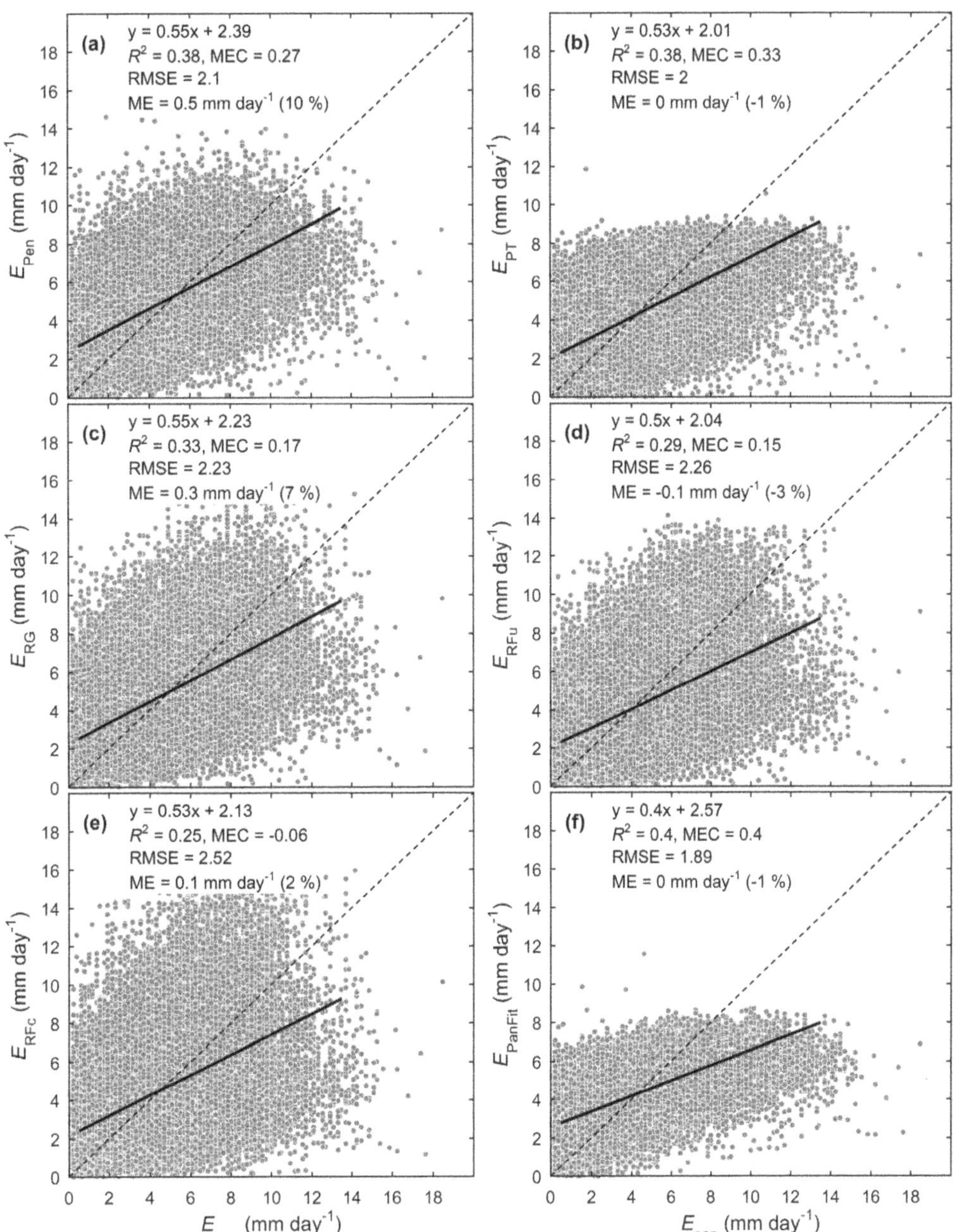

Figure 2. Comparison between observations of daily total pan evaporation (E_{Pan}) (after applying the pan coefficient of 0.77) and models of potential evaporation based on: (**a**) Penman's model (E_{Pen}); (**b**) the Priestley–Taylor model (E_{PT}); (**c**) reference rate for grassland based on the Penman–Monteith model (**d**) reference rate for forests canopies that are moderately coupled to the atmosphere (E_{RFu}); (**e**) reference rates for forest canopies that are well coupled to the atmosphere (E_{RFc}); (**f**) linear mixed-effects model fit to daily pan evaporation measurements (E_{PanFit}). Linear curves indicate the ordinary least squares fits. Statistics include: coefficient of determination (R^2), model efficiency coefficient (MEC, values of 0.0 or <0.0 indicate equivalent, or less predictive power than applying the mean of observations, respectively, while a value of 1.0 indicates a perfect predictor), root mean squared error (RMSE, mm·day^{-1}), and mean error (ME, mm·day^{-1} or %).

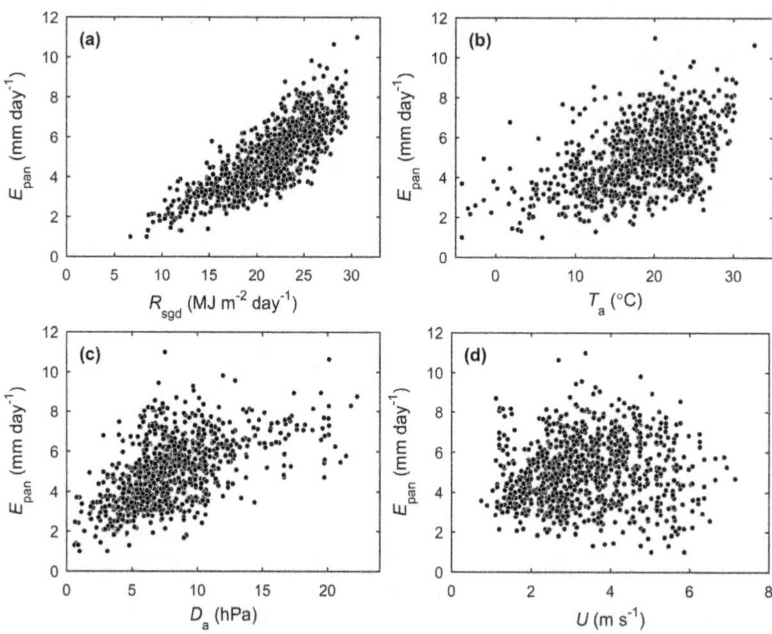

Figure 3. Relationships between monthly mean pan evaporation (E_{Pan}) and climate variables, including incident global shortwave radiation (R_{sgd}), air temperature (T_a), vapour pressure deficit (D_a), and wind speed (U).

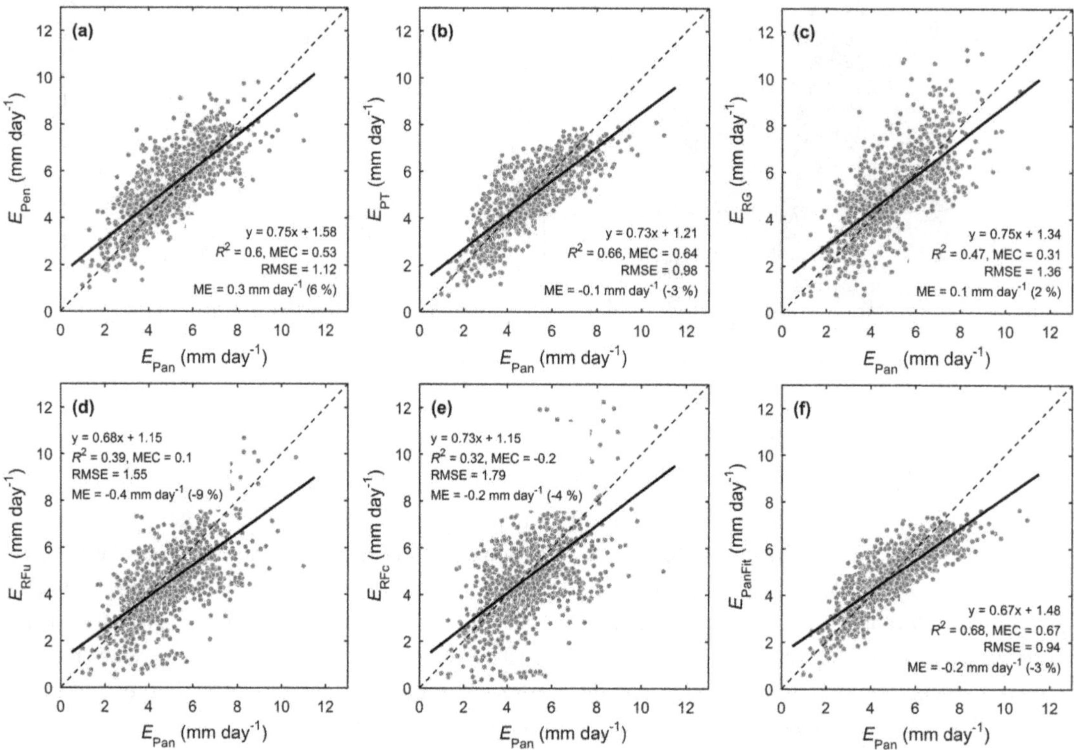

Figure 4. Comparison between observations of monthly mean pan evaporation (E_{Pan}) and model predictions of potential evaporation (PE) based on: (**a**) Penman's model; (**b**) the Priestley–Taylor model; (**c**) reference rates for grasslands based on the Penman–Monteith model; (**d**) reference rates for forests based on the Penman–Monteith model assuming the canopy is moderately coupled to the atmosphere; (**e**) reference rates for forests based on the Penman–Monteith model assuming the canopy is well coupled to the atmosphere; (**f**) linear mixed-effects model fit to monthly pan evaporation (E_{PanFit}).

3.3. Attribution of Variance in PE to Meteorological Drivers

In decreasing order of sensitivity, daily variability of E_{Pan} was controlled by R_{sgd}, T_a, D_a, and U (Figure 5). On a monthly time scale, the controls of T_a and D_a on E_{Pan} dissipated and were compensated by increases in sensitivity to R_{sgd}, while control by U was no longer significant.

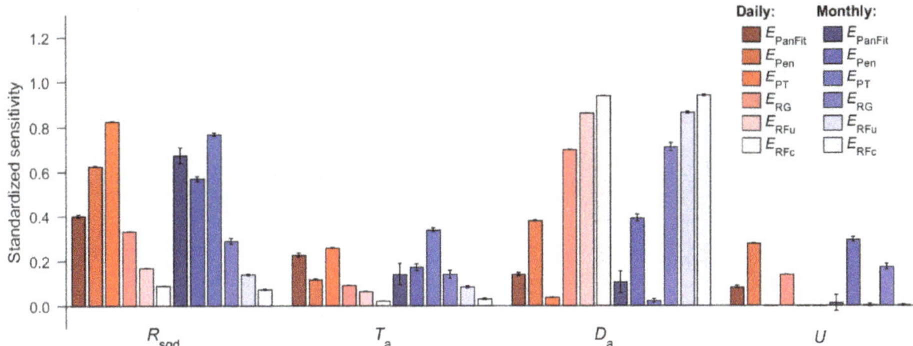

Figure 5. The sensitivity of potential evaporation (PE) to driving variables on a daily and monthly time scale. Left-to-right (red bars): Fixed effects from linear mixed-effects model fits to daily pan evaporation (E_{PanFit}); fixed effects from model fits to daily predictions of PE from Penman's model; fixed effects from model fits to daily predictions of PE from the Priestley–Taylor model; fits to daily reference evaporation from grasslands based on the Penman–Monteith model; fits to daily reference evaporation from forest canopies that are moderately coupled to the atmosphere based on the Penman–Monteith model; fits to daily reference evaporation from forest canopies that are well coupled to the atmosphere based on the Penman–Monteith model. Blue bars then repeat through the same estimates, only for fits to monthly instead of daily time scale.

On a daily time scale, E_{Pen} was most strongly driven by R_{sgd} and secondarily by D_a and U, while T_a exerted a low positive effect. There were no significant differences in sensitivity between daily and monthly time scale for E_{Pen}. E_{PT} exhibited the strongest sensitivity to R_{sgd}, with secondary dependence on T_a. On a monthly time scale, control on E_{PT} shifted slightly from R_{sgd} to T_a. Despite differences in aerodyne mic, the behaviour of E_{RFu} and E_{RFc} only showed modest differences in sensitivity to D_a. Both exhibited strong dependence on D_a, followed by R_{sgd} and weak positive dependence on T_a. Hence, none of the models perfectly matched the variance attribution of monthly E_{PanFit}; sensitivity of monthly E_{PT} to T_a was too high, while sensitivities of monthly E_{Pen}, E_{RFu}, and E_{RFc} to D_a were too high.

3.4. Spatial and Temporal Variability

Differences in the spatial variation of PE were evident from visual inspection of the mean warm-season values over the 1971–2000 base period (Figure 6). Patterns for Penman's model and the reference grassland PE were similar, with the exception that E_{Pen} showed slightly greater values of PE across regions of southern Canada (Figure 6a,c). The Priestley–Taylor model indicated relatively low PE in arid regions of the southwest U.S. compared to other models and also showed slightly greater levels of PE across southern Canada relative to E_{RG} (Figure 6b). Reference forest values showed more prominent zonal gradients and the differences between spatial patterns of PE for moderately- and well-coupled canopies were much less than that between reference forest rates and other models (Figure 6d,e).

All models indicated decreases in warm-season mean daily PE as latitude increased from 25 to 80 degrees north (Figure 7). Zonal gradients were greatest for estimates of reference evaporation using the Penman–Monteith equation, and similar among other models. At the scale of ecozones, historical values of warm-season mean daily PE ranged from 2.4 to 4.5 mm·day^{-1} (Table 3). Mean historical values of warm-season PE ranged from 2.9 mm·day^{-1} (E_{RG}) to 3.8 mm·day^{-1} (E_{PanFit}) (Table 3).

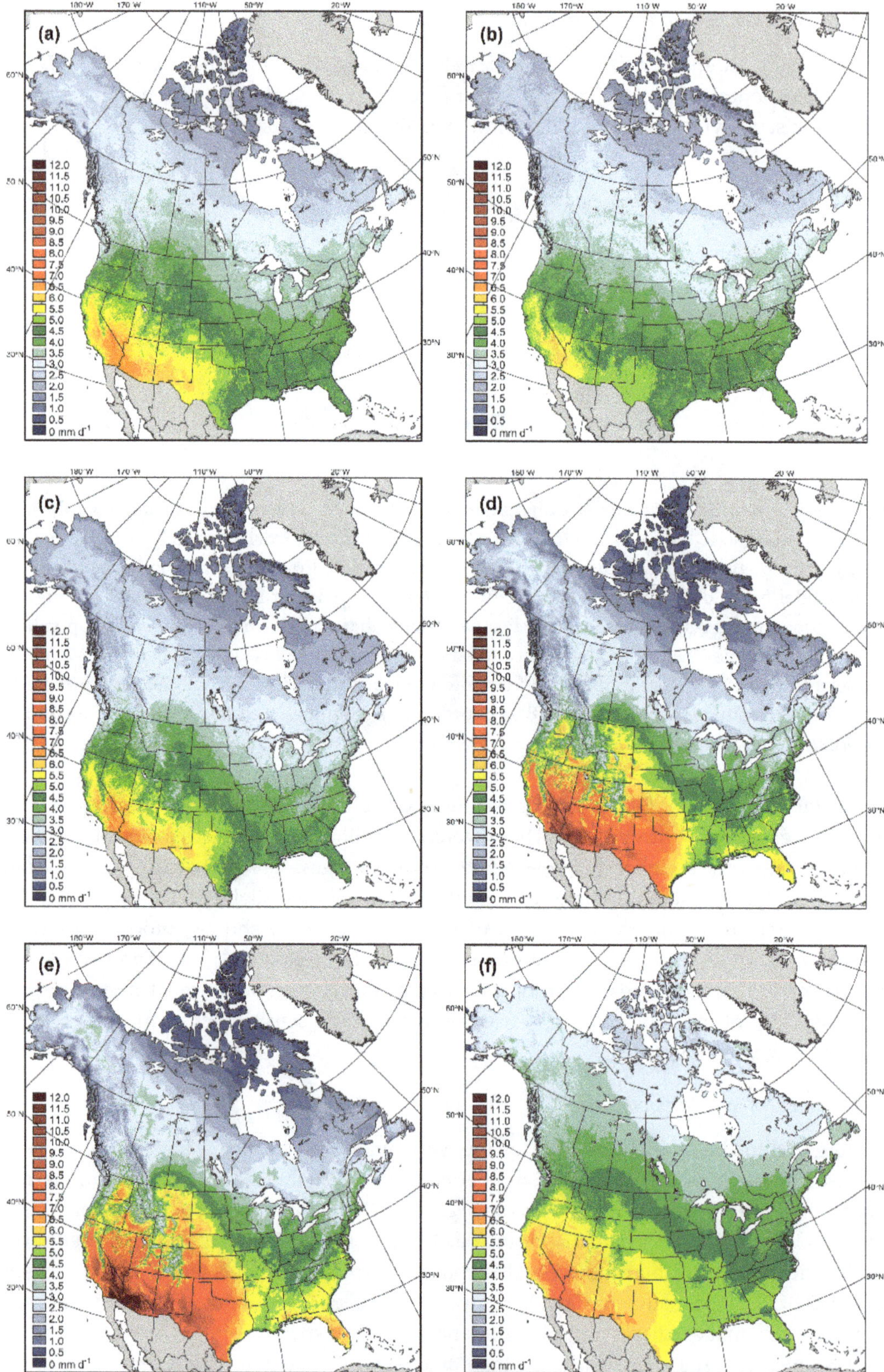

Figure 6. Comparison between long-term (1971–2000) mean daily potential evaporation (PE) over the warm season (May–September). Units of PE are expressed as mm·day^{-1}: (**a**) Penman's model (E_{Pen}); (**b**) Priestley–Taylor model (E_{PT}); (**c**) reference grassland (E_{RG}) based on the Penman–Monteith model; (**d**) reference forest with moderately coupled canopy (E_{RFu}) based on the Penman–Monteith model; (**e**) reference forest with coupled canopy (E_{RFc}) based on the Penman–Monteith model; (**f**) linear mixed-effects model fit to monthly mean pan evaporation (E_{PanFit}).

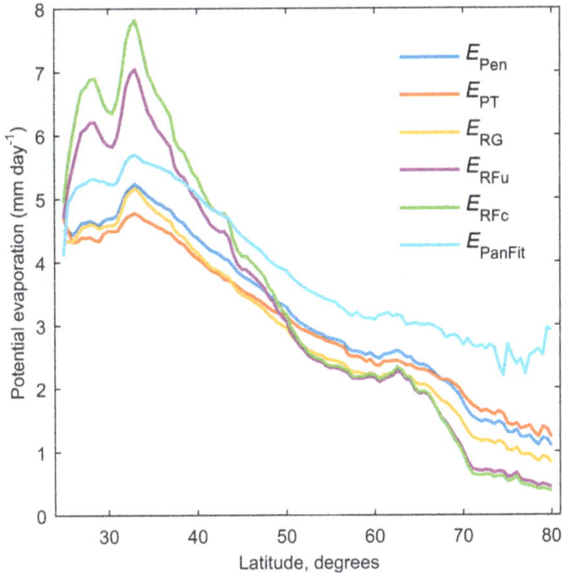

Figure 7. Zonal mean warm-season (May–September) potential evaporation from various models over the land surface of North America.

Table 3. Ecozone-average mean warm-season (May–September) variables over the period 1951–2014: Incident global solar radiation, R_{sgd} (MJ·m^{-2}·day^{-1}), air temperature, T_a (°C), vapour pressure, e_a (hPa), vapour pressure deficit, D_a (hPa), Penman's model, E_{Pen} (mm·day^{-1}), the Priestley–Taylor model, E_{PT} (mm·day^{-1}), the Penman–Monteith model for a reference grassland, E_{RG} (mm·day^{-1}), the Penman–Monteith model for a reference forest with moderately coupled canopy, E_{RFu} (mm·day^{-1}), and reference forest with well coupled canopy, E_{RFc} (mm·day^{-1}), and statistical fits to pan evaporation measurements E_{PanFit} (mm·day^{-1}).

	R_{sgd}	T_a	e_a	D_a	E_{Pen}	E_{PT}	E_{RG}	E_{RFu}	E_{RFc}	E_{PanFit}
Boreal, west	16.0	9.6	8.0	4.3	2.7	2.5	2.4	2.6	2.7	3.2
Boreal Plains	17.7	11.7	9.4	4.7	3.1	3.0	2.8	2.8	2.9	3.6
Boreal, east	17.2	11.6	10.4	3.8	2.9	2.9	2.5	2.5	2.4	3.4
Temperate, west	18.1	12.2	10.8	4.0	3.1	3.1	2.7	2.5	2.5	3.6
Montane Cordillera	21.4	11.8	8.0	6.3	4.0	3.8	3.6	3.7	3.9	4.2
Temperate, east	19.5	20.7	19.1	6.2	3.9	4.0	3.6	4.1	3.8	4.5
Mean	18.3	12.9	11.0	4.9	3.3	3.2	2.9	3.0	3.1	3.8

On average over forest ecozones, there was little change in R_{sgd} from 1951–2014 (Table 4). According to NCEP/NCAR global reanalysis, R_{sgd} increased considerably over the boreal forest ecozones, and decreased considerably over eastern temperate forest ecozone. The station-based reconstruction of warm-season air temperature suggested an increase of 1.21 °C from 1951–2014 (Table 4). A large positive effect of warming on saturation vapour pressure led to a net increase in vapour pressure deficit of 0.81 hPa despite concomitant increase in vapour pressure of 0.38 hPa. No attempt was made to analyze trends in U.

Continuous high-resolution reconstructions of warm-season (May–September) mean PE based on values of E_{Pen}, E_{PT}, E_{RG}, E_{RFu}, E_{RFc}, and E_{PanFit} suggested a range between negative and positive time trends from 1951–2014 (Figure 8). The magnitude of time trends were irregularly distributed, but similarities in the pattern were apparent among models. Negative trends were consistently concentrated over the east-central U.S. and pockets of industrialized areas across Canada and Alaska. On average across forest ecozones, the long-term change in estimates of potential evaporation (inferred from least-squares fits of potential evaporation against time for each model) ranged from just 0.08 mm·day^{-1} for the Priestley–Taylor model, to 0.45 mm·day^{-1} for the reference forest that is well coupled to the atmosphere (Table 4). Average long-term changes were intermediate for Penman's model and the LME fit to pan evaporation measurements at 0.16 and 0.14 mm·day^{-1} respectively, slightly higher for the reference grassland (0.19 mm·day^{-1}), and high for reference forests (0.37 and 0.45 mm·day^{-1} for moderately- and well-coupled canopies, respectively).

Based on a compilation of the time trends in warm-season PE for reference forests, using a range of aerodynamic and surface conductance, a general relationship was produced from regression analysis (Figure 9). The analysis confirmed that time trends increased strongly with increasing aerodynamic conductance and weakly with increasing surface conductance.

Table 4. Ecozone-average long-term changes in warm-season (May–September) mean daily variables over the period 1951–2014: Incident global solar radiation, R_{sgd} (MJ·m^{-2}·day^{-1}), air temperature, T_a (°C), vapour pressure deficit, D_a (hPa), Penman's model, E_{Pen} (mm·day^{-1}), the Priestley–Taylor model, E_{PT} (mm·day^{-1}), the Penman–Monteith model for a reference grassland, E_{RG} (mm·day^{-1}), the Penman–Monteith model for a reference forest with moderately-coupled canopy, E_{RFu} (mm·day^{-1}), and reference forest with well-coupled canopy, E_{RFc} (mm·day^{-1}), and statistical fits to pan evaporation measurements E_{PanFit} (mm·day^{-1}).

	R_{sgd}	T_a	e_a	D_a	E_{Pen}	E_{PT}	E_{RG}	E_{RFu}	E_{RFc}	E_{PanFit}
Boreal, west	0.09	1.30	0.43	0.63	0.17	0.10	0.19	0.31	0.36	0.15
Boreal Plains	0.08	1.33	0.20	1.01	0.22	0.10	0.26	0.48	0.57	0.18
Boreal, east	0.40	1.49	0.72	0.65	0.23	0.18	0.24	0.34	0.38	0.21
Temperate, west	0.09	1.11	0.37	0.74	0.17	0.10	0.20	0.34	0.41	0.14
Montane Cordillera	−0.04	1.14	0.19	0.95	0.18	0.09	0.22	0.43	0.52	0.14
Temperate, east	−0.58	0.86	0.35	0.89	0.00	−0.10	0.06	0.31	0.44	0.04
Mean	0.01	1.21	0.38	0.81	0.16	0.08	0.19	0.37	0.45	0.14

4. Discussion

Ecological studies commonly make use of basic expressions of evaporative demand. Here, we explored the behavior of several common approaches to calculating PE. While such basic expressions of PE may have strong practical value in studies that attempt to predict climate change impacts on terrestrial ecosystems across large spatial and temporal domains, further work is needed to improve the representation of spatial variation in meteorological conditions above forests, canopy structure and physiology, and ultimately the prediction of actual evapotranspiration rather than PE or reference rates [52,53].

We considered measurements of pan evaporation as one independent indicator of PE. While there are likely key differences between PE above forests and evaporation pans, the comparison helped place the attribution of variance and spatial and temporal variability of PE predictions into context. Upon compiling the pan evaporation data from Canada and the U.S., we discovered that records were only rarely accompanied by on-site meteorological variables, so considerable effort was placed on deriving best available values from nearby climate stations, which constitutes an important source of error. Although varying conditions at the evaporation pans, and errors associated with site, wind, heating, splashing, likely led to different pan coefficients, we chose to restrict the scope of the study by applying a constant pan coefficient and including a random effect on the intercept coefficient at the scale of individual pan evaporation stations in the LME models. Applying a mid-range value of the pan coefficient, $k = 0.77$, led to a mean E_{PanFit} of 3.8 mm·day^{-1} across North American forests. As this value was 0.5 mm·day^{-1} greater than the next highest value (E_{Pen}), any future consideration of the dataset should likely consider reducing k.

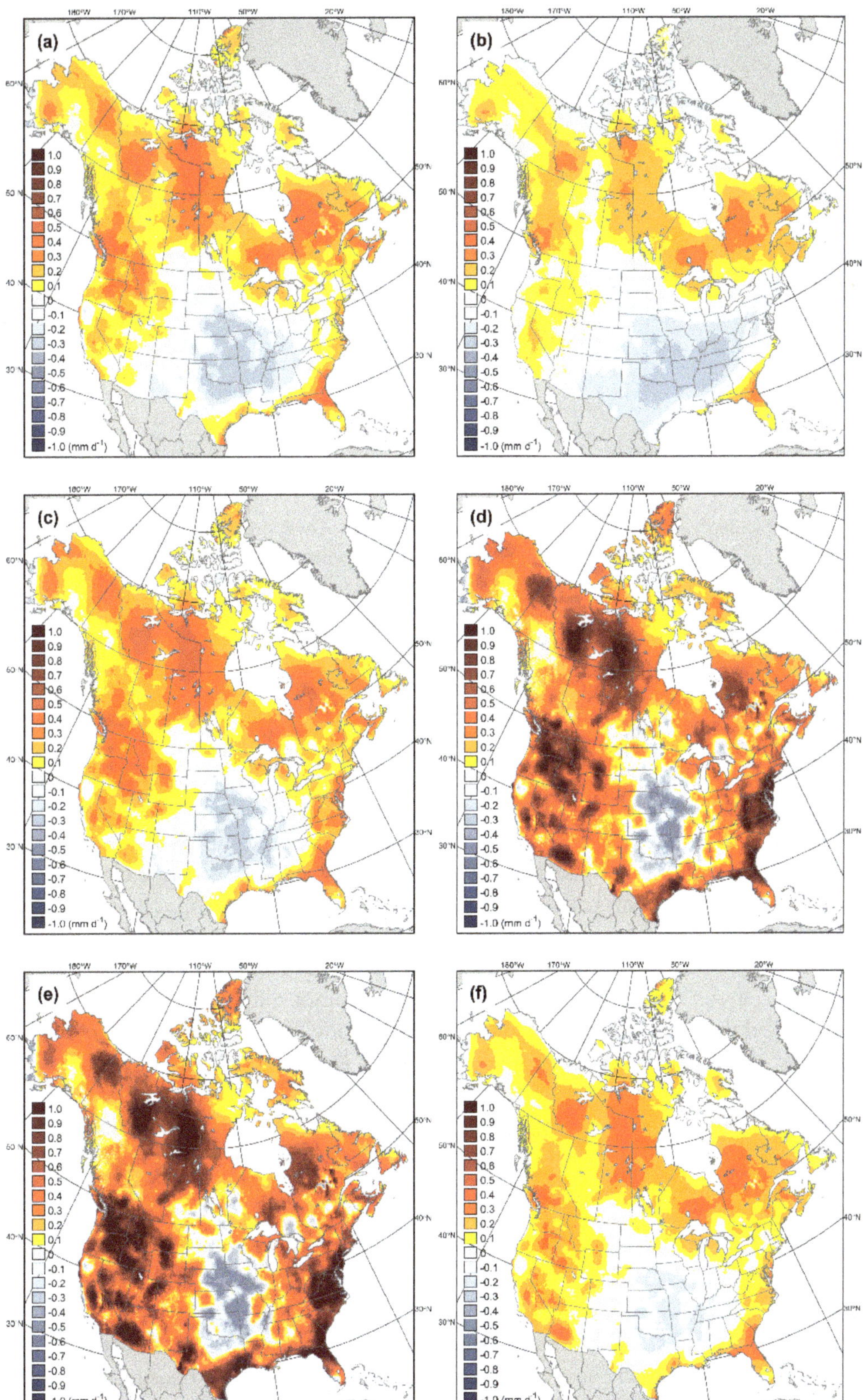

Figure 8. Long-term change in warm-season (May–September) mean potential evaporation (PE) from 1951–2014 (mm·day^{-1}): (**a**) Penman's model (E_{Pen}); (**b**) Priestley–Taylor model (E_{PT}); (**c**) reference grassland (E_{RG}); (**d**) reference forest with moderately coupled canopy (E_{RFu}); (**e**) reference forest with well coupled canopy (E_{RFc}); (**f**) linear mixed-effects model fit to monthly mean pan evaporation (E_{PanFit}). Values were calculated by multiplying the slope coefficient from regression relationship between PE and time by the length of the study period (64 years).

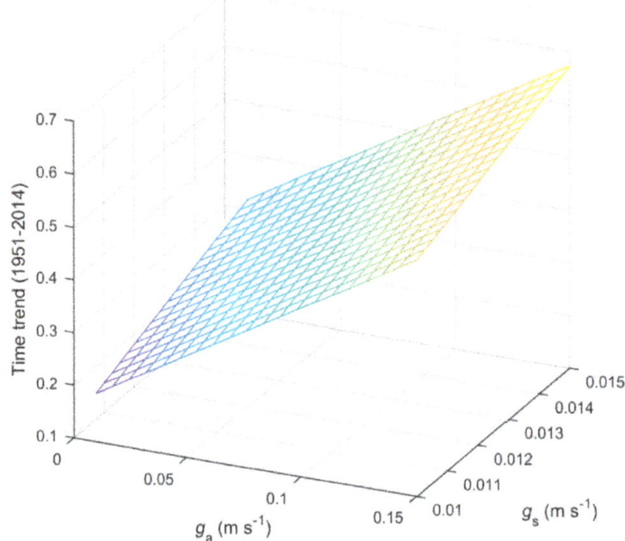

Figure 9. Dependence of long-term changes in reference rates of evaporation on levels of aerodynamic conductance (g_a) and surface conductance (g_s) set in the Penman–Monteith equation. Trends are representative of mean warm-season daily reference evaporation over forest regions of North America (z-axis indicates change in mm·day^{-1} from 1951–2014).

While plenty of surface water balance research is performed at a sub-monthly time interval, it remains far more practical to simulate the water balance at a monthly time step in large-scale forest ecosystem modelling. Analysis of variance attribution suggested that controls on E_{Pan} were scale dependent. On a daily time scale, E_{Pan} was significantly related to R_{sgd}, T_a, D_a, and U, whereas on a monthly time scale E_{Pan} was only significantly related to R_{sgd}, T_a, and D_a. We interpreted this difference to mean that systematic effects of advection (as an energy source over pans) and ground heat flux (as an energy sink) diminish as time scale increases. That is, on any given day, advection and ground heat flux can enhance or suppress PE, while beyond this temporal scale, it cancels out. The lack of effect of U at the monthly time scale may suggest that accuracy in large-scale analysis of spatial and temporal variation (including trend analysis) may not suffer from failure to represent wind.

Varying sensitivity to D_a was a defining difference among the most popular models of PE. The Priestley–Taylor model and the Penman–Monteith model parameterized for conditions above reference forest canopies reflect estimates with absent D_a-sensitivity and strong D_a-sensitivity, respectively. Any inferences about change in surface hydrology and consequent socio-economic implications over time are, therefore, going to strongly depend on which model is applied. The differentiation in long-term change among models was exacerbated by strong apparent positive trend in historical D_a across North America [85]. Interestingly, the different dependence on D_a among models also shaped the zonal variation in PE (Figure 7), where estimates of PE derived from reference forest rates actually exceeded estimates of PE with infinite surface conductance in Penman's model. Specifically, estimates of E_{RFu} and E_{RFc} start to exceed E_{Pen} when $D_a > 18.5$ hPa.

On average across forests of North America, results indicate that PE increased from 1951–2014. If this is correct, increasing PE must increase the tendency for drought stress. Confidence in the claim of a positive trend is gained from the fact that all models consistently indicated positive trends. However, the claim of positive trends remains uncertain due to reliance on irradiance from NCEP/NCAR reanalysis, which has poorly constrained uncertainty, and by the inability to account for trends in wind speed. These sources of uncertainty will affect all models similarly. If we ignore uncertainty in irradiance and wind trends, it must still be recognized that the range of trend estimates among models of PE includes low estimates, which may have been exceeded by increasing trends in precipitation [86]. If we exclude consideration of E_{PT} on the basis that it does not include dependence on D_a, we estimate average long-term changes in warm season PE ranging from 0.14 to 0.45 mm·day^{-1}. Hence, there is still large uncertainty in how much evaporative demand increased.

Greater increases in E_{RFu} and E_{RFc}, owing to strong dependence on increasing D_a (and the general pattern of increasing trends with increasing g_a and g_s) must impose substantial impacts on the level of water stress imposed upon plants, either through a preemptive regulation of gas exchange to avoid desiccation at the expense of carbon supply, or lax regulation of gas exchange at the expense of water transport efficiency and risk of hydraulic failure. Yet either way, the magnitude of those increases in evaporative demand (i.e., the forcing imposed on guard cells or hydraulic conductivity) is not supported by other models of PE, including a statistical fit to E_{Pan}. That is not to say, however, that the more modest changes in evaporative demand implied by E_{PT} cannot impose severe water stress, as the sensitivity of plant health and productivity to changes in E_{PT}, specifically, remains largely unexplored.

At the regional scale, all trends in PE were positive with the exception of trends in the eastern temperate ecozone, where E_{PT} was negative and E_{Pen} was zero. Interestingly, this could be the signature of high tropospheric aerosol forcing on surface climates over the east-central conterminous U.S. (Figure 8). Similar patterns exist for pockets of industrial activity across the continent. Further research is needed to understand the quantitative relationship between PE and tropospheric aerosols. If there is a link, it stands to reason that continued trends in PE may be strongly affected by the onset of aerosol emission regulations that began late in our study period (i.e., 1990's), and conversely by any new or growing emission sources.

5. Conclusions

We investigated spatial and temporal variation of PE across forest regions of North America. While records of E_{Pan} were not long enough to perform direct trend analysis, we instead chose to produce a statistical model, E_{PanFit}, that helped to place the magnitude of model predictions in context and evaluate differences in the relative effect of the driving meteorological factors. Consistent with previous model inter-comparison studies [1,87,88], we found substantial differences in PE among models. Large-scale (zonal) differences were clearly demonstrated across North America. Long-term change in PE from 1951–2014 ranged from 0.08 to 0.45 mm·day^{-1} depending on whether it was calculated with the Priestley–Taylor model, or with the Penman–Monteith model set for a reference forest by making the common assumption that the canopy is well coupled to the atmosphere. A general analysis of variance attribution among the approaches helped to put the underlying reasons for divergent behavior in quantitative terms. We conclude that until a consensus emerges on which expression is more representative of reality in forests, studies that adopt such basic expressions of PE should consider the range of different model predictions to represent uncertainty in trend magnitude. The need to forecast impacts of a changing climate forests, including wildfire risk, insect outbreak risk, drought-related mortality, and health decline syndromes, indicates a need to improve monitoring of the key drivers of PE, including R_{sgd} and D_a. This source of high resolution, continuous monthly variation of PE from various models can facilitate research on a wide range of applications that require knowledge of the surface water balance.

Acknowledgments: This research was supported by the Pacific Institute for Climate Solutions.

Author Contributions: R.H., N.C., and D.S. designed the study. R.H. produced and analyzed results and wrote the first draft. R.H., N.C. and D.S. provided editorial comments and suggestions.

Conflicts of Interest: The authors declare no conflict of interest.

References

1. Fisher, J.B.; Whittaker, R.J.; Malhi, Y. ET come home: Potential evapotranspiration in geographical ecology. *Glob. Ecol. Biogeogr.* **2011**, *20*, 1–18. [CrossRef]
2. Webb, W.L.; Lauenroth, W.K.; Szarek, S.R.; Kinerson, R.S. Primary production and abiotic controls in forests, grasslands, and desert ecosystems in the united states. *Ecology* **1983**, *64*, 134–151. [CrossRef]

3. Stephenson, N.L. Actual evapotranspiration and deficit: biologically meaningful correlates of vegetation distribution across spatial scales. *J. Biogeogr.* **1998**, *25*, 855–870. [CrossRef]
4. Sperry, J.S.; Tyree, M.T. Water-stress-induced xylem embolism in three species of conifers. *Plant Cell Environ.* **1990**, *13*, 427–436. [CrossRef]
5. Tyree, M.T.; Sperry, J.S. Do woody plants operate near the point of catastrophic xylem dysfunction caused by dynamic water stress? Answers from a model. *Plant Physiol.* **1988**, *88*, 574–580. [CrossRef] [PubMed]
6. Choat, B.; Jansen, S.; Brodribb, T.J.; Cochard, H.; Delzon, S.; Bhaskar, R.; Bucci, S.J.; Feild, T.S.; Gleason, S.M.; Hacke, U.G.; et al. Global convergence in the vulnerability of forests to drought. *Nature* **2012**, *491*, 752–755. [CrossRef] [PubMed]
7. Hember, R.A.; Kurz, W.A.; Coops, N.C. Relationships between individual-tree mortality and water-balance variables indicate positive trends in water stress-induced tree mortality across North America. *Glob. Change Biol.* **2016**. [CrossRef] [PubMed]
8. Millar, C.I.; Stephenson, N.L. Temperate forest health in an era of emerging megadisturbance. *Science* **2015**, *349*, 823–826. [CrossRef] [PubMed]
9. McDowell, N.G.; Williams, A.P.; Xu, C.; Pockman, W.T.; Dickman, L.T.; Sevanto, S.; Pangle, R.; Limousin, J.; Plaut, J.; Mackay, D.S.; et al. Multi-scale predictions of massive conifer mortality due to chronic temperature rise. *Nat. Clim. Change* **2016**, *6*, 295–300. [CrossRef]
10. Martínez-Vilalta, J.; Lloret, F. Drought-induced vegetation shifts in terrestrial ecosystems: The key role of regeneration dynamics. *Glob. Planet. Change* **2016**, *144*, 94–108. [CrossRef]
11. Clark, J.S.; Iverson, L.; Woodall, C.W.; Allen, C.D.; Bell, D.M.; Bragg, D.C.; D'Amato, A.W.; Davis, F.W.; Hersh, M.H.; Ibanez, I.; et al. The impacts of increasing drought on forest dynamics, structure, and biodiversity in the United States. *Glob. Change Biol.* **2016**. [CrossRef] [PubMed]
12. Sucoff, E. Water potential in red pine: Soil moisture, evapotranspiration, crown position. *Ecology* **1972**, *53*, 681–686. [CrossRef]
13. Sharkey, T.D. Transpiration-induced changes in the photosynthetic capacity of leaves. *Planta* **1984**, *160*, 143–150. [CrossRef] [PubMed]
14. Novick, K.A.; Ficklin, D.L.; Stoy, P.C.; Williams, C.A.; Bohrer, G.; Oishi, A.C.; Papuga, S.A.; Blanken, P.D.; Noormets, A.; Sulman, B.N.; et al. The increasing importance of atmospheric demand for ecosystem water and carbon fluxes. *Nat. Clim. Change* **2016**. [CrossRef]
15. Sanderson, M. Drought in the Canadian Northwest. *Geogr. Rev.* **1948**, *38*, 289. [CrossRef]
16. Thornthwaite, C.W. An approach toward a rational classification of climate. *Geogr. Rev.* **1948**, *38*, 55–94. [CrossRef]
17. McMahon, T.A.; Peel, M.C.; Lowe, L.; Srikanthan, R.; McVicar, T.R. Estimating actual, potential, reference crop and pan evaporation using standard meteorological data: A pragmatic synthesis. *Hydrol. Earth Syst. Sci.* **2013**, *17*, 1331–1363. [CrossRef]
18. Oke, T.R. *Boundary Layer Climates*, 2nd ed.; Routledge: Cambridge, UK, 1987.
19. Rosenberg, N.; Blad, B.; Verma, S. *Microclimate, the Biological Environment*, 2nd ed.; Jon Wiley and Sons Inc.: Toronto, ON, Canada, 1983.
20. Slatyer, R.; McIlroy, I. Practical Microclimatology. *Commonw. Sci. Ind. Res. Organ.* **1961**. [CrossRef]
21. Mather, J.R.; Ambroziak, R.A. A Search for Understanding Potential Evapotranspiration. *Geogr. Rev.* **1986**, *76*, 355–370. [CrossRef]
22. Ferguson, H.L.; O'Neill, A.D.J.; Cork, H.F. Mean Evaporation over Canada. *Water Resour. Res.* **1970**, *6*, 1618–1633. [CrossRef]
23. Kohler, M.A.; Nordenson, T.J.; Fox, W.E. *Evaporation from Pans and Lakes*; US Department of Commerce: Washington, DC, USA, 1955.
24. Thornthwaite, C.W.; Mather, J.R. *The Water Balance*; Laboratory of Climatology: Centerton, NJ, USA, 1955.
25. Hargreaves, G.H.; Allen, R.G. History and Evaluation of Hargreaves Evapotranspiration Equation. *J. Irrig. Drain. Eng.* **2003**, *129*, 53–63. [CrossRef]
26. Christiansen, J.E. *Estimating Evaporation and Evpoatranspiration from Climate Data*; Utah Water Research Laboratory, Utah State University: Logan, UT, USA, 1966.
27. Penman, H.L. Natural evaporation from open water, bare soil and grass. *Proc. R. Soc. Lond.* **1948**, *193*, 120–145. [CrossRef]
28. Monteith, J.L. Evaporation and environment. *Symp. Soc. Exp. Biol.* **1965**, *19*, 205–234. [PubMed]

29. Priestley, C.H.B.; Taylor, R.J. On the assessment of surface heat flux and evaporation using large-scale parameters. *Mon. Weather Rev.* **1972**, *100*, 81–92. [CrossRef]
30. Black, T.A. Evapotranspiration from douglas-fir stands exposed to soil-water deficits. *Water Resour. Res.* **1979**, *15*, 164–170. [CrossRef]
31. Brümmer, C.; Black, T.A.; Jassal, R.S.; Grant, N.J.; Spittlehouse, D.L.; Chen, B.; Nesic, Z.; Amiro, B.D.; Arain, M.A.; Barr, A.G.; et al. How climate and vegetation type influence evapotranspiration and water use efficiency in Canadian forest, peatland and grassland ecosystems. *Agric. For. Meteorol.* **2012**, *153*, 14–30. [CrossRef]
32. Greve, P.; Seneviratne, S.I. Assessment of future changes in water availability and aridity. *Geophys. Res. Lett.* **2015**, *42*. [CrossRef] [PubMed]
33. Hogg, E.H. Temporal scaling of moisture and the forest-grassland boundary in western Canada. *Agric. For. Meteorol.* **1997**, *84*, 115–122. [CrossRef]
34. Willmott, C.J.; Rowe, C.M.; Mintz, Y. Climatology of the terrestrial seasonal water cycle. *J. Climatol.* **1985**, *5*, 589–606. [CrossRef]
35. Allen, R.G.; Pereira, L.S.; Raes, D.; Smith, M. *Crop Evapotranspiration: Guidelines for Computing Crop Water Requirements, FAO Irrigation and Drainage Paper 56*; Food and Agriculture Organization of the United Nations: Rome, Italy, 1998.
36. Cleugh, H.A.; Leuning, R.; Mu, Q.Z.; Running, S.W. Regional evaporation estimates from flux tower and MODIS satellite data. *Remote Sens. Environ.* **2007**, *106*, 285–304. [CrossRef]
37. Spittlehouse, D.L. Water availability, climate change and the growth of Douglas-fir in the Georgia Basin. *Can. Water Resour. J.* **2003**, *28*, 673–687. [CrossRef]
38. Coops, N.; Coggins, S.; Kurz, W. Mapping the environmental limitations to growth of coastal Douglas-fir stands on Vancouver Island, British Columbia. *Tree Physiol.* **2007**, *27*, 805–815. [CrossRef] [PubMed]
39. Seely, B.; Welham, C.; Scoullar, K. Application of a hybrid forest growth model to evaluate climate change impacts on productivity, nutrient cycling and mortality in a montane forest ecosystem. *PLoS ONE* **2015**, *10*, e0135034. [CrossRef] [PubMed]
40. King, D.A.; Bachelet, D.M.; Symstad, A.J.; Ferschweiler, K.; Hobbins, M. Estimation of potential evapotranspiration from extraterrestrial radiation, air temperature and humidity to assess future climate change effects on the vegetation of the Northern Great Plains, USA. *Ecol. Model.* **2015**, *297*, 86–97. [CrossRef]
41. Hogg, E.H.; Barr, A.G.; Black, T.A. A simple soil moisture index for representing multi-year drought impacts on aspen productivity in the western Canadian interior. *Agric. For. Meteorol.* **2013**, *178–179*, 173–182. [CrossRef]
42. Wang, Y.; Hogg, E.H.; Price, D.T.; Edwards, J.; Williamson, T. Past and projected future changes in moisture conditions in the Canadian boreal forest. *For. Chron.* **2014**, *90*, 678–691. [CrossRef]
43. Chattopadhyay, N.; Hulme, M. Evaporation and potential evapotranspiration in India under conditions of recent and future climate change. *Agric. For. Meteorol.* **1997**, *87*, 55–73. [CrossRef]
44. Hobbins, M.; Ramirez, J.; Brown, T. Trends in pan evaporation and actual evapotranspiration across the conterminous US: Paradoxical or complementary? *Geophys. Res. Lett.* **2004**, *31*. [CrossRef]
45. Linacre, E.T. Estimating U.S. Class A pan evaporation from few climate data. *Water Int.* **1994**, *19*, 5–14. [CrossRef]
46. Hobbins, M.T.; Dai, A.; Roderick, M.L.; Farquhar, G.D. Revisiting the parameterization of potential evaporation as a driver of long-term water balance trends. *Geophys. Res. Lett.* **2008**, *35*, L12403. [CrossRef]
47. Kingston, D.G.; Todd, M.C.; Taylor, R.G.; Thompson, J.R.; Arnell, N.W. Uncertainty in the estimation of potential evapotranspiration under climate change. *Geophys. Res. Lett.* **2009**, *36*, L20403. [CrossRef]
48. Scheff, J.; Frierson, D.M.W. Scaling potential evapotranspiration with greenhouse warming. *J. Clim.* **2013**, *27*, 1539–1558. [CrossRef]
49. Dai, A. Increasing drought under global warming in observations and models. *Nat. Clim. Change* **2013**, *3*, 52–58. [CrossRef]
50. Dai, A. Characteristics and trends in various forms of the Palmer Drought Severity Index during 1900–2008. *J. Geophys. Res. Atmospheres* **2011**, *116*. [CrossRef]
51. Sheffield, J.; Wood, E.F.; Roderick, M.L. Little change in global drought over the past 60 years. *Nature* **2012**, *491*, 435–438. [CrossRef] [PubMed]

52. Amatya, D.; Tian, S.; Dai, Z.; Sun, G. Long-term potential and actual evapotranspiration of two different forests on the Atlantic Coastal Plain. *Trans. ASABE* **2016**, *59*, 647–660.
53. Brauman, K.A.; Freyberg, D.L.; Daily, G.C. Potential evapotranspiration from forest and pasture in the tropics: A case study in Kona, Hawai'i. *J. Hydrol.* **2012**, *440–441*, 52–61. [CrossRef]
54. NREL 2013 US Department of Energy, National Renewable Energy Laboratory. National Solar Radiation Data Base. Available online: http://rredc.nrel.gov/solar/old_data/nsrdb/ (accessed on 26 April 2016).
55. GSOD 2013 National Oceanic and Atmospheric Administration, Global Summary of the Day. Available online: https://data.noaa.gov/dataset/global-surface-summary-of-the-day-gsod (accessed on 31 October 2013).
56. Brouwer, C.; Heibloem, M. *Irrigation Water Management: Irrigation Water Needs*; Food and Agriculture Orgainzation of the United Nations: Rome, Italy, 1985.
57. Lu, J.; Sun, G.; McNulty, S.G.; Amatya, D. A comparison of six potential evapotranspiration methods for regional use in the Southeastern United States. *J. Am. Water Resour. Assoc.* **2005**, *41*, 621–633. [CrossRef]
58. Kelliher, F.; Leuning, R.; Schulze, E. Evaporation and canopy characteristics of coniferous forests and grasslands. *Oecologia* **1993**, *95*, 153–163. [CrossRef]
59. Landsberg, J.; Sands, P. *Physiological Ecology of Forest Production: Principles, Processes and Models*, 1st ed.; Elsevier Inc.: Boston, MA, USA, 2011.
60. Pereira, A.R. The Priestley–Taylor parameter and the decoupling factor for estimating reference evapotranspiration. *Agric. For. Meteorol.* **2004**, *125*, 305–313. [CrossRef]
61. Bladon, K.D.; Silins, U.; Landhäusser, S.M.; Lieffers, V.J. Differential transpiration by three boreal tree species in response to increased evaporative demand after variable retention harvesting. *Agric. For. Meteorol.* **2006**, *138*, 104–119. [CrossRef]
62. Humphreys, E.; Black, T.; Ethier, G.; Drewitt, G.; Spittlehouse, D.; Jork, E.; Nesic, Z.; Livingston, N. Annual and seasonal variability of sensible and latent heat fluxes above a coastal Douglas-fir forest, British Columbia, Canada. *Agric. For. Meteorol.* **2003**, *115*, 109–125. [CrossRef]
63. Martin, T.A.; Hinckley, T.M.; Meinzer, F.C.; Sprugel, D.G. Boundary layer conductance, leaf temperature and transpiration of Abies amabilis branches. *Tree Physiol.* **1999**, *19*, 435–443. [CrossRef] [PubMed]
64. Jarvis, P.G.; McNaughton, K.G. Stomatal Control of Transpiration: Scaling Up from Leaf to Region. In *Advances in Ecological Research*; Elsevier: Amsterdam, The Netherlands, 1986; Volume 15, pp. 1–49.
65. Running, S.; Coughlan, J. A General-Model of Forest Ecosystem Processes for Regional Applications. 1. Hydrologic Balance, Canopy Gas-Exchange and Primary Production Processes. *Ecol. Model.* **1988**, *42*, 125–154. [CrossRef]
66. Bates, D.M. *Lme4: Mixed-Effects Modeling with R*; Springer: New York, NY, USA, 2010.
67. Kalnay, E.; Kanamitsu, M.; Kistler, R.; Collins, W.; Deaven, D.; Gandin, L.; Iredell, M.; Saha, S.; White, G.; Woollen, J.; et al. The NCEP/NCAR 40-year reanalysis project. *Bull. Am. Meteorol. Soc.* **1996**, *77*, 437–471. [CrossRef]
68. Schroeder, T.; Hember, R.; Coops, N.; Liang, S. Validation of solar radiation surfaces from modis and reanalysis data over topographically complex terrain. *J. Appl. Meteorol. Climatol.* **2009**, *48*, 2441–2458. [CrossRef]
69. NARR 2013 National Oceanic and Atmospheric Administration, Earth System Research Laboratory. North American Regional Reanalysis. Available online: www.esrl.noaa.gov/psd/data/gridded (accessed on 18 December 2013).
70. ESRI. ArcGIS 2013 ArcGIS 9.2 Desktop Help. Available online: webhelp.esri.com/arcgisdesktop/9.2/index.cfm?TopicName=Area_Solar_Radiation (accessed on 23 July 2013).
71. Cogley, J.G. Greenland accumulation: An error model. *J. Geophys. Res. Atmospheres* **2004**, *109*. [CrossRef]
72. Press, W.; Teukolsky, S.; Vetterling, W.; Flannery, B. *Numerical Recipes: The Art of Scientific Computing*, 3rd ed.; Cambridge Universiy Press: New York, NY, USA, 2007.
73. Wang, T.; Hamann, A.; Spittlehouse, D.; Aitken, S. Development of scale-free climate data for western Canada for use in resource management. *Int. J. Climatol.* **2006**, *26*, 383–397. [CrossRef]
74. Wang, T.; Hamann, A.; Spittlehouse, D.; Carroll, C. Locally downscaled and spatially customizable climate data for historical and future periods for north america. *PLoS ONE* **2016**, *11*, e0156720. [CrossRef] [PubMed]
75. EC-NCDIA 2013 Environment Canada, National Climate Data and Information Archive. 2013. Available online: http://climate.weather.gc.ca/index_e.html (accessed on 5 January 2013).

76. Menne, M.J.; Williams, C.N.; Vose, R.S. *United States Historical Climatology Network (USHCN)*; Version 2.5 Serial Monthly Dataset; Oak Ridge National Laboratory, Carbon Dioxide Information Analysis Center: Oak Ridge, TN, USA.
77. EC-AHCCD 2013 Environment Canada, Adjusted and Homogenized Canadian Climate Data. 2013. Available online: http://www.ec.gc.ca/dccha-ahccd/ (accessed on 15 July 2013).
78. CRU31 2013 UK Met Office, Climate Research Unit. *Climate Database*, Version 3.1; Available online: http://badc.nerc.ac.uk/view/badc.nerc.ac.uk__ATOM__dataent_1256223773328276 (accessed on 25 July 2013).
79. Kimball, J.; Running, S.; Nemani, R. An improved method for estimating surface humidity from daily minimum temperature. *Agric. For. Meteorol.* **1997**, *85*, 87–98. [CrossRef]
80. Thornton, P.; Running, S.; White, M. Generating surfaces of daily meteorological variables over large regions of complex terrain. *J. Hydrol.* **1997**, *190*, 214–251. [CrossRef]
81. USGS National Atlas of the United States. Available online: https://nationalmap.gov/small_scale/ (accessed on 13 March 2015).
82. Coakley, J.A. Reflectance and albedo, surface. In *Encyclopedia of the Atmospheric Sciences*; Holton, J.R., Curry, J.A., Pyle, J.A., Eds.; Academic Press: Cambridge, MA, USA, 2002; pp. 1914–1923.
83. Oak Ridge National Laboratory Distributed Active Archive Center (ORNL DAAC). *FLUXNET Web Page*; ORNL DAAC: Oak Ridge, TN, USA, 2015.
84. CEC 1997 Commission for Environmental Cooperation (CEC): Ecological Regions of North America—Towards a Common Perspective. GIS Data (shapefiles, metadata and symbology). Available online: http://www.cec.org/tools-and-resources/map-files/terrestrial-ecoregions-level-i (accessed on 29 November 2009).
85. Isaac, V.; van Wijngaarden, W.A. Surface water vapor pressure and temperature trends in North America during 1948–2010. *J. Clim.* **2012**, *25*, 3599–3609. [CrossRef]
86. Vincent, L.A.; Zhang, X.; Brown, R.D.; Feng, Y.; Mekis, E.; Milewska, E.J.; Wan, H.; Wang, X.L. Observed Trends in Canada's Climate and Influence of Low-Frequency Variability Modes. *J. Clim.* **2015**, *28*, 4545–4560. [CrossRef]
87. Vörösmarty, C.J.; Federer, C.A.; Schloss, A.L. Potential evaporation functions compared on US watersheds: Possible implications for global-scale water balance and terrestrial ecosystem modeling. *J. Hydrol.* **1998**, *207*, 147–169. [CrossRef]
88. Yao, H. Long-term study of lake evaporation and evaluation of seven estimation methods: Results from Dickie Lake, south-central Ontario, Canada. *J. Water Resour. Prot.* **2009**, *1*, 59–77. [CrossRef]

4

Change in Future Rainfall Characteristics in the Mekrou Catchment (Benin), from an Ensemble of 3 RCMs (MPI-REMO, DMI-HIRHAM5 and SMHI-RCA4)

Ezéchiel Obada [1,2,*], Eric Adéchina Alamou [2], Josué Zandagba [1,2], Amédée Chabi [1,2] and Abel Afouda [2,3]

1. International Chair in Mathematical Physics and Applications (ICMPA), University of Abomey-Calavi (UAC), Cotonou 072 B.P 50, Benin; zjosua@yahoo.fr (J.Z.); amedees2005@yahoo.fr (A.C.)
2. Laboratory of Applied Hydrology, University of Abomey-Calavi (UAC), Cotonou 01 BP 4521, Benin; ericalamou@yahoo.fr (E.A.A.); aafouda@yahoo.fr (A.A.)
3. West African Science Service Center on Climate Change and Adapted Land Use (WASCAL), GRP Climate Change and Water Resources, University of Abomey-Calavi (UAC), Abomey-Calavi BP 2008, Benin

* Correspondence: e.obada83@yahoo.fr

Abstract: This study analyzes the impact of climate change on several characteristics of rainfall in the Mekrou catchment for the twenty-first century. To this end, a multi-model ensemble based on regional climate model experiments considering two Representative Concentration Pathways (RCP4.5 and RCP8.5) is used. The results indicate a wider range of precipitation uncertainty (roughly between −10% and 10%), a decrease in the number of wet days (about 10%), an increase (about 10%) of the total intensity of precipitation for very wet days, and changes in the length of the dry spell period, as well as the onset and end of the rainy season. The maximum rainfall amounts of consecutive 24 h, 48 h and 72 h will experience increases of about 50% of the reference period. This change in rate compared to the reference period may cause an exacerbation of extreme events (droughts and floods) in the Mekrou basin, especially at the end of the century and under the RCP8.5 scenario. To cope with the challenges posed by the projected climate change for the Mekrou watershed, strong governmental policies are needed to help design response options.

Keywords: climate change; precipitation; future projections; trends; extreme events

---

1. Introduction

These last decades, climate change and variability issues have been the main focus of scientists and policy makers around the world. Projected temperatures and precipitation under different scenarios show that climate change will have different impacts in different regions of the globe, with spatio-temporal changes in the occurrence and amounts of rainfall, but usually with an increase in temperature [1–6] in unequal proportions since the Paleolithic period [7,8]. The increase in temperature will affect rainfall and its variability, in particular, droughts and floods [9,10]. The impacts of climate change will vary across regions and populations, through space and time, depending on multiple factors, including non-climate stressors and the extent of mitigation and adaptation [11]. According to [12], Africa is one of the most vulnerable continents to climate disturbances due to the diversity of effects of multiple stress and the low adaptive capacity. Recent work in the West African region [13,14], under Representative Concentration Pathway (RCP4.5 and RCP8.5) scenarios of *climate change* projections indicate that continued warming (1–6.5 °C increase in temperature), and great uncertainty in rainfall (between −10% and 10%) will be observed in the Sahel until 2100. According to these authors, the remaining part of West Africa will also experience a more intense extreme climate in the future, but to a lesser extent. It is also reported by previous studies that the rainy and agricultural seasons will become shorter [15,16] while the torrid, arid and semi-arid climate will extend to Sahel [17,18]. Such conditions can have important constraints on agricultural activities, water resource management, etc.

Given the alarming regional predictions, it is important to know the precipitation forecasts at local scales in order to propose adaptation and mitigation measures falling within the regional context but with local specificities. It is in this vein that this paper was initiated and aims at studying trends in annual precipitation parameters (annual precipitation amounts, maximum amounts of rainfall in 24, 48 and 72 consecutive hours, the beginning and end dates of the rainy season, the length of the rainy season, the number of wet days in the year, the length of the dry spell period and total intensity of rainfall for very wet days) in the Mekrou basin over the period 1981–2100.

2. Study Area, Data and Method

Our study area is the Mekrou watershed in Kompongou. Covering an area of 5670 km^2, it is located in the North of Benin between 1°30′ and 2°15′ East Longitude and 10°20′ and 11°30′ North Latitude. With an elongated shape, it covers three main cities, i.e., Kérou, Kouandé and Péhunco. This watershed belongs to the Benin side of the Niger Basin. The highest point of watershed is Kampuya (639 m) around Kouandé, while the lowest point (266 m) is located around Kérou and is precisely in the bed of the Mekrou River [19]. The average slope of the stream bed is approximately 2.47%. The soil and the vegetation types encountered in the basin are ferruginous soils on crystalline bedrock, histosols, swamps and fertile gallery forests [20,21]. The rainfall amounts between 1981 and 2014 show that the months of July, August and September are the wettest (Figure 1). The discharge of the Mekrou River at Kompongou varies from 250 m^3·s^{-1} in September to 0 m^3·s^{-1} from December to April. The annual mean of flow is about 21 m^3·s^{-1}. High flows occur mostly during August to October.

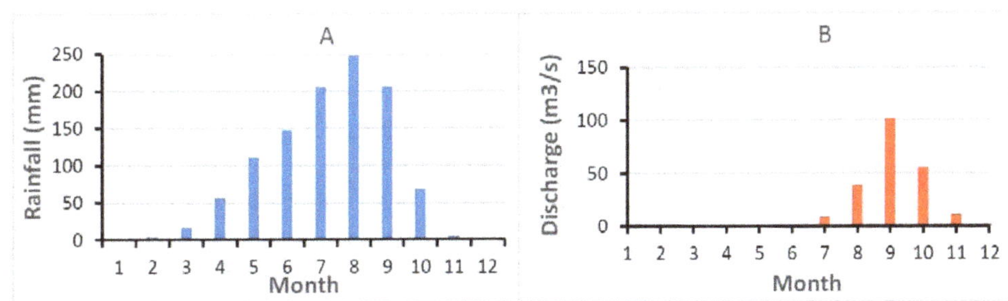

Figure 1. Seasonal variability of rainfall and discharge in the basin (**A**: Monthly rainfall amounts over 1981–2014; **B**: Monthly mean of discharge over 2006–2012).

The data used in this study are of two types: observed data and simulated data from Regional Climate Models (RCMs). The two types of data are daily rainfall data. The first were obtained from the National Directorate of Meteorology of Benin. Across the whole watershed, 02 rain gauges are available: these are the Kouandé and Kérou gauges. In addition to these 02 gauges, 12 rain gauges distributed around the basin are also available (Figure 2), resulting in a total of 14 rainfall stations. Over the period 1981–2010, all precipitation stations were functional.

The simulated data are future (RCP4.5 and RCP8.5 scenarios) rainfall projection data of three regional models (SMHI-RCA4, MPI-REMO, DMI-HIRHAM5) obtained from the CORDEX Africa project. Their characteristics are shown in Table 1. The future projections, i.e., the RCP4.5 and RCP8.5 scenarios, are considered over the period from 2011 to 2100. The data bias was corrected using the Empirical Quantile Mapping Method [22].

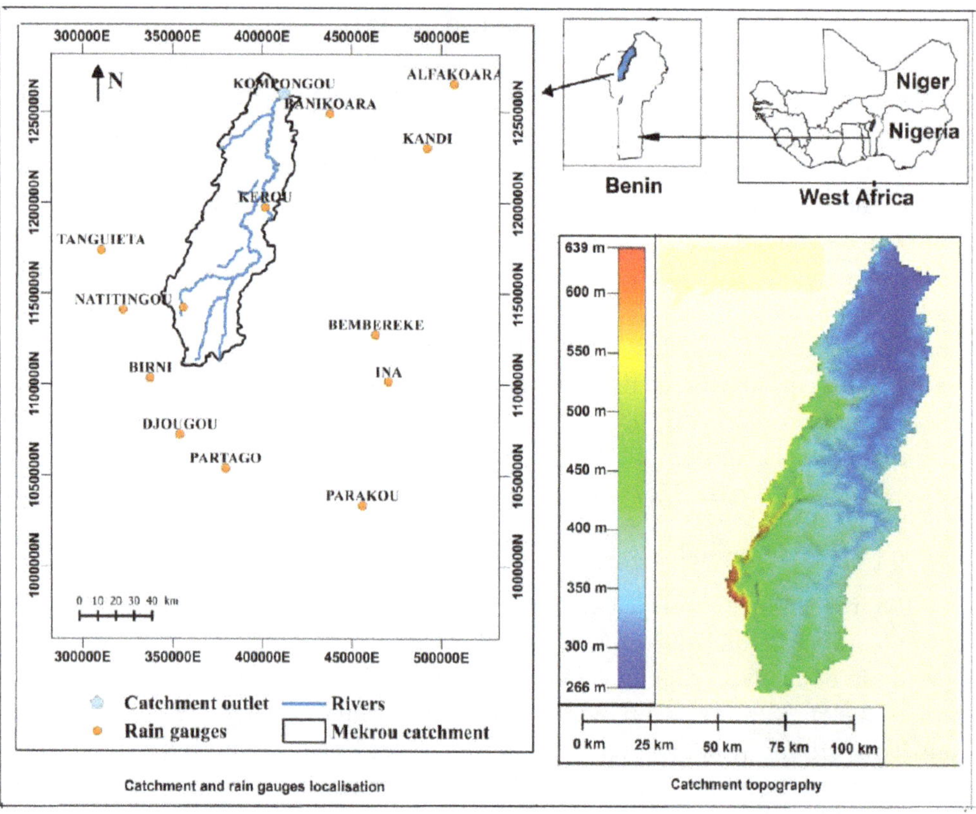

Figure 2. Study area.

Table 1. Main characteristics of the RCMs.

Model (RCM)	Institution	Driving GCM	Horizontal Resolution	No. of Vertical Levels	Simulation Period	Reference
HIRHAM5	DMI	GFDL-ESM2M	50 km	31	1951–2100	[23]
REMO	CSC	MPI-ESM-LR	50 km	27	1951–2100	[24]
RCA4	SMHI	EC-EARTH	50 km	40	1951–2100	[25]

The precipitation data from the three RCMs are used as an ensemble that represents the average of the three models. [22] corrected the bias of the precipitation simulated by each of these models and their ensemble with several bias correction methods and concluded that the ensemble of the 3 RCMs better simulated the rainfall than each RCM over the historical period for the same study area. Moreover, according to [11], in Oust Africa, climate models do not converge for the prediction of precipitation. For this, it is better to use a set of climate models in order to reduce the uncertainties of predictions.

The rates of change were calculated by considering four (04) different periods. The first period is the reference period (1981–2010). The three other periods are the projected periods (2011–2040, 2041–2070 and 2071–2100). For each parameter of precipitation and the rainfall stations, the rate of change was calculated using Equation (1).

$$Change\ rate = \frac{\overline{X}_p - \overline{X}_r}{\overline{X}_r} \times 100, \tag{1}$$

where $\overline{X}_p$ is the mean of a parameter over the considered projected period, and $\overline{X}_r$ is the mean of the same parameter over the reference period.

The interpolation was performed using the Kriging method. The first step is to construct a spatial structure of each change rate using a semivariogram. The spherical model $\gamma_{mod}(h)$ (Equation (2)), where h is the distance between two rain gauges, was adopted to adjust the sample semivariogram. A regular grid point was adopted, and ordinary Kriging, which assumes an unknown mean as well as a second-order stationary process, was implemented.

$$\gamma_{mod}(h) = S\left[3h/2a - 0.5(h/a)^3\right], \quad (2)$$

where S is the sill and a is the range of the semivariogram.

3. Results

3.1. Mean Values of Study Parameters for Reference Period (1981–2010) in the Mekrou Basin at Kompongou

The 1970s were characterized by a severe drought in West Africa [26–33]. Over the whole of West Africa, rainfall decreased by 180 mm compared to the previous period [30]. The following decades were characterized by a recovery of rainfall amounts. According to [15], in the last two decades, not only the total annual precipitation increased, but rainy days occurred more frequently, which led to the partial recovery of precipitation amounts. An increasing trend of annual precipitation was thus observed [13,34–36]. This recovery of precipitation amounts was mainly due to the direct influence of higher levels of anthropogenic greenhouse gases in the atmosphere, as well as changes in the emissions of anthropogenic aerosol precursors [37–40]. Above all, natural variability could also play a significant role in this recovery [41].

In order to quantify future changes, some characteristic parameters of rainfall in the Mekrou basin at the outlet of Kompongou were evaluated for the period from 1981 to 2010 and are considered as reference data for evaluating future rainfall trends. The parameters studied were: the annual rainfall amounts, maximum amounts of rainfall in consecutive 24, 48 and 72 h that can be used as extreme event indicators, the dates of the beginning and end of the rainy season (determined by the method of "anomalous accumulation" proposed by [42], the length of the rainy season, the number of wet days in the year, the length of the dry spell period (maximum number of consecutive days without precipitation or precipitation less than 1 mm) that characterizes the occurrence of droughts [43] and total precipitation intensity of very wet days in a year (wet days are considered as the days with precipitation amounts greater than or equal to 1 mm, and very wet days are days with precipitation amounts above the 95th percentile), which measures the occurrence of floods [44].

Despite the low density of rain gauges in the study area, errors related to the interpolation of the different parameters are relatively low and vary between 0 and 0.1 (Figure 3k). These low values of the variance show that the number of rain gauges used are sufficient.

The mean annual precipitation amounts over the reference period varies from 900 to 1200 mm in the basin (Figure 3a). The spatial distribution of annual rainfall amounts shows a reverse direction of change with respect to latitude. There are 3 major areas of precipitation. An area of heavy rainfall (average rainfall of 1100–1200 mm per year) between 10° N and 10.5° N, a medium rainfall zone in the basin (1000–1100 mm) between 10.5° N and 11° N, and an area of low annual rainfall amounts (below 1000 mm) between 11° N and 11.5° N. Numbers of wet days in the basin have a spatial distribution almost identical to that of the rainfall amounts. Indeed, in general, the average number of wet days in the basin ranges from 55 to 75 days per year (Figure 3b). Note the existence of three areas with identical variations to those of rainfall amounts. The length of the dry spell period ranges from 140 to 170 days. The length of the drought period also has a spatial distribution that shows the existence of three zones of different drought periods identical to those previously identified (Figure 3c). Here, however, the numbers of consecutive days without rain move in the same direction as the latitude. The mean of the onset dates of the rainy season vary between the 125th day in the year (4 May) and the 141st day (20 May) (Figure 3d). The spatial distribution of the onset of the rainy season also shows an evolution depending on the latitude. The more the latitude increases, the more the rainy seasons are slow to start. By contrast, the ends of the rainy season vary little in the basin (274th to 281st) (Figure 3e) and correspond generally to the first week of October. Regarding the lengths of the rainy seasons, they vary between 135 and 155 (Figure 3f). They decrease with increasing latitude. Therefore, the seasons are longer in the South Basin than in the North. The maximum precipitation amounts in consecutive 24 h, 48 h and 72 h decrease as the latitude increases; they vary between 64 and 80 mm, 75 mm and 95 mm and 85 and 110 mm, respectively. The percentages of total precipitation amounts of very wet days compared to total precipitation vary from 57%–64% in the basin. These rates increase with latitude, indicating that the northern basin is more exposed to extreme events such as floods.

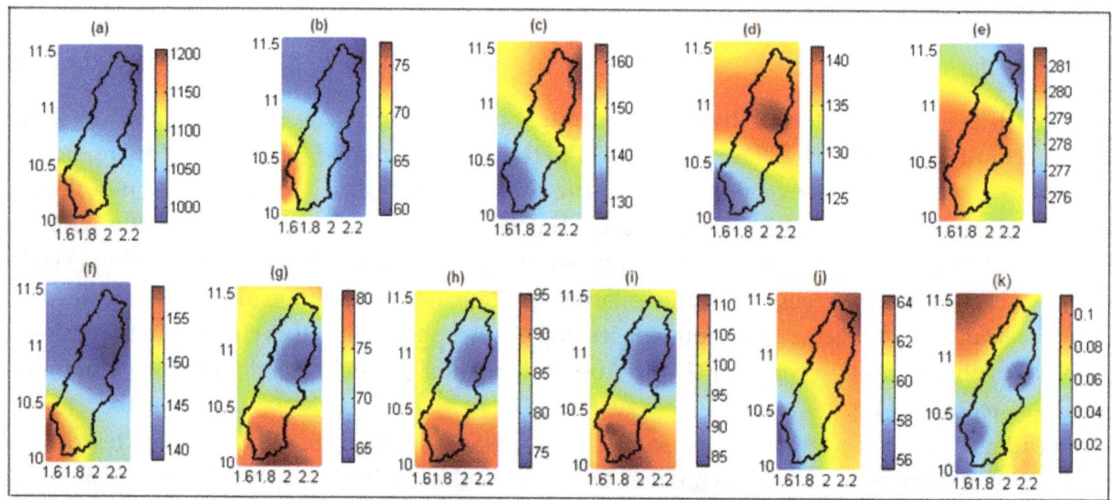

Figure 3. Mean values of rainfall characteristics over the period 1981–2010 (**a**: annual precipitation amounts (mm); **b**: number of wet days (day); **c**: dry spell length (day); **d**: onset of rainy seasons (Nth day of year); **e**: end of rainy seasons (Nth day of year); **f**: length of rainy seasons (day); **g**: maximum precipitation amounts during 24 consecutive hours (mm); **h**: maximum precipitation amounts during 48 consecutive hours (mm); **i**: maximum precipitation amounts during 72 consecutive hours (mm); **j**: total precipitation intensity of very wet days (%)) **k**: Kriging variance.

3.2. Future Changes in Precipitation Characteristics

3.2.1. Annual Precipitation Amounts

Changes in future annual rainfall amounts compared to the reference period are either positive or negative depending on the periods and the scenarios of the greenhouse gas emissions considered. In all cases, these changes are within the range predicted, which is between −30% and 30% of the reference period for West Africa [44], indicating that the projected rainfalls are very uncertain in the region. Changes in the Mekrou basin over the period from 2011 to 2040 show a decrease of 0%–5% of the precipitation amounts compared to the reference period for the RCP4.5 scenario (Figure 4a) and a decrease of 5%–12% for the RCP8.5 scenario (Figure 4e).

The period 2041–2070 is characterized by a net decrease in precipitation amount (3%–6%) inside the basin for the RCP4.5 scenario (Figure 4b) and a variation between −4% and 5% for the RCP8.5 scenario (Figure 4f). For both scenarios considered, future projections show an increase in precipitation amounts over the period 2071–2100. These increases vary from 0 to 6% for the RCP4.5 scenario (Figure 4c) and from 4%–12% for the RCP8.5 scenario. The general mean (2011–2100) indicates a decrease in precipitation amounts of 0-4% for the RCP4.5 scenario (Figure 4d) versus a variation of −4% to 4% for the RCP8.5 scenario (Figure 4h). These results are consistent with those of [13], who predicted rates of change of −10% to 10% for future rainfall in West Africa and those of [15], who stated that precipitation projections are uncertain in West Africa. The spatial variation of projected rainfall indicates a deficit of rainfall in the northern half of the basin regardless of the scenario and the period considered (2071–2100), with a growth rate of about 4%.

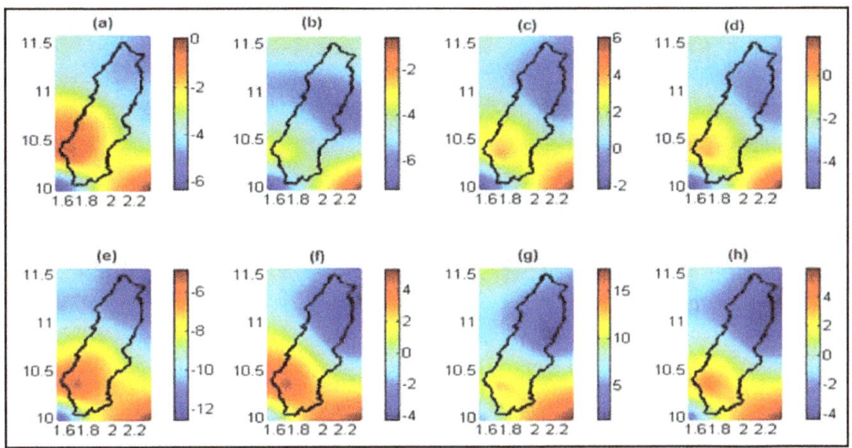

Figure 4. Changes in annual precipitation amounts (%) compared to the reference period: **a**: RCP4.5-2011–2040; **b**: RCP4.5-2041–2070; **c**: RCP4.5-2071–2100; **d**: RCP4.5-2011–2100; **e**: RCP8.5-2011–2040; **f**: RCP8.5-2041–2070; **g**: RCP8.5-2071–2100; **h**: RCP8.5-2011–2100.

3.2.2. Number of Wet Days

The average number of wet days according to the RCP4.5 scenario will decrease by about 0%–3% in the South of the basin versus a slight increase of about 0%–1% toward the outlet of the basin for the period 2011–2040 (Figure 5a). However, for the periods of 2041 to 2070 (Figure 5b) and 2071 to 2100 (Figure 5c), the average number of wet days will decrease throughout the basin, by between 4%–11%. This indicates a downward trend of wet days in the basin over the period 2011–2100 with an average decrease between 2%–7% (Figure 5d). According to the RCP8.5 scenario, there will be a downward trend in the number of wet days in the watershed regardless of the considered period. Moreover, this decrease would be greater than the one projected by the RCP4.5 scenario. In fact, over the period from 2011 to 2040, the decrease ranges from 5% to 10% (Figure 5e). The same range of variation is observed over the period from 2041 to 2070 (Figure 5f). The decrease of number of wet days will be more pronounced over the period from 2071 to 2100 with a decrease between 6% to 13% (Figure 5g) compared to the reference period (1981–2010). Over the period from 2011 to 2100, there will be a decrease of 4%–10% (Figure 5h) with respect to the period 1981-2010. We noted that regardless of the period considered and the scenario, the decrease was greater in the northern part of the basin than in the southern part. These results confirm those of [15], with decreases of 7%, 3.1% and 2.1% for rainy days using the CCLM, RCA and REMO models, respectively, over the period from 2021–2050 in Burkina-Faso.

3.2.3. Onset of the Rainy Seasons

The onset dates of the rainy seasons will also experience changes in the basin. Depending on the scenario and the period considered, the future evolution compared to the period of 1981 to 2010 can be positive, negative or both. Over the period from 2011 to 2040, the scenarios RCP4.5 and RCP8.5 (Figure 6a,e) yield a decrease of 0%–5% of the onset of the rainy seasons (rainy season will start one week early) while over the period 2041–2070, the variation compared to the reference period ranges from −8% to 0% for the RCP4.5 scenario (Figure 6b) and −6% to 5% for the RCP8.5 scenario (Figure 6f). Over the period 2071–2100, the RCP4.5 scenario yields an increase of 0%–10% for the onset dates of the rainy season compared to the reference period (Figure 6c), while the scenario RCP8.5 yields a variation of −5% to 10% for the dates of the onset of the rainy seasons compared to the period of reference (Figure 6g). For the period 2011 to 2100, both scenarios project changes varying from −5%

to 5% (an uncertainty of one week) for the onset dates of the rainy season compared to the reference period (Figure 6d,h). These results indicate that no trend has emerged with respect to the evolution of the onset of rainy seasons but depending on the period, the scenario and the portion of the basin considered, the rainy seasons will begin one week earlier or two weeks later. Similar results were obtained by [15] using 5 different RCMs.

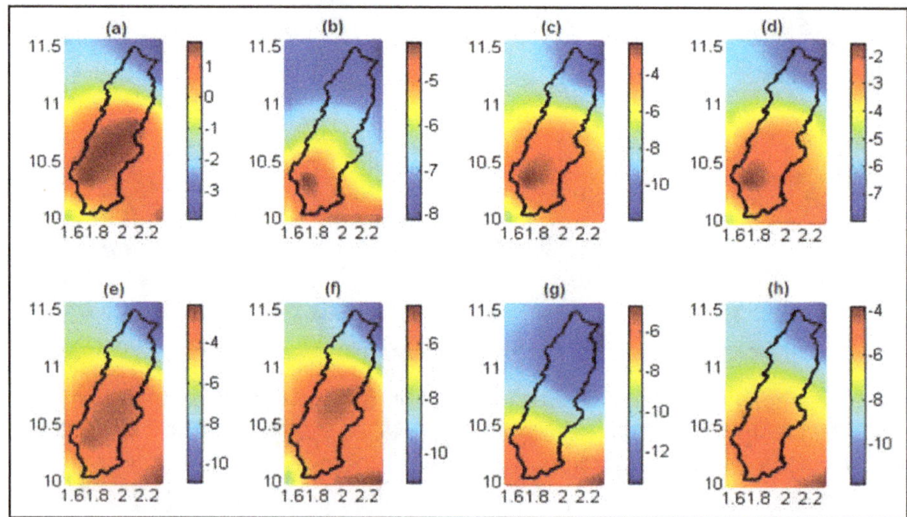

Figure 5. Changes in the number of wet days (%) compared to the reference period **a**: RCP4.5-2011–2040; **b**: RCP4.5-2041–2070; **c**: RCP4.5-2071–2100; **d**: RCP4.5-2011–2100; **e**: RCP8.5-2011–2040; **f**: RCP8.5-2041–2070; **g**: RCP8.5-2071–2100; **h**: RCP8.5-2011–2100.

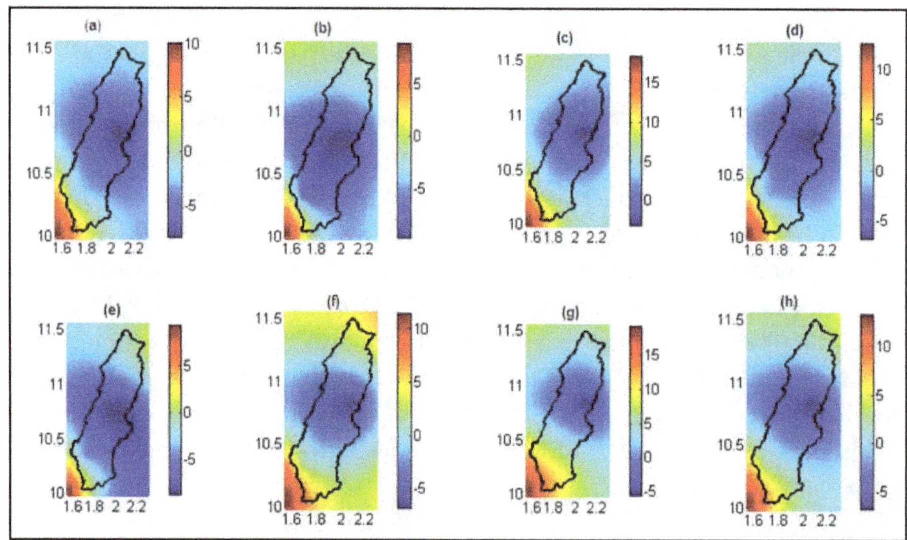

Figure 6. Changes in the onset of the rainy season (%) compared to the reference period **a**: RCP4.5-2011–2040; **b**: RCP4.5-2041–2070; **c**: RCP4.5-2071–2100; **d**: RCP4.5-2011–2100; **e**: RCP8.5-2011–2040; **f**: RCP8.5-2041–2070; **g**: RCP8.5-2071–2100; **h**: RCP8.5-2011–2100.

3.2.4. End of the Rainy Seasons

The end dates of the rainy seasons will be earlier compared to reference period, regardless of the scenario and the future period considered. However, these decreases are not important and are 1%–2.2% for the period from 2011 to 2040 under both the RCP4.5 and RCP8.5 scenarios (Figure 7a,e). Over the period 2041–2070, the rates of decrease are in the order of 0.4%–1% for the RCP4.5 scenario (Figure 7b) and 1.5%–2.5% for the RCP8.5 scenario (Figure 7f). Over the period 2071–2100, the RCP4.5

scenario projects a declining rate that ranges between 0% and 2% (Figure 7c) versus the rate of decrease of 0.5%–3% for the RCP8.5 scenario compared to the values of the reference period (Figure 7g). Over the period from 2011 to 2100, the RCP4.5 scenario predicts earlier end datesof the rainy season, which range from 0.6%–1.6% (Figure 7d), whereas for the RCP8.5 scenario, these decreases are between 1% and 2.7% (Figure 7h) with respect to the reference period values. These results indicate that the season will end earlier and will probably decrease the length of the rainy seasons. These results are in contrast to those of [15], who predicted an later end date of the rainy season for five models used in Burkina-Faso.

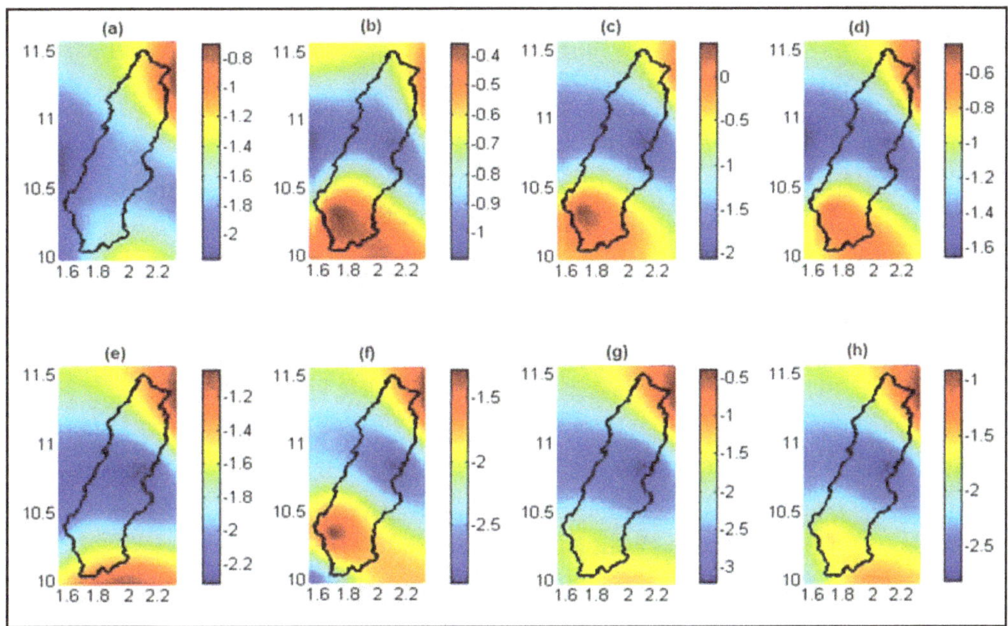

Figure 7. Changes in the end of the rainy season (%) compared to the reference period **a**: RCP4.5-2011–2040; **b**: RCP4.5-2041–2070; **c**: RCP4.5-2071–2100; **d**: RCP4.5-2011–2100; **e**: RCP8.5-2011–2040; **f**: RCP8.5-2041–2070; **g**: RCP8.5-2071–2100; **h**: RCP8.5-2011–2100.

3.2.5. Length of the Rainy Seasons

The length of the rainy seasons will experience changes in the future. These changes will vary from one period to another, and their changes compared to the reference period appear to be related to the greenhouse gas emission scenarios (Figure 8). Over the period from 2011 to 2040, both the RCP4.5 and RCP8.5 scenarios predict changes that are both negative and positive in the lengths of the rainy seasons compared to the reference period (Figure 8a,e). These changes are almost identical for both scenarios and generally range from −5% to 3%. Over the period 2041–2070, the RCP4.5 scenario projects an increase of 0%–6% in the lengths of the rainy seasons (Figure 8b) compared to the reference period, while the RCP8.5 scenario projects a decrease in the length of the rainy seasons (between 0%–8%; Figure 8f). Large decreases in the length of the rainy seasons will be observed in the period from 2071 to 2100. These decreases vary from 2% to 10% for the RCP4.5 scenario (Figure 8c). They are more pronounced for the RCP8.5 scenario and are between 3% and 15% (Figure 8g) with respect to the reference period values. Generally, the RCP4.5 scenario shows variations ranging from −6% to 3% (Figure 8d) for the length of the rainy seasons in the period from 2011 to 2100 compared to the reference period, while the scenario RCP8.5 indicates decreases of 0%–8% (Figure 8h). For the period 2011–2040, there is uncertainty in the trend of the change in the length of the rainy seasons in the basin; it is clear that for the following periods there will be a net decrease in the length of the rainy seasons, especially for RCP8.5 scenario.

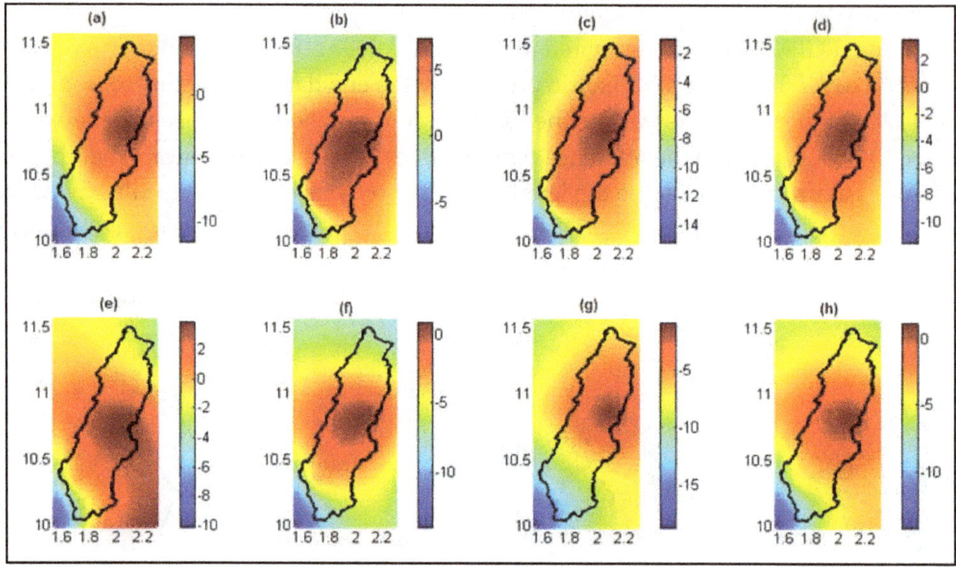

Figure 8. Changes in the length of the rainy season (%) compared to the reference period: **a**: RCP4.5-2011–2040; **b**: RCP4.5-2041–2070; **c**: RCP4.5-2071–2100; **d**: RCP4.5-2011–2100; **e**: RCP8.5-2011–2040; **f**: RCP8.5-2041–2070; **g**: RCP8.5-2071–2100; **h**: RCP8.5-2011–2100.

3.2.6. Length of Dry Spells

Figure 9 shows that there will be an increase in the length of dry spells irrespective of the period and the scenario considered. In fact, both the RCP4.5 and RCP8.5 scenarios predict an increase of 0%–10% in the length of the dry spells in the period from 2011 to 2040 compared to the reference period (Figure 9a,e). In the period 2041–2070, there will be an accentuation of the increase in the length of the dry spells compared to the reference period. This increase ranges between 4% and 12% for the RCP4.5 scenario (Figure 9b) versus 4% to 15% for the RCP8.5 scenario (Figure 9f). In the period from 2070 to 2100, these increases range from 5% to 9% for the RCP4.5 scenario (Figure 9c) versus 0% to 10% for the RCP8.5 scenario (Figure 9g). These results are in line with those of other authors who also predict an increasing trend in the lengths of dry spells in the West Africa area [13,15].

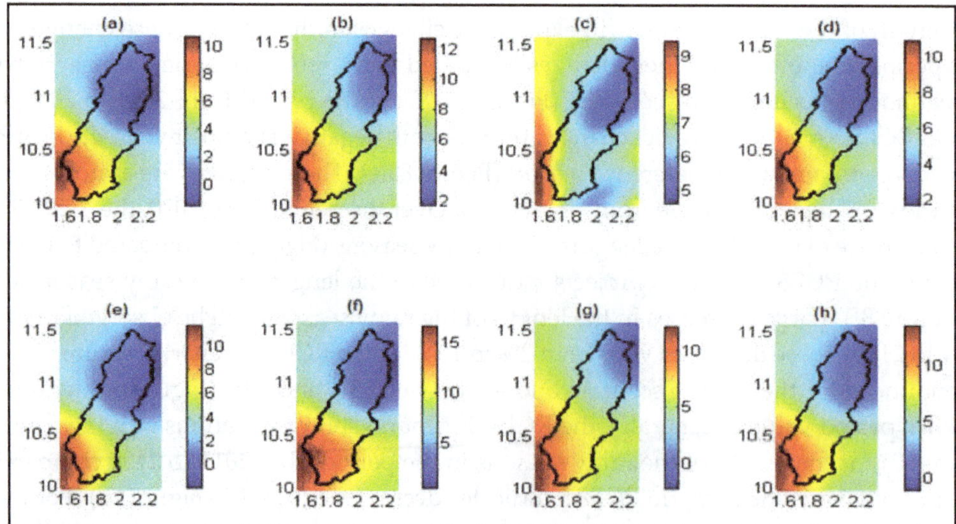

Figure 9. Changes in the lengths of the dry spells (%) compared to the reference period: **a**: RCP4.5-2011–2040; **b**: RCP4.5-2041–2070; **c**: RCP4.5-2071–2100; **d**: RCP4.5-2011–2100; **e**: RCP8.5-2011–2040; **f**: RCP8.5-2041–2070; **g**: RCP8.5-2071–2100; **h**: RCP8.5-2011-2100.

3.2.7. Total Precipitation Intensity of Very Wet Days

The total precipitation intensity of very wet days will also experience changes in future years depending on the scenario considered. In general (period of 2011–2100), the RCP4.5 scenario projects increases of 3%–5% (Figure 10d) for the total precipitation intensity of very wet days compared to the reference period. These increases are more important under the RCP8.5 scenario and vary from 6%–9% for the same period (Figure 10h). However, when one considers the different sub-periods, we notice that over the period from 2011 to 2040 and for the RCP4.5 scenario, the rate of change in the total precipitation intensity of the very wet days is very low and ranges from −1% to 1% (Figure 10a). For the same period, the RCP8.5 scenario projects an increase of 2%–5% compared to the reference period (Figure 10e). In the next period (2041–2070), there will be an increase in the total precipitation intensity of very wet days compared to the previous period. These intensifications result in an increase of 5%–7% (Figure 10b) of total precipitation intensity of very wet days relative to the reference period for the RCP4.5 scenario and 7%–10% for the RCP8.5 scenario (Figure 10f). The intensity of total precipitation for very wet days in the period 2071–2100 will be the highest in the entire period from 1981–2100. In fact, the RCP4.5 scenario projects growth rates of 4%–9% (Figure 10C) compared to the reference period, while the RCP8.5 scenario projects rates of 8%–12%, which are the highest across periods and scenarios. The increase in the total precipitation intensity of very wet days confirms the trend of exacerbation of extreme events in West Africa announced by [8].

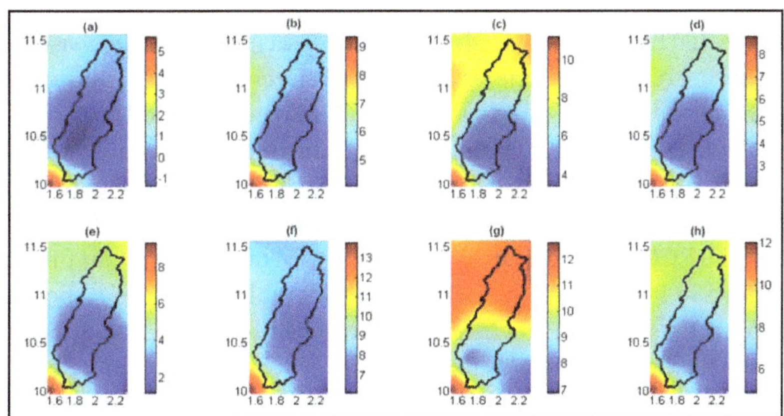

Figure 10. Changes in the total precipitation intensity of very wet days (%) compared to the reference period: **a**: RCP4.5-2011–2040; **b**: RCP4.5-2041–2070; **c**: RCP4.5-2071–2100; **d**: RCP4.5-2011–2100; **e**: RCP8.5-2011–2040; **f**: RCP8.5-2041–2070; **g**: RCP8.5-2071–2100; **h**: RCP8.5-2011–2100.

With the objective of assessing the occurrence and intensity of extreme events, changes in the maximum precipitation amounts in 24, 48 and 72 consecutive hours per year were studied.

3.2.8. Maximum Precipitation Amounts During 24 Consecutive Hours

In general, the two scenarios project an upward trend of daily maximum precipitation amounts. The average maximum daily precipitation amounts over the period of 2011–2100 show increases ranging from 5 to 15% compared to the reference period for the RCP4.5 scenario (Figure 11d) and 22%–30% for the RCP8.5 scenario (Figure 11h). Irrespective of the scenario considered, the maximum daily precipitation increases over time, i.e, 0%–7% for the period 2011–2040 (Figure 11a), 0%–15% for the period 2041–2070 (Figure 11b) and 10%–25% for the period 2071–2100 (Figure 11c) compared to the period 1981–2010 for RCP4.5 scenario. For the RCP8.5 scenario, the increase in maximum daily precipitation amount compared to the reference period is higher than that projected by the RCP4.5 scenario and is about 4%–11% for the period 2011–2040 (Figure 11e), 20%–35% for the period 2041–2070 (Figure 11f) and 35%–45% for the period 2071–2100 (Figure 11g). This indicates an intensification of extreme rainy events in the basin, which may increase the occurrence of floods.

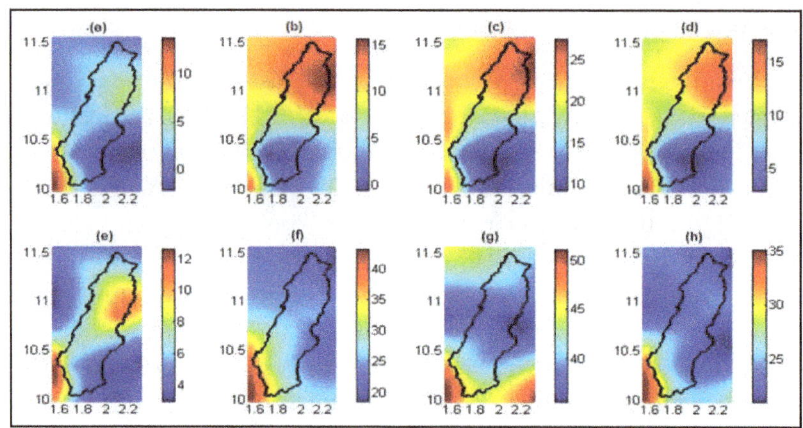

Figure 11. Changes in the maximum precipitation amounts during 24 consecutive hours (%) compared to the reference period: **a**: RCP4.5-2011–2040; **b**: RCP4.5-2041–2070; **c**: RCP4.5-2071–2100; **d**: RCP4.5-2011–2100; **e**: RCP8.5-2011–2040; **f**: RCP8.5-2041–2070; **g**: RCP8.5-2071-2100; **h**: RCP8.5-2011–2100.

3.2.9. Maximum Precipitation Amounts During 48 Consecutive Hours

The maximum precipitation amounts during 48 consecutive hours also showed an upward trend over the period 2011–2100 for both the RCP4.5 and RCP8.5 scenarios. The increase rates are 13%–31% for the RCP4.5 scenario compared to the values of the reference period (Figure 12d). For scenario RCP8.5, the increase rates vary from 34% to 41% (Figure 12h). For the periods 2011–2040 and 2041–2070, the change rate in the maximum precipitation amounts during 48 consecutive ours projected by the RCP4.5 scenario are identical and represent an increase of about 8%–25% (Figure 12a,b) compared to the reference period, but for the period 2071–2100, these rates reach 22%–42% (Figure 12c). With respect to the projections of the RCP8.5 scenario, the rates of increase of the maximum precipitation amounts during 48 h compared to the reference period are considerably higher than those projected by the RCP4.5 scenario. In fact, values of 12%–28% will be reached for the period 2011–2040 (Figure 12e); the projections of the maximum precipitation amounts during 48 consecutive hours reach increase rates of 30%–50% (for 2041–2070) (Figure 12f) and 50%–60% (for 2070–2100) (Figure 12g) relative to the reference period. It appears from these results that an increased frequency of flooding could occur in the basin.

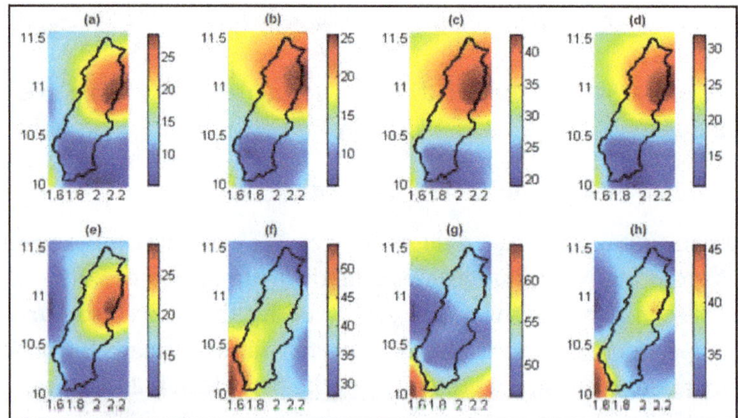

Figure 12. Changes in the maximum precipitation amounts during 48 consecutive hours (%) compared to the reference period: **a**: RCP4.5-2011–2040; **b**: RCP4.5-2041–2070; **c**: RCP4.5-2071–2100; **d**: RCP4.5-2011–2100; **e**: RCP8.5-2011–2040; **f**: RCP8.5-2041–2070; **g**: RCP8.5-2071–2100; **h**: RCP8.5-2011–2100.

3.2.10. Maximum Precipitation Amounts during 72 Consecutive Hours

Similar to the maximum precipitation amounts for 24 and 48 consecutive hours, projections of precipitation amounts during 72 consecutive hours also show an increasing trend over the period 2011–2100. This trend is characterized by an increase in maximum precipitation amounts for 72 consecutive hours of 5%–25% for the medium scenario of greenhouse gas emissions (Figure 13d) and of 22%–34% for the high scenario of greenhouse gas emissions (Figure 13h) compared to the reference period. Over the period 2011–2040, the two scenarios (RCP4.5 and RCP8.5) indicate an increase of about 3%–20% compared to the reference period (Figure 13a,e). The northern area of the basin is the most exposed to these changes, with rates reaching 20%. In the period 2041–2100, a large difference is observed between the projections of the two scenarios. The RCP4.5 scenario indicates changes of −5% to 20% over the period from 2041 to 2070 (Figure 13b) with a decrease trend in the southern part of the basin versus an increasing trend towards the North. For the same scenario, the projections of 2071–2100 indicate high increases (20%–41%) of maximum precipitation amounts of 72 consecutive hours compared to the reference period (Figure 13c). The northern area of the basin also has the highest rates of increase, reaching 40%. The RCP8.5 scenario projections show alarming increases for the maximum precipitation amounts during 72 consecutive h during the period from 2041 to 2100 throughout the watershed. These increases are in the range of 25%–35% over the period 2041–2070 (Figure 13f) and 38%–50% for the 2071–2100 period. These strong increase rates have certainly resulted in an increase in flooding frequency.

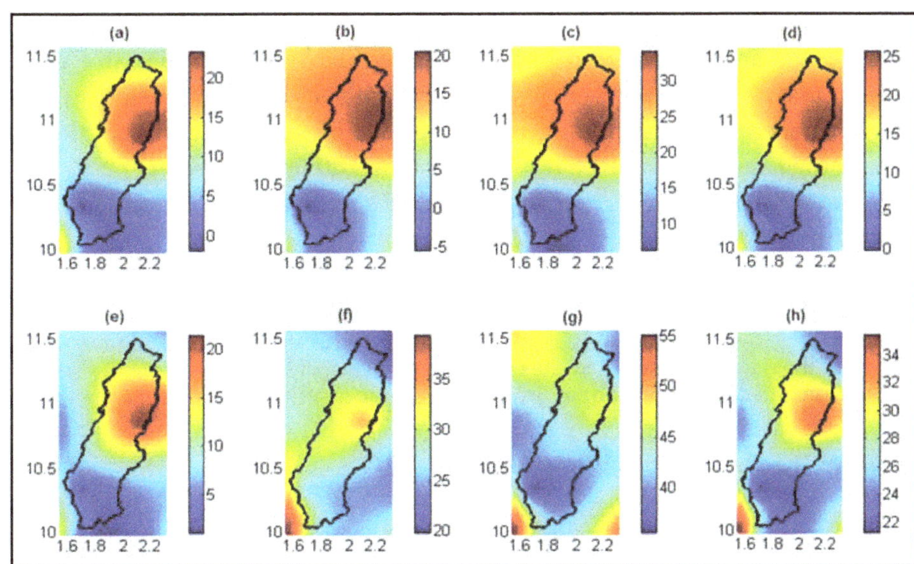

Figure 13. Changes in the maximum precipitation amounts during 72 consecutive hours (%) compared to the reference period: **a**: RCP4.5-2011–2040; **b**: RCP4.5-2041–2070; **c**: RCP4.5-2071–2100; **d**: RCP4.5-2011–2100; **e**: RCP8.5-2011–2040; **f**: RCP8.5-2041–2070; **g**: RCP8.5-2071–2100; **h**: RCP8.5-2011–2100.

4. Discussion

The structure of future projections of rainy seasons based a set of 3 Regional Climate Models (MPI-REMO, DMI-HIRHAM5 and SMHI-RCA4) for RCP4.5 and RCP8.5 scenarios was studied in this paper through a set of ten (10) characteristics of precipitation. The impacts of climate change on the annual precipitation amounts in the basin are quite varied depending on the period and the scenario, similar to many studies in the West African region [13–15,45–47]. For the two (2) scenarios and the period 2011–2070, a decrease in annual rainfall amounts reaching 12% is projected versus an increase of around 10% for the 2071–2100 period. This indicates that in the coming decades, the basin could be exposed to a long dry period that can lead to declining agricultural production, food insecurity, famine, rural migration, etc.

The number of wet days and the rainy season length gradually decrease from 2011 to 2100. These simultaneous decreases in the two (02) parameters will certainly have great consequences for agriculture, which is essentially dependent on rainfall in the region. As for the occurrence and intensification of extreme events in the basin, the simultaneous increase trends related to total precipitation intensity of very wet days and maximum precipitation amounts for 24, 48 and 72 consecutive hours constitute evidence of a possible increase in the frequency and magnitude of floods in the future. Otherwise, the increase trend in the length of dry spells is also an index of exacerbation of drought in the basin. In short, the risk of exacerbation of extreme events exists in the future (drought, flood, food insecurity, rural migration, etc.). This would be the result of future anthropogenic emissions of greenhouse gases [16,48–52]. It has been established that the aggravation of extreme events intensifies local hydrological cycles [53–55]. As established for the West African region, the Mekrou basin will experience changes in climate that will impact agriculture, water resources, and ecosystems, which will affect the rural population, as well as those with urban lifestyles.

5. Conclusions

West Africa is already facing the consequences of climate change. In this paper, future trends of some precipitation characteristics of multi-RCMs under projections (2011–2100) of the RCP4.5 and RCP8.5 scenarios are examined in relation to the observations of the period 1981–2010 in the Mekrou basin at the Kompongou outlet. The results confirm the expected projections in the West African region with regards to annual rainfall amounts (ranging from −10% to 10%), the number of wet days (decrease of up to 10%), total precipitation intensity of very wet days (increase of around 10%), the length of the dry spells (increase up to 10%) and the dates of onset and ending of the rainy seasons. The maximum rainfall amounts during 24, 48 and 72 consecutive hours will experience increases of about 50% compared to the reference period. These rates of change may result in exacerbation of extreme events (droughts and floods) in the Mekrou basin.

Author Contributions: Ezéchiel Obada, Eric Adéchina Alamou, Josué Zandagba, Amédée Chabi and Abel Afouda designed the study, developed the methodology and wrote the manuscript. Ezéchiel Obada performed the field work, collected the data and conducted the computer analysis with Josué Zandagba and Amédée Chabi. Eric Adéchina Alamou and Abel Afouda supervised this part of the work.

Conflicts of Interest: The authors declare no conflict of interest.

References

1. Barrios, S.; Ouattara, B.; Strobl, E. The impact of climatic change on agricultural production: Is it different for Africa? *Food Policy* **2008**, *33*, 287–298. [CrossRef]
2. Dang, H.; Gillett, N.P.; Weaver, A.J.; Zwiers, F.W. Climate change detection over different land surface vegetation classes. *Int. J. Climatol.* **2007**, *27*, 211–220. [CrossRef]
3. IPCC. *Climate Change 2001: The Climate Change. Contribution of the Working Group I to the Third Asseesment Report of the Intergovmmental Panel on Climate Change*; Cambridge University Press: Cambridge, UK, 2001.
4. Meissner, K.; Weaver, A.; Matthews, H.; Cox, P. The role of land surface dynamics in glacial inception: A study with the UVic Earth System Model. *Clim. Dyn.* **2003**, *21*, 515–537. [CrossRef]
5. Snyder, P.; Delire, C.; Foley, J. Evaluating the influence of different vegetation biomes on the global climate. *Clim. Dyn.* **2004**, *23*, 279–302. [CrossRef]
6. Tarhule, A. Damaging rainfall and flooding: The other Sahel hazards. *Clim. Change* **2005**, *72*, 355–377. [CrossRef]
7. Karl, T.R.; Arguez, A.; Huang, B.; Lawrimore, J.H.; Menne, M.J.; Peterson, T.C.; Vose, R.S.; Zhang, H.M. Possible artifacts of data biases in the recent global surface warming hiatus. *Science* **2015**, *348*, 1469–1472. [CrossRef] [PubMed]

8. IPCC. *Climate Change 2013: The Physical Science Basis. Contribution of Working Group I to the Fifth Assessment Report of the Intergovernmental Panel on Climate Change*; Cambridge University Press: Cambridge, UK; New York, NY, USA, 2013.
9. Zwiers, F.W.; Alexander, L.V.; Hegerl, G.C.; Knutson, T.R.; Kossin, J.; Naveau, P.; Nicholls, N.; Schär, C.; Seneviratne, S.I.; Zhang, X. Challenges in estimating and understanding recent changes in the frequency and intensity of extreme climate and weather events. In *Climate Science for Serving Society: Research, Modeling and Prediction Priorities*; Springer: Berlin, Germany, 2013.
10. Giorgi, F.; Coppola, E.; Raffaele, F. A consistent picture of the hydroclimatic response to global warming from multiple indices: Models and observations. *J. Geophys. Res. Atmos.* **2014**, *119*. [CrossRef]
11. IPCC. Summary for Policymakers. In *Climate Change 2014: Impacts, Adaptation, and Vulnerability. Contribution of Working Group II to the Fifth Assessment Report of the Intergovernmental Panel on Climate Change*; Cambridge University Press: Cambridge, UK; New York, NY, USA, 2014.
12. IPCC. *Climate Change 2007—Synthesis Report. Intergovernmental Panel on Climate Change*; Cambridge University Press: Cambridge, UK, 2007.
13. Sylla, M.B.; Nikiema, P.M.; Gibba, P.; Kebe, I.; Klutse, N.A.B. Climate Change over West Africa: Recent Trends and Future Projections. In *Adaptation to Climate Change and Variability in Rural West Africa*; Springer: Basel, Switzerland, 2016.
14. Tall, M.; Sylla, M.B.; Diallo, I.; Pal, J.S.; Faye, A.; Mbaye, M.L.; Gaye, A.T. Projected impact of climate change in the hydroclimatology of Senegal with a focus over the Lake of Guiers for the twenty-first century. *Theor. Appl. Clim.* **2016**. [CrossRef]
15. Ibrahim, B.; Karambiri, H.; Polcher, J.; Yacouba, H.; Ribstein, P. Changes in rainfall regime over Burkina Faso under the climate change conditions simulated by 5 regional climate models. *Clim. Dyn.* **2014**, *42*, 1363–1381. [CrossRef]
16. Cook, K.H.; Vizy, E.K. Impact of climate change on mid-twenty-first century growing seasons in Africa. *Clim. Dyn.* **2012**, *39*, 2937–2955. [CrossRef]
17. Elguindi, N.; Grundstein, A.; Bernardes, S.; Turuncoglu, U.; Feddema, J. Assessment of CMIP5 global model simulations and climate change projections for the 21st Century using a modified Thornthwaite climate classification. *Clim. Chang.* **2014**, *122*, 523–538. [CrossRef]
18. Sylla, M.B.; Elguindi, N.; Giorgi, F.; Wisser, D. Projected robust shift of climate zones over West Africa in response to anthropogenic climate change for the late 21st century. *Clim. Chang.* **2016**, *134*, 241–253. [CrossRef]
19. Gaba, O.U.C.; Biao, I.E.; Alamou, A.E.; Afouda, A. An Ensemble Approach Modelling to Assess Water Resources in the Mékrou Basin, Benin. *IJCET* **2015**, *3*, 22–32.
20. GLEauBe. *Etude Portant État Des Lieux et Gestion de L'information sur les Ressources en eau Dans le Bassin de la Mékrou*; Rapport Technique: Cotonou, Benin, 2002; p. 104.
21. Benoit, M. *Statut et usages du sol en périphérie du parc national du "w" du Niger. Tome 1 Contribution à l'étude du milieu naturel et des ressources végétales du canton de Tamou et du Parc du "W"*; ORSTOM: Bondy, France, 1998.
22. Obada, E.; Alamou, A.E.; Zandagba, E.J.; Biao, I.E.; Chabi, A.; Afouda, A. Comparative study of seven bias correction methods applied to three Regional Climate Models in Mekrou Catchment (Benin, West Africa). *IJCET* **2016**, *6*, 1831–1840.
23. Christensen, O.B.; Drews, M.; Christensen, J.H. The HIRHAM regional climate model version 5. Available online: http://orbit.dtu.dk/fedora/objects/orbit:118724/datastreams/file_8c69af6e-acfb-4d1a-aa53-73188c001d36/content (accessed on 15 February 2017).
24. Jacob, D.; Bärring, L.; Christensen, O.B.; Christensen, J.H.; Hagemann, S.; Hirschi, M.; Kjellström, E.; Lenderink, G.; Rockel, B.; Schär, C.; et al. An inter-comparison of regional climate models for Europe: Design of the experiments and model performance. *Clim. Chang.* **2007**, *81*, 31–52. [CrossRef]
25. Samuelsson, P.; Jones, C.G.; Wille´n, U.; Ullerstig, A.; Gollvik, S.; Hansson, U.; Kjellström, E.; Nikulin, G.; Wyser, K. The Rossby Centre regional climate model RCA3: model description and performance. *Tellus. A* **2011**, *63*, 4–23. [CrossRef]
26. Servat, E.; Paturel, J.E.; Kouamé, B.; Lubès-Niel, H.; Ouedraogo, M.; Masson, J.M. Climatic variability in humid Africa along the Gulfe of Guinea. Part I: Detailed analysis of the phenomenon in Côte d'Ivoire. *J. Hydrol.* **1997**, *191*, 1–15. [CrossRef]

27. Servat, E.; Paturel, J.E.; Kouamé, B.; Travaglio, M.; Ouédraogo, M.; Boyer, J.F.; Lubès-Niel, H.; Fritsch, J.M.; Masson, J.M.; Marieu, B. Identification, caractérisation et conséquences d'une variabilité hydrologique en Afrique de l'Ouest et centrale. *IAHS Publ.* **1998**, *252*, 323–337.
28. Servat, E.; Paturel, J.E.; Lubès-Niel, H.; Kouamé, B.; Masson, J.M.; Travaglio, M.; Marieu, B. De différents aspects de la variabilité de la pluviométrie en Afrique de l'ouest et centrale non sahélienne. *Rev. Sci. Eau* **1999**, *12*, 363–387. [CrossRef]
29. Ardoin, S.; Lubès-Niel, H.; Servat, E.; Dezetter, A.; Boyer, J.F. Analyse de la persistance de la sécheresse en Afrique de l'Ouest: Caractérisation de la situation de la décennie 1990. *IAHS Publi.* **2003**, *278*, 223–228.
30. Vissin, E.W.; Boko, M.; Perard, J.; Houndenou, C. Recherche de ruptures dans les séries pluviométriques et hydrologiques du bassin béninois du fleuve Niger (Bénin, Afrique de l'Ouest). *Assoc. Int. Clim.* **2003**, *15*, 368–376.
31. Ardoin, S. Variabilité hydroclimatique et impacts sur les ressources en eau de grands bassins hydrographiques en zone soudano-sahélienne. Thèse de Doctorat, Université de Montpellier II, Monpellier, France, 2004.
32. Kouassi, A.M.; Kouamé, K.F.; Goula, B.T.A.; Lasm, T.; Paturel, J.E.; Biémi, J. Influence de la variabilité climatique et de la modification de l'occupation du sol sur la relation pluie-débit à partir d'une modélisation globale du bassin versant du N'zi (Bandama) en Côte d'Ivoire. *Revue Ivoirienne. des Sciences et Technologie* **2008**, *11*, 207–229.
33. Tapsobat, D. Caractérisation événementielle des régimes pluviométriques ouest-africains et de leur récent changement. Thèse de Doctorat, Université Paris XI (ORSAY), Paris, France, 1997.
34. Nicholson, S. On the question of the "recovery" of the rains in the West African Sahel. *J. Arid Environ.* **2005**, *63*, 615–641. [CrossRef]
35. Mahe, G.; Paturel, J.E. 1896–2006 Sahelian annual rainfall variability and runoff increase of Sahelian Rivers. *C.R. Geosci.* **2009**, *341*, 538–546. [CrossRef]
36. Riede, J.O.; Posada, R.; Fink, A.H.; Kaspar, F. What's on the 5th IPCC Report for West Africa? In *Adaptation to Climate Change and Variability in Rural West Africa*; Springer: Basel, Switzerland, 2016.
37. Haarsma, R.J.; Selten, F.M.; Weber, S.L.; Kliphuis, M. Sahel rainfall variability and response to greenhouse warming. *Geophys. Res. Lett.* **2005**, *32*. [CrossRef]
38. Ackerley, D.; Booth, B.B.B.; Knight, S.H.E.; Highwood, E.J.; Frame, D.J.; Allen, M.R.; Rowell, D.P. Sensitivity of twentieth-century sahel rainfall to sulfate aerosol and CO2 forcing. *J. Clim.* **2011**, *24*, 499–5014. [CrossRef]
39. Biasutti, M. Forced Sahel rainfall trends in the CMIP5 archive. *J. Geophys. Res.: Atmos.* **2013**, *118*, 1613–1623. [CrossRef]
40. Dong, B.W.; Sutton, R. Dominant role of greenhouse gas forcing in the recovery of Sahel rainfall. *Nat. Clim. Chang.* **2015**, *5*, 757–760. [CrossRef]
41. Mohino, E.; Janicot, S.; Bader, J. Sahel rainfall and decadal to multi-decadal sea surface temperature variability. *Clim. Dyn.* **2011**, *37*, 419–440. [CrossRef]
42. Liebmann, B. A definition for onset and end of the rainy season. 2006. Available online: www.eol.ucar.edu/projects/cppa/meetings/200608/posters/brant.pdf (accessed on 12 February 2007).
43. WMO. *Report on Drought and Countries Affected by Drought during 1974–1985*; World Climate Programme 118; World Meteorological Organization: Geneva, Switzerland, 1986.
44. WMO. Statistical distributions for flood frequency analysis. *Oper. Hydrol. Rep.* **1989**, *33*. WMO no 718.
45. Dai, A. Increasing drought under global warming in observations and models. *Nat. Clim. Chang.* **2013**, *3*, 52–58. [CrossRef]
46. Johns, T.; Gregory, J.; Ingram, W.; Johnson, C.; Jones, A.; Lowe, J.; Mitchell, J.; Roberts, D.; Sexton, D.; Stevenson, D.; et al. Anthropogenic climate change for 1860 to 2100 simulated with the HadCm3 model under updated emissions scenarios. *Clim. Dyn.* **2003**, *20*, 583–612.
47. Sylla, M.B.; Giorgi, F.; Coppola, E.; Mariotti, L. Uncertainties in daily rainfall over Africa: Assessment of observation products and evaluation of a regional climate model simulation. *Int. J. Climatol.* **2013**, *33*, 1805–1817. [CrossRef]
48. Sylla, M.B.; Dell'Aquila, A.; Ruti, P.M.; Giorgi, F. Simulation of the intraseasonal and the interannual variability of rainfall over West Africa with RegCM3 during the monsoon period. *Int. J. Climatol.* **2010**, *30*, 1865–1883. [CrossRef]

49. Lintner, B.R.; Biasutti, M.; Diffenbaugh, N.S.; Lee, J.E.; Niznik, M.J.; Findell, K.L. Amplification of wet and dry month occurrence over tropical land regions in response to global warming. *J. Geophys. Res.* **2012**, *117*, D11106. [CrossRef]
50. Scoccimarro, E.; Gualdi, S.; Bellucci, A.; Zampieri, M.; Navarra, A. Heavy precipitation events In a warmer climate: results from CMIP5 models. *J. Clim.* **2013**, *26*, 7902–7911. [CrossRef]
51. Abiodun, B.J.; Lawal, K.A.; Salami, A.T.; Abatan, A.A. Potential influences of global warming on future climate and extreme events in Nigeria. *Reg. Environ. Chang.* **2013**, 477–491. [CrossRef]
52. Giorgi, F.; Coppola, E.; Raffaele, F.; Diro, G.T.; Fuentes-Franco, R.; Giuliani, G.; Mamgain, A.; Llopart, M.P.; Mariotti, L.; Torma, C. Changes in extremes and hydroclimatic regimes in the CREMA ensemble projections. *Clim. Chang.* **2014**, *125*, 39–51. [CrossRef]
53. Giorgi, F.; Im, E.S.; Coppola, E.; Diffenbaugh, N.S.; Gao, X.J.; Mariotti, L.; Shi, Y. Higher hydroclimatic intensity with global warming. *J. Clim.* **2011**, *24*, 5309–5324. [CrossRef]
54. Sylla, M.B.; Gaye, A.T.; Jenkins, G.S. On the fine-scale topography regulating changes in atmospheric hydrological cycle and extreme rainfall over West Africa in a regional climate model projections. *Int. J. Geophys.* **2012**, 981649. [CrossRef]
55. Sylla, M.B.; Giorgi, F.; Pal, J.S.; Gibba, P.; Kebe, I.; Nikiema, M. Projected changes in the annual cycle of high intensity precipitation events over West Africa for the late 21st century. *J. Clim.* **2015**, *28*, 6475–6488. [CrossRef]

5

Hydrological Modelling Using a Rainfall Simulator over an Experimental Hillslope Plot

Arpit Chouksey [1,*], Vinit Lambey [1], Bhaskar R. Nikam [1], Shiv Prasad Aggarwal [1] and Subashisa Dutta [2]

1 Water Resources Department, Indian Institute of Remote Sensing, Indian Space Research Organisation, 4 Kalidas Road, Dehradun-248001, Uttarakhand, India; vinitlambey39@gmail.com (V.L.); bhaskarnikam@iirs.gov.in (B.R.N.); spa@iirs.gov.in (S.P.A.)

2 Civil Engineering Department, Indian Institute of Technology Guwahati, Guwahati-781039, Assam, India; subashisa@iitg.ernet.in

* Correspondence: arpit@iirs.gov.in

Abstract: Hydrological processes are complex to compute in hilly areas when compared to plain areas. The governing processes behind runoff generation on hillslopes are subsurface storm flow, saturation excess flow, overland flow, return flow and pipe storage. The simulations of the above processes in the soil matrix require detailed hillslope hydrological modelling. In the present study, a hillslope experimental plot has been designed to study the runoff generation processes on the plot scale. The setup is designed keeping in view the natural hillslope conditions prevailing in the Northwestern Himalayas, India where high intensity rainfall events occur frequently. A rainfall simulator was installed over the experimental hillslope plot to generate rainfall with an intensity of 100 mm/h, which represents the dominating rainfall intensity range in the region. Soil moisture sensors were also installed at variable depths from 100 to 1000 mm at different locations of the plot to observe the soil moisture regime. From the experimental observations it was found that once the soil is saturated, it remains at field capacity for the next 24–36 h. Such antecedent moisture conditions are most favorable for the generation of rapid stormflow from hillslopes. A dye infiltration test was performed on the undisturbed soil column to observe the macropore fraction variability over the vegetated hillslopes. The estimated macropore fractions are used as essential input for the hillslope hydrological model. The main objective of the present study was to develop and test a method for estimating runoff responses from natural rainfall over hillslopes of the Northwestern Himalayas using a portable rainfall simulator. Using the experimental data and the developed conceptual model, the overland flow and the subsurface flow through a macropore-dominated area have been estimated/analyzed. The surface and subsurface runoff estimated using the developed hillslope hydrological model compared well with the observed surface runoff for a rainfall intensity of 100 mm/h. The surface runoff hydrograph was very well predicted by the model, with correlation coefficient (R^2) and Nash–Sutcliffe efficiency coefficient (E) as 0.95 and 0.91, respectively. The observed soil/macropore storage component was estimated with the help of water balance equation and compared with the model predicted macropore storage. The error in computing the soil/macropore storage was estimated as 0.38 mm i.e., 13%.

Keywords: hydrological processes; hillslope hydrological modeling; rainfall simulator; macropores; subsurface flow processes

---

1. Introduction

A rainfall simulator is an important principal apparatus for the study of infiltration, soil erosion, surface runoff and sediment transport, as it allows rainfall-runoff generation under controlled and repeatable conditions. A rainfall simulator permits generation of the rainfall at a known depth and intensity in controlled manner. In country like India, which has an agriculture-dominated economy and where the increasing population is constantly exerting pressure on the land and water resources, this type of field experiment is very useful for the understanding of complex water resources systems, especially in the hilly terrain of the Himalayas. Quantification of hydrological process on the hilly terrain is much more complex than in plain areas [1]. The hillslope hydrology is mainly controlled by subsurface storm flow, saturation excess flow, overland flow, return flow and pipe storage. Field experiments using rainfall simulators for estimating these parameters are scant, specifically over hilly terrain of the Northwestern Himalayas.

In the study of rainfall-runoff response by using rainfall simulator, the main query that arises is whether runoff response generated with a rainfall simulator matches with the natural storms. Researchers worldwide have focused on relative results such as the fraction of surface and subsurface runoff. Two problems occur in relating the runoff losses from simulated and natural rainstorms: reproducing the kinetic energy of a specific storm with a simulator, and scaling-up the results from a small simulator to a watershed/basin. The best way to reproduce the kinetic energy of natural storms is to replicate the natural rainfall duration and intensity with the simulator. For this, one requires the historical rainfall climatology records to select the desirable rainfall duration and intensity for the simulation. A well-adopted procedure is to choose a defined precipitation intensity, then run the simulator till steady state runoff is achieved or for a specified time [2]. Moreover, in hillslope hydrological studies, the additional problem of the incorporation of macropore-dominated processes has to be tackled. Macropores are actually large soil pores that are usually greater than 0.08 mm in diameter. Macropores allow free movement of air and water by gravity [3]. Macropores provide habitat for soil organisms and plant roots can grow into them. The process of infiltration into macroporous soils is primarily controlled by the network, density, connectivity and depth-wise distribution of macropores. It is important to quantify soil macroporosity and trace the dominating flow paths within continuous soil macropores to interpret the underlying flow mechanisms. The experimentally-derived quantitative data of soil macroporosity can have wide range of applications in various study domains such as water quality monitoring and groundwater pollution assessment due to preferential leaching of solutes and pesticides, study of soil structural properties and infiltration behavior of soils, investigation of flash floods in rivers, and hydrological modelling of the watersheds [4]. To understand the flow behavior of infiltrated water in active macropore structures of saturated undisturbed soil columns, dye tracing experiments and subsequent digital image processing exercises were carried out for the experimental plot in the present study. The method provided quantitative information about average fraction of macropores and volume density in terms of stained path width with depth as a descriptive variable. There are field and modelling studies that have attempted to understand and incorporate the spatially dynamic nature of the macropore flow system [5]. In spite of this progress in conceptualizing and modelling the macropore flow at the larger hillslope or watershed scale, little has been done to examine the details of accurate flow networks at the scale of individual macropores or soil pipes.

Rainfall simulators have been used successfully in research on many aspects of water resources over the last 70 years. Adams et al. [6] performed an analysis of surface runoff generation source by using large-scale rainfall simulator experiments. They used field experiments results to calibrate the hydrological model and interpret the runoff mechanisms. Sheridan et al. [7] used a rainfall simulator to obtain a modified erodibility index which could be used to predict annual erosion rates for forest roads. Arnaez et al. [8] used a rainfall simulator to compare runoff and sediment production under distinct rainfall intensities in a vineyard plantation in Spain. Verbist et al. [9] obtained soil loss values in 10 plots with bare soil in the Coquimbo Region using rainfall simulator setup. It has been shown in prior studies that rainfall simulators should have the ability to produce controlled and reproducible artificial rainfall which represents natural conditions at a given location. Rainfall simulators were proven as a useful tool in representing the natural rainfall events with fast data acquisition and controllable spatio-temporal variability of intensity, duration and kinetic energy.

Several studies have been performed on rainfall simulators so far and same have been referred for the selection of the design of rainfall simulator and the type of nozzle, in the present study. There are basically two types of rainfall simulators described in the literature: (1) nozzle type; and (2) tube type. Nozzle-type rain drop producing devices are much more common compared to the tube-type drop formers due to portability, capability of producing variable intensity and wide distribution range. Humphry et al. [10] designed a rainfall simulator which is easy to operate and transport while maintaining the intensity, distribution and energy characteristics of the natural rainfall. A single 50WSQ nozzle was used, producing rainfall with a kinetic energy of 25 $J/mm{\cdot}m^2$, which is 87% of that of natural rainfall, and a drop size of 1.8 mm diameter, with an intensity of 70 mm/h. In this design, the usage of water was also less due to use of single nozzle. Sousa and Siqueira [11] developed a cost-efficient rainfall simulator for urban hydrology studies. The developed rainfall simulator simulated the rainfall events with raindrops of median diameter (D_{50}) of 2.12 mm and kinetic energy (KE) of 22.53 $J/mm{\cdot}m^2$. The designed rainfall simulator was able to simulate rainfall intensities from 40 mm/h to 182 mm/h with Christiansen's uniformity coefficient (CUC), ranging from 68.3% to 82.2%. Pe' rez Latorre et al. [12] designed two different rainfall simulators using full-cone jet nozzles (RS1) and plane-jet nozzles (RS2) to obtain different rainfall intensities with drop sizes and energies similar to natural rainfall. It was observed that the design using plane jets (RS2) provided a more realistic drop size distribution and lower cost than that using full-cone jet nozzle (RS1) for lower rainfall intensity experiments. Abudi et al. [13] designed a high accuracy rainfall simulator for runoff and soil erosion studies. The mean drop size was found to be 1.5 mm and energy flux was 76% of the energy flux expected for natural rainfall of same intensity. Bubenzer [14] and Meyer [15] used Veejet 80100 nozzles in rainfall simulators, which have the ability to simulate physical characteristics of natural rainfall with 80% accuracy. The main drawbacks for this design were its complexity and time-consuming mechanisms. Later on, Swanson [16], Foster et al. [17] and Moore et al. [18] developed a rotating boom type of rainfall spray simulator with the same Veejet nozzles. Improved designs with Veejet nozzles achieved a wide range of intensities, up to 130 mm/h. Shelton et al. [19], and Miller [20] enhanced the capability of rainfall simulators by introducing spraying systems fulljet cone nozzles that provided continuous application, a wide angle, and variable intensities by adjusting the valve openings. In spite of many differences in the design and capabilities of the simulators and nozzle types described above, there are certain characteristics anticipated in a rainfall simulator. Therefore, on the basis of vast application of rainfall simulators, Meyer [21] prepared a list of desirable rainfall characteristics such as fall velocity, drop-size distribution, kinetic energy, uniformity, and intensity. However, the desirable characteristics and rainfall simulator design largely depend on the operational requirements, plot sizes, portability and the cost [18].

The main objective of the study was to develop and test a hillslope hydrological model with integration of hillslope-dominated runoff generation processes such as subsurface storm flow, saturation excess flow, overland flow, return flow and pipe storage using rainfall simulator experiments. In this study, a rainfall simulator was installed in a hillslope experimental plot of an area measuring 50 m^2 and used to generate rainfall event of variable intensities. Kinetic energy and rain drop size was then calculated so as to match with the natural rainfall events. The experimental rainfall response was observed over a hillslope plot. A hillslope hydrological model was developed and tested for surface, subsurface runoff and other water balance components.

2. Materials and Methods

2.1. Study Area

The experimental hillslope site is located in the campus of Indian Institute of Remote Sensing, Dehradun, India at an elevation of 435 m above mean sea level. The average annual rainfall received in the area is around 2000 mm. The monsoon starts from June and continues until September. Around 70%–80% of annual rainfall occurs in the period of monsoon months and majority of this rain gets converted in to runoff due to terrain governed rainfall-runoff generation processes. To simulate rainfall-runoff generation processes, an experimental plot of 5 m × 10 m was developed. A detailed topographic survey of the experimental plot at a uniform grid of 0.05 m^2 has been done using a total station. The undisturbed soil samples were collected from the experimental plot (hillslope site) from top layer (0–10 cm) and bottom layer (10–30 cm). Bulk density test and soil texture analysis were done in the laboratory. The study area has loamy sand texture on top surface. Table 1 shows the soil characteristics of the experimental plot. The hard rock layer was observed at a varying depths starting from a depth of 0.5–1.2 m below the ground surface. Such an area can be highly conducive for quick subsurface flow generation under saturated conditions. A surface sealing effect is not applicable in these type of soils, as the infiltration rate is high and texture is loamy sand [22]. The experimental plot is covered with small shrubs throughout the year. The degree of vegetation depends on the climatic conditions. Figure 1(a-b) shows the location and image of experimental setup of hillslope plot.

Table 1. Soil profile of plot area.

Soil Layer	Texture	% Sand	% Silt	% Clay	Bulk Density (g/cm^3)
Top Layer (0–10 cm)	Loamy Sand	78.68	12.74	8.66	1.450
Bottom Layer (10–30 cm)	Loam	49.12	43.44	7.44	1.527

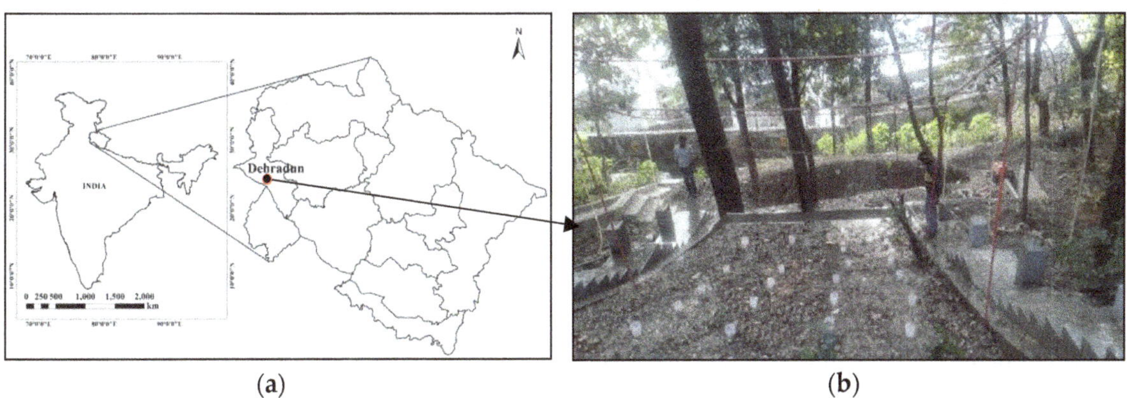

(a) (b)

Figure 1. (**a**) Location of the study area; (**b**) Experimental hillslope plot setup.

2.2. Rainfall Simulator Design

Our main purpose for designing a rainfall simulator was to enhance the uniformity and controllability over the previously designed rainfall simulators that would allow us to perform runoff studies in remote and difficult hilly terrain with more ease and less manpower. To make sure that designed rainfall simulator met the desirable technical requirements, the objective was fixed to develop the simulator that produces near natural rainfall characteristics such as median drop size, velocity, kinetic energy, intensity ranges, uniformity and continuity of flow.

On the basis of information available in peered review literature regarding size, type of nozzle and working mechanism of rainfall simulator, a portable rainfall simulator was fabricated at our laboratory, which is a continuous sprinkler system. The frame of the simulator was constructed from the 1.25-inch-diameter steel pipe and was installed on the experimental plot size of 5 m × 10 m. The simulator frame consisted of six legs made of steel pipe, three on each side. The legs were inserted into the ground to a depth of 30 cm and hinged with surrounded trees for the vertical support. Above the frame, four parallel pipes measuring 0.5 inches in diameter, with uniform spacing between them and each consisting of two nozzles and a pressure gauge, were connected. The parallel pipes are connected with the water supply pipes from both sides. The working setup of rainfall simulator is shown in Figure 1b.

The water supply system consists of two storage tanks, the first one that acts as primary tank, is of 10,000 L capacity. The water from this primary tanks is supplied to a small tank with a capacity of 400 L through gravity. This small tank is connected to a pump which supplies the water to both the pipes which are connected to the parallel pipes over the rainfall simulator frame. The nozzle system used on the simulator was the Spraying Systems Fulljet 1/2HH 50WSQ solid cone nozzle. Two nozzles were placed at the top of the frame on each 0.5-inch pipe, almost 3 m above the ground surface. The nozzle was threaded directly into the 0.5-inch pipe. The selection of the nozzle was done based on literature review for best possible replication of natural rainfall on the plot.

2.3. Experimental Setup and Investigations

For conducting the simulated rainfall experiments on the plot scale level, the experimental setup was designed keeping in view the natural hillslope conditions which are dominant in the Northwestern Himalayas. Keeping the above conditions in mind, an overland flow and subsurface flow collection system was set up at the experimental site. The intensity of rainfall was controlled with the help of the pressure gauges and pressure regulators installed at each parallel pipe on which nozzles were fixed. The simulator was designed in such a way that it covered the entire area of the test plot and distributed the rainfall equally and uniformly over the whole test plot area. Soil moisture sensors were also installed at different depths and locations to monitor the soil moisture variability before, during and after the rainfall simulation experiments.

To collect the runoff, a collecting channel with a gentle slope of 2% was constructed with H flume at the outlet of channel. The runoff water was diverted to the stilling basin where the digital water level recorder (DWLR) was installed to record the water level variations. Small holes were kept in the wall of the channel along the side of the plot and connected to collection tank through pipes to record the subsurface flow. A pump was used to feed the water from the upstream storage tank to the water distribution pipe network of a rainfall simulator. For uniform sprinkling of water through nozzles, proper pressure was maintained at each pressure gauge located at each top pipe. A schematic diagram of the experimental plot and digital elevation model of the experimental plot is shown in Figure 2. The digital elevation map has been presented in the paper to represent the hilly slope profile of the experimental plot. A portable rain gauge was also installed at the experimental site to continuously measure the intensity of the rainfall from simulator.

2.3.1. Raindrop Size Estimation

Raindrop characteristics play important role in many scientific, commercial and industrial applications [23]. The rainfall intensity is also related to the median raindrops diameter by a power function and hence knowledge of raindrop size becomes essential for understanding and modeling the hydrological process in the experimental plot. In this exercise, the drop-size distribution of simulated rainfall was determined by using the flour pellet method described by Hudson [24]. A tray of flour was exposed to simulated rainfall for a period of around 2 s. The flour was then dried for 24 h at room temperature (~28 °C) and the pellets formed as shown in Figure 3 were passed through a series of sieves (4.75, 3.35, 2.36, 1.18 and 0.85 mm). The distributed pellets were then dried for 24 h at 105 °C, weighed and measured. Drops smaller than 1 mm in diameter could not be produced, while those larger than 6 mm could not be determined accurately by this method. The results revealed that the rain drop size follows the gamma distribution and size varies between 1 mm to 5 mm for rainfall intensity 100 mm/h, while the 80% raindrop size lies between 2 to 4 mm, which is very noteworthy and assumed to be a replication of the natural rain storms.

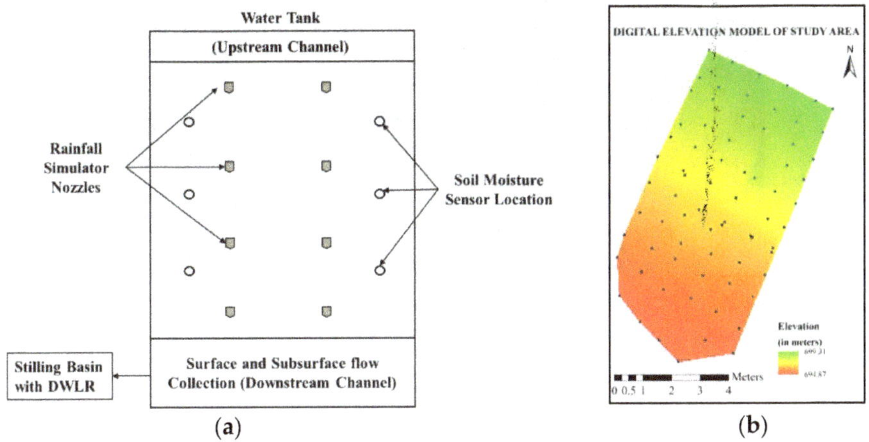

Figure 2. (**a**) Schematic diagram of the experimental setup; (**b**) Digital elevation model (DEM) of the experimental hillslope plot. DWLR: Digital water level recorder.

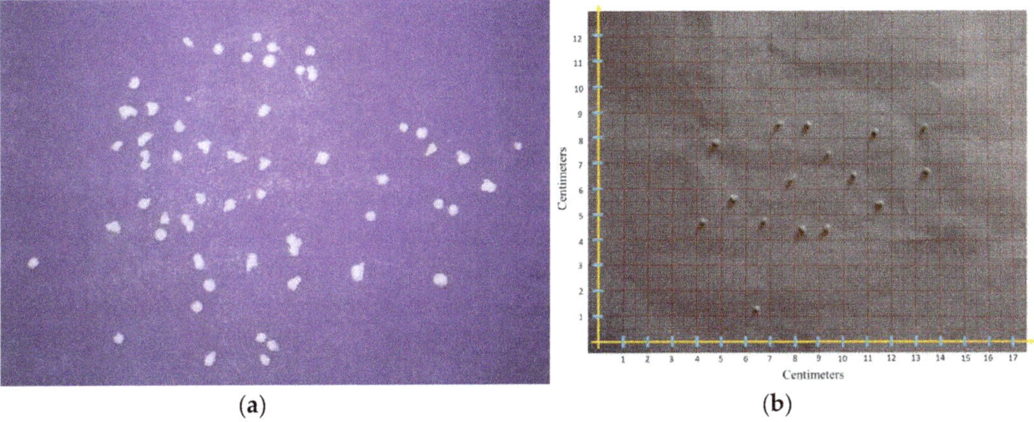

Figure 3. (**a**) Raindrop flour pellet after oven drying; (**b**) Measurement of the raindrop size by flour pellet method.

2.3.2. Performance Evaluation of Rainfall Simulator

The uniformity of rainfall application to the field with rainfall simulator is usually reported as either the distribution uniformity (DU) or Christiansen's uniformity coefficient (CUC) [25]. CUC is a widely used method of calculating the uniformity of water application from rainfall simulators and expressed in percentage as:

$$CUC = 100\;(1 - \frac{Average\ deviation\ from\ average\ depth\ of\ application}{Overall\ average\ depth\ of\ application})\%$$

$$CUC = 100\;(1 - (\textstyle\sum|x - \overline{x}|/n\overline{x})) \quad (1)$$

where, n = no. of sample points; x = rainfall (mm) at each point; and $\overline{x}$ = mean of x values.

Rainfall experiments were performed three times for 30 min with intensity 100 mm/h to check the uniformity of rainfall simulator. The simulated rainfall was collected in the 36 containers placed over the experimental plot at uniform grid of 1 m apart from each other. The capacity of each container was 600 mL. Figure 4 shows the setup of experiment conducted for estimation of coefficient of uniformity on the plot. The average CUC calculated was 79%, which is quite satisfactory on the plot of bigger size [26].

Figure 4. Rainfall experiment for calculation of coefficient of uniformity.

2.3.3. Estimation of Kinetic Energy and Velocity of Rainfall

The kinetic energy of rainfall is a widely used indicator of the potential ability of rain to detach soil. The empirical equation given by Wischmeier and Smith [27] for the kinetic energy of the raindrop is given as:

$$e = 11.897 + 8.73 \log_{10} \mathrm{I} \tag{2}$$

$$\mathrm{E} = \sum_{i=0}^{n} e \times P \tag{3}$$

where, e is the kinetic energy (J/mm·m^2); I is the rainfall intensity (mm/h); P is the rainfall amount (mm); E is the kinetic energy (J/m^2); and n is the number of rainfall periods.

Basically, the kinetic energy of rainfall is estimated from the kinetic energy of each individual raindrop that strikes the ground surface. The drop-size distribution measurements along with fall velocity measurements or empirical laws linking fall velocity (v) and drop diameter (d), let one calculate the rain kinetic energy. Raindrop velocity is based on the raindrop diameter, which can be estimated by modified Newton's equation.

$$\mathrm{v} = (17.20 - 0.84\mathrm{d}) \times (\mathrm{d}) \times 0.5 \tag{4}$$

where, d is diameter of the raindrop.

The relation between drop velocity and drop size when compared with its terminal velocity given by Lows and Parson [28] is given in Table 2.

Table 2. Comparison of measured raindrop velocity and terminal velocity with respect to drop diameter (Lows and Parson, 1943).

Drop Diameter (mm)	Terminal Velocity (m/s)	Velocity (m/s) Measured
1	4	3.3
1.5	5.3	4
2	6.5	5
2.5	7.2	5.7
3	8	6.2
3.5	8.5	6
more	≤9	-

2.3.4. Soil Macropore Characteristics

In order to quantify the macropore structures in the hillslopes, an undisturbed soil column was obtained from the study site. The soil column has a circular dimension with a diameter of 53 cm, which was extracted from the field plot using a steel ring of the same diameter and a depth of 30 cm.

The initial wet conditions of the soil columns were obtained by continuously supplying the water for 1 h with a constant ponding depth of 3 cm. Then, the soil column was left for 4–5 h to attain field capacity. The dye test experiment was done after the soil attained the field capacity. The dye was applied for 1 h with a constant ponding depth of 3 cm. Then, the soil column was left for another 4–5 h for proper distribution of dye in soil column. Then, the steel ring was removed and soil column was sliced horizontally by a sharp edged thin plate at 2 cm intervals from the top to analyze the dye distribution patterns. The distortions in dye patterns due to slicing were negligible as the soils were not cohesive and had low water-holding capacity. A graduated frame was placed on the soil surface to provide a reference for the image analysis.

Digital images (nadir photographs) of each horizontal slice were taken using a digital SLR camera (Canon EOS 400D, 10 Mega Pixels resolution) for digital image analysis of the dye patterns. For accurate measurement of stained areas, the images were color corrected, digitally rectified, and scaled to a resolution of one square millimeter per pixel [29]. These corrected images were analyzed to determine the characteristics of macropore flow for the soils collected from the hillslope plot. The stained paths indicated flow paths with continuous macropore connectivity. The unstained pores that are visible on a horizontal slice are the macropores for which the connectivity was disturbed. Therefore, in relation to active macropore flow, the main area of interest was to quantify the stained path width and their distribution in both horizontal and vertical faces. After every image analysis, a classification report was generated to find the percentage coverage area of the dye in the soil slices. The image analysis of soil photographs provided the useful detail of the depth of the dye penetration. The dye penetration was visible up to the last sliced part of the soil. This indicates that the continuous macropores are present in the soil throughout the depth. This type of macropore connectivity is generally observed in the soils with densely vegetated roots. Figure 5 shows the subset images of the sliced soil columns, which were further used for digital image analysis to find out macropore characteristics of experimental plot. Fraction of macropore present at different soil depths were used as input and to calibrate the hillslope hydrological model.

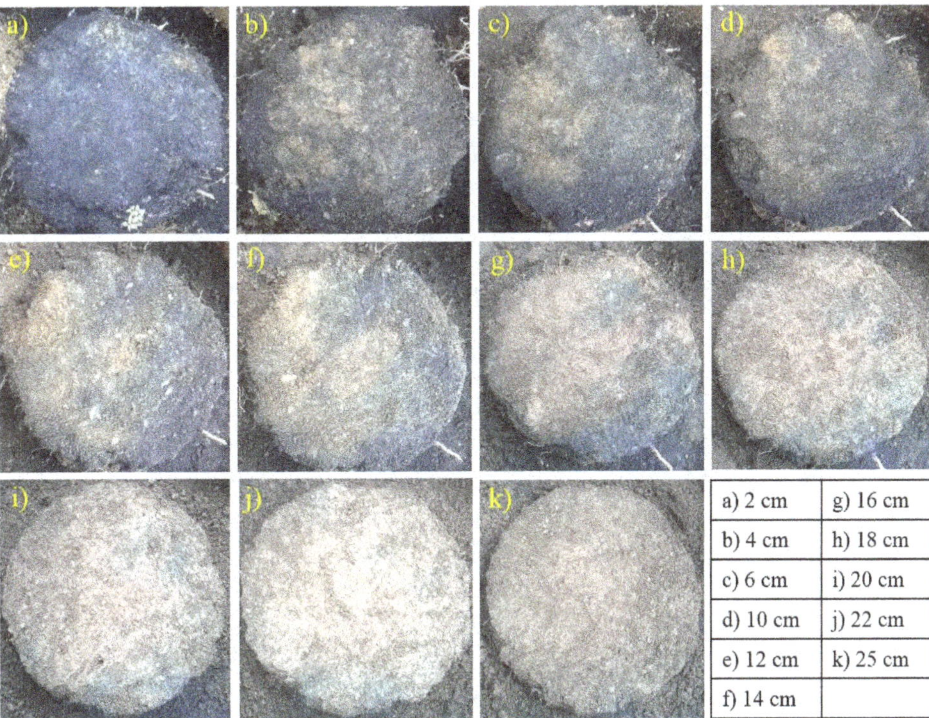

Figure 5. (**a–k**) Subset digitally pre-processed images of soil samples used in macropore analysis of the hillslope at different depths mentioned in the figure.

3. Hydrological Modeling

This part describes the hydrological model for the hillslope areas. The study area is conceptually divided into four hydrological similarity classes (HSC), namely, vegetated hillslope, agricultural field, settlement area and bare soil. The developed model framework is based on the physical processes on the hillslopes, where topmost surface layer interacts with the rainfall. At a hillslope site, surface runoff can be generated by any of three (1) infiltration excess; (2) saturation excess; or (3) variable source area processes [4]. If the soil is saturated, then saturation excess overland flow or retention excess flow occurs from the vegetated hillslope areas and the agriculture fields, most common near the toe of the slopes where the accumulated water from the entire hillslopes is enormous in volume [30]. The excess saturation excess water directly contributes to the channel flow and if the soil of the above two classes is partially saturated or un-saturated, then the rain water will infiltrate into soil and enter the macropore area causing the occurrence of macropore-dominated processes like subsurface flow, return flow (water that not consumed by soil root zone and throughflow but returned back to the surface due to excess pressure), throughflow (a subcomponent of interflow, is the lateral unsaturated flow of water in the soil zone) and pipe storage. Agnese et al. [31] derived a simple storage-based hillslope hydrological response model by assuming the quick runoff by surface runoff generation and compared the results with Horton's theory of runoff generation. They argued that difficulties and complexity in estimating the local parameters cause model prediction errors. The overland flow is not a source of runoff generation in the forested watersheds [32]. Therefore, insight relating to macropore flow processes is most desirable in the recent scenario. Subsurface movement of water largely contributes to the storm runoff generation.

From macropore storage, surface water enters in the root zone storage from where it will be a part of return flow or throughflow depending on the root zone storage capacity. As name suggests, return flow directly contributes to the channel flow and the water from the throughflow adds to subsurface flow processes. The hydrological processes on the other two classes i.e., settlement area and the bare soil are based on the Hortonian infiltration excess runoff generation theory [33]. If the rainfall intensity is less than the infiltration capacity of the soil, then the water will be added to the subsurface flow while for the reverse case, infiltration excess flow will occur, which contributes to the channel flow. Total subsurface flow was calculated by adding the throughflow and the flow from Hortonian infiltration and the total channel flow was estimated by adding return flow, retention excess overland flow and infiltration excess overland flow. Freeze [34] discussed the importance of subsurface flow in generation of surface runoff in highland areas. Figure 6 shows the conceptual flow chart of hydrological processes and their interdependence in the hilly slope areas/hilly watersheds. Anderson et al. [32] and Dunne [35] performed several experiments to identify the dominant flow pathways and discuss the condition of occurrence of these flow paths.

Mathematical Framework of Hillslope Hydrological Model

The mechanism of infiltration in the hillslope area/hilly watershed is mainly controlled by macropores. Macropores initiate the subsurface stormflow after saturating the surrounding soil matrix. The water flow and balance equation for this condition is given by Kroes et al. [36] as:

$$\mathrm{S}^{\mathrm{t}} - \mathrm{S}^{\mathrm{t_0}} = \int_{t0}^{t} (\boldsymbol{Ipr} + \boldsymbol{Iru} + \boldsymbol{Qli} - \boldsymbol{Qlu} - \boldsymbol{Qls}) \tag{5}$$

where,

$$\mathrm{Q_{li}} = \int_{z,if,bot}^{z,if,top} Q_{li};\ \ \mathrm{Q_{lu}} = \int_{z,uns,bot}^{z,=0} Q_{lu};\ \ \mathrm{Q_{ls}} = \int_{z,prof,bot}^{z,uns,bot} Q_{ls}$$

The +ve terms represent infiltration into soil matrix, while the −ve term shows exfiltration from soil matrix. Depths *z,if,top*, *z,if,bot*, *z,uns,bot and z,prof,bot* (cm) refer to top and bottom of interflow zone, and bottom of unsaturated zone and soil profile, respectively and:

(1) Storage of water in the main bypass domain of macropore S_{mb} (cm);
(2) Infiltration of water into macropores at the soil surface, by precipitation, irrigation and snowmelt water falling directly into macropores I_{pr} and by overland flow (runoff) into the macropores I_{ru} (cm/d);
(3) Lateral infiltration into the unsaturated soil matrix Q_{lu} (cm/d);
(4) Lateral exfiltration out of the saturated soil matrix Q_{ls} (cm/d);
(5) Lateral exfiltration out of the saturated soil matrix by interflow out of a zone with perched groundwater Q_{li} (cm/d).

The rate of precipitation I_{pr}, irrigation and snowmelt water routed directly into the macropores at the soil surface at a given precipitation/irrigation/snowmelt intensity P (cm/d) is calculated as:

$$I_{pr} = A_{mp} \times P \tag{6}$$

where, A_{mp} (cm^2/cm^2) is the horizontal macropore area fraction at the soil surface which equals V_{mpo} (cm^3/cm^3), the total macropore volume fraction at soil surface.

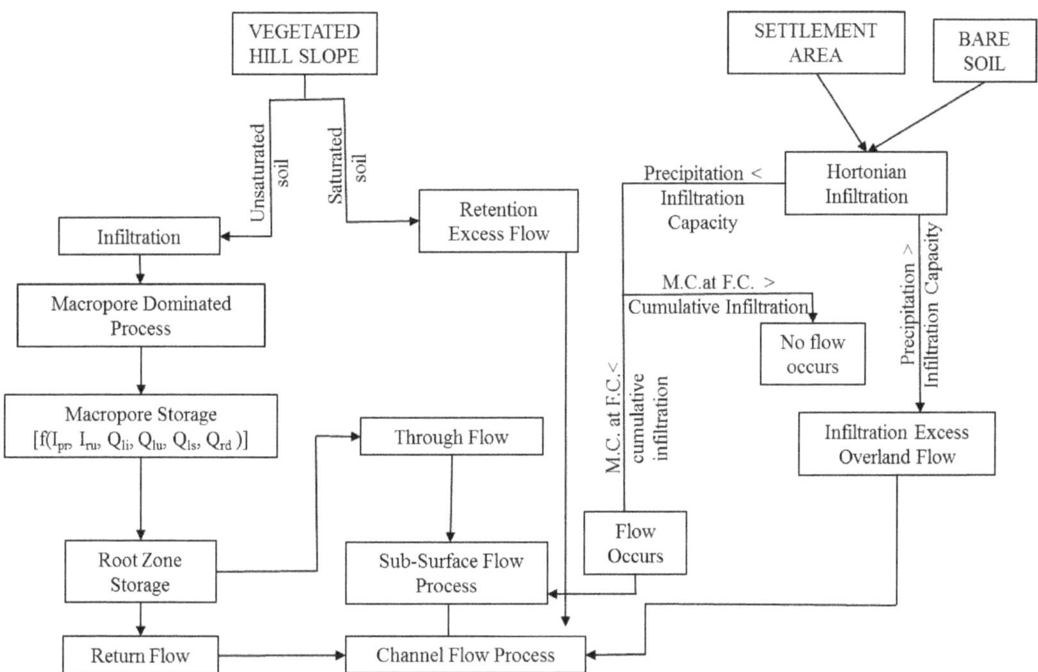

Figure 6. Conceptual flow chart of the hillslope hydrological model. M.C.: Moisture content, F.C.: Field capacity.

Ponding occurs when the total of precipitation, irrigation and inundation intensity exceed soil matrix infiltration capacity and subsequently, overland flow occurs. Infiltration rate I_{ru} due to overland flow or surface runoff is numerically given as:

$$I_{ru} = h_o / Y_{Iru} \tag{7}$$

where, h_o is the pressure head at the soil surface which is equal to the ponding height in cm and Y_{Iru} is the resistance of macropore inflow at soil surface. Lateral infiltration of macropore water into the unsaturated soil matrix (Q_{lu}) takes place strictly over the depth where stored macropore water is in contact with the unsaturated matrix. Absorption is the dominant mechanism at low soil moisture content. It will be negligible under wet conditions even when there is a large pressure head gradient and for this condition, Darcy flow will be dominant. Darcy flow is very small under dry conditions because of very low hydraulic conductivities.

Using a one-dimensional flow equation combined with Darcy's equation [37], the lateral infiltration can be computed as:

$$Q_{lu} = S_r/2t^{-1/2} \quad (8)$$

where, S_r is the sorptivity, which is computed by using Young's estimation formula as:

$$S_r = 6.3\,(\theta - \theta_r)^{0.5}\, K_{sat}{}^{0.25} \quad (9)$$

where, θ is the moisture content at the present time, θ_r is the residual moisture content i.e., moisture content at wilting point, and K_{sat} is the saturated hydraulic conductivity.

Lateral exfiltration out of saturated soil matrix water into the macropores (Q_{ls}), only concerns static macropores below the groundwater table, since in the present concept in case of the saturated condition the soil is assumed to be swollen to its maximum volume. The lateral exfiltration rate per unit of depth $Q_{ls,}$ (cm/cm.d) in the case of water-filled macropores (Pressure head (h_{mp}) > 0) is described by Darcy flow:

$$Q_{ls} = f_{shp} \times 8 \times K_{sat} \times (h_{mp} - h_{mt})/(d^2{}_{pol}) \quad (10)$$

where, h_{mp} and h_{mt} are the pressure head in the water-filled macropores and in the unsaturated soil matrix, respectively.

Parameter f_{shp} is a shape factor to account for the uncertainties in the theoretical description of lateral infiltration by Darcy flow originating from uncertainties in the exact shape of the soil matrix polygons. Theoretically, the value of f_{shp} lies between 1 and 2. Infiltration occurs if $h_{mp} > h_{mt}$ and exfiltration if $h_{mp} < h_{mt}$.

Lateral exfiltration out of the saturated matrix as interflow ($Q^*{}_{li}$) is a special case of exfiltration of soil water from the saturated zone into the macropores and is described as:

$$Q^*{}_{li} = -(f_{shp} \times 8 \times K_{sat} \times (h_{mp} - h_{mt})/(d^2{}_{pol})) \quad (11)$$

If $h_{mp} > h_{mt}$, infiltration into the saturated matrix in the perched groundwater zone occurs. Here, perched groundwater is defined as the subsurface water that forms a saturated horizon within porous media at an elevation higher than the local or regional groundwater table, d_{pol} is the effective diameter of soil polygon which is nothing but the macropore diameter and is given by:

$$d_{pol} = d_{p,min} + (d_{p,max} - d_{p,min}) \times (1 - M) \quad (12)$$

where, $d_{p,min}$ and $d_{p,max}$ are the minimum and maximum diameter of the macropores, M is the relative macropore density which is the ratio of the static macropore volume to the static macropore volume at surface.

Retention excess flow (Q_{re}) occurs if the soil is already saturated. The main factor affecting this flow is the availability of soil moisture content. The retention excess flow can be calculated as:

$$Q_{re} = P - I \quad (13)$$

where, P is the rainfall intensity in mm/h and I is the infiltration capacity in mm/h.

Return flow (saturation overland flow) occurs where the soil is completely saturated and no additional water can be accepted into soil. This type of flow is most common near the toe of the slopes where the accumulated water from the entire hillslopes is enormous in volume. This is a time-dependent condition i.e., the longer the rainfall occurs, the more water will be in the soil layers, and hence a greater area will be subjected to saturation. This flow returns to the land surface after flowing a short distance in the upper soil horizon. Return flow per unit length at the hillslope can be calculated using the following equation:

$$Q_{return} = H_o\, V_{lat}\, (L - L_s) \tag{14}$$

where, Q_{return} is the return flow (mm/day), and H_o is the saturated thickness normal at the hillslope outlet expressed as function of total thickness (mm/mm). V_{lat} is the velocity of the flow at the outlet (mm/day) which can be defined as:

$$V_{lat} = K_s \sin(\alpha) \tag{15}$$

where, α is the hillslope angle.

Throughflow is the downslope flow of water occurring physically within soil surface under unsaturated condition. Throughflow can maintain both low flows (baseflow) in rivers by low subsurface drainage and also contribute to high peak flows (stormflow) through its role in generating saturation excess overland flow. In this study, throughflow is calculated as the difference between the root zone storage and the return flow.

$$\text{Through flow} = S_{rz} - \text{Return flow} \tag{16}$$

where, S_{rz} is the root zone storage.

The above process will occur in the areas with vegetated hillslopes and agriculture land. For the areas covered with settlement and the bare soil, two scenarios occur. The first refers to when the rainfall intensity is less than infiltration capacity. In this case, the water directly infiltrates into the soil and contributes to the subsurface flow. In the second case i.e., when rainfall intensity is higher than infiltration capacity, infiltration excess overland flow occurs. This flow is also known as Hortonian flow and occurs mainly in irrigated areas, urban areas and generally during the storms with very high intensity of rainfall. The infiltration excess overland flow is calculated by Horton's equation which is given as:

$$f_p = f_c + (f_0 + f_c)\, e^{-kt} \tag{17}$$

where, f_p is the infiltration capacity (depth/time) at some time t, k is constant depending on soil characteristics and vegetative cover, f_c is a final or equilibrium capacity, and f_0 is the initial infiltration capacity.

Rainfall event was simulated for the rainfall intensity 100 mm/h by maintaining the nozzle pressure to around 82 kPa through pressure regulators installed in the water distribution network of the rainfall simulator [38]. Each of the experiments was repeated three times for 30 min. Hence, a total of nine rainfall simulations were conducted in three months. Each set of simulation was started only after ensuring that the surface soil moisture in the experimental plot was equal to the soil moisture of the nearby area outside the experimental plot (to resemble with the natural condition). The surface runoff was collected in the channel and measured with the help of digital water level recorder installed at the end of channel. Subsurface flow was collected through the runoff collection mechanisms as discussed in Section 2.

The observed rainfall and soil characteristics were given as input parameters into the developed hillslope hydrological model to estimate the surface, subsurface flow and other macropore flow components. Then, the simulated and observed values of surface and subsurface flow for the experimental plot were compared and performance of the model was tested.

4. Results and Discussion

4.1. Field Observations

The typical characteristics of the hillslopes, climate, and hydro-geologic conditions prevailing in the Northwestern Himalayas have been elaborated in previous section in order to justify the adoption of the experimental techniques. A detailed description of the experimental setup and the methodologies for the in-situ observations were also provided previously along with description of instrumentation used in the hillslope plot and the data captured from artificial storm events. The observed field data and the results obtained from the simulated rainfall event with an intensity of 100 mm/h have been discussed in detail to draw suitable inferences about the hydrological response of the hillslope plot.

4.1.1. Study of Soil Moisture Profile in the Hillslope Plot

Temporal variations of soil moisture profile within the hillslope plot before, during, and after the rainfall-runoff experiments were precisely monitored using the profile probe soil moisture sensors. Interesting observations could be made from the soil moisture patterns in the hillslope soil. The graphs in Figure 7a–d clearly show that once a wet antecedent moisture condition is attained, the moisture content of soils at different depths remains almost constant during the rainfall-runoff events as well as after the cessation of surface runoff. It can be observed that the constant soil moisture conditions over the plot have been attained very quickly during the runoff event. This is a very important finding related to the present investigation. Figure 7a–d shows temporal variations in soil moisture content at different depths (below ground level) in the hillslope plot during and immediately after the runoff experiment.

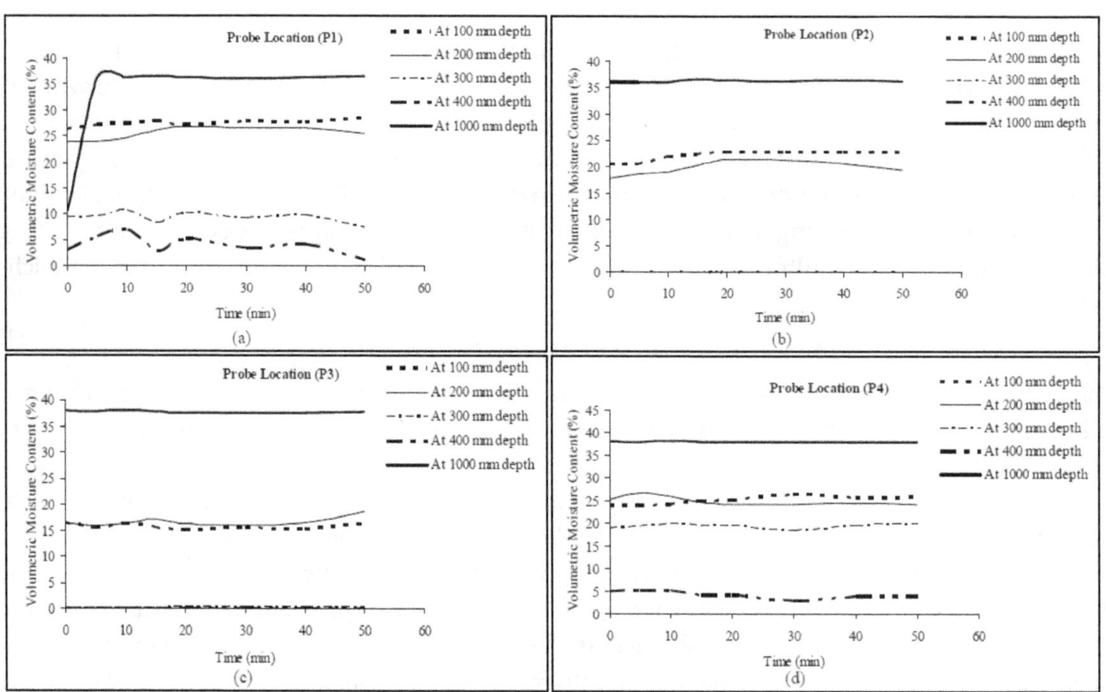

Figure 7. (**a–d**) Soil moisture variations at experimental hillslope plot.

Figure 7a–d clearly depicts that after a wet antecedent condition has been established, the moisture content in the top soil layer does not vary. Rainfall experiments were started after ensuring the resemblance of soil moisture to the natural condition outside the experimental plot. The soil moisture variations for the middle layer (300 mm and 400 mm) for probe location P2 and P3 could not be captured due to a technical fault at the sensor level. However, the middle layer soil (300–400 mm) for probe location P1 and P4 shows relatively low but stable moisture content. Temporal variations of the soil moisture profiles at the probe location P1 at a depth of 1000 mm indicate that the infiltrated water bypasses this layer to reach the bottom layer where the buildup of water table takes place over the impermeable bed and causes lateral diversion of water in the form of subsurface stormflow. Such bypassing flow patterns within the soil combined with rapid buildup and recession of water table in the hillslope soil profile strongly indicates the existence of highly active lateral preferential pathways in the subsoil. However, the quick rise of soil moisture at probe location P2, P3 and P4 might be due to the vertical drainage along the installed soil moisture probe.

4.1.2. Results of Dye Pattern Analysis

The horizontal dye patterns provide detailed information about the maximum depth of dye penetration and percentage dye coverage of the sections. Percentage dye coverage versus depth was plotted for the soil column. Figure 8 shows the depth-wise distribution of percentage dye coverage for the hillslope soil column. The soil column had maximum dye coverage of 7.41% at a 2-cm·depth and an average of 3.08%. In the soil column, the color dye penetration was clearly visible up to the last soil layer. This indicates the presence of continuous macropores throughout the soil column. Such distribution of macropores can be expected from densely vegetated undisturbed hillslope soils where growth of plant roots provides connectivity to preferential pathways for water movement. The occurrence of maximum dye coverage within a 14-cm·depth also represents higher root density and activity of soil fauna and flora in the top soil layer. From the average dye coverage, it can be noted that most of the flow pathways were concentrated to 3.08% of total area in soil column. These results justify the demand of dedicated hydrological model for hilly-slope areas having capability of handling macropore-dominant processes as is done by the model implemented in the present study.

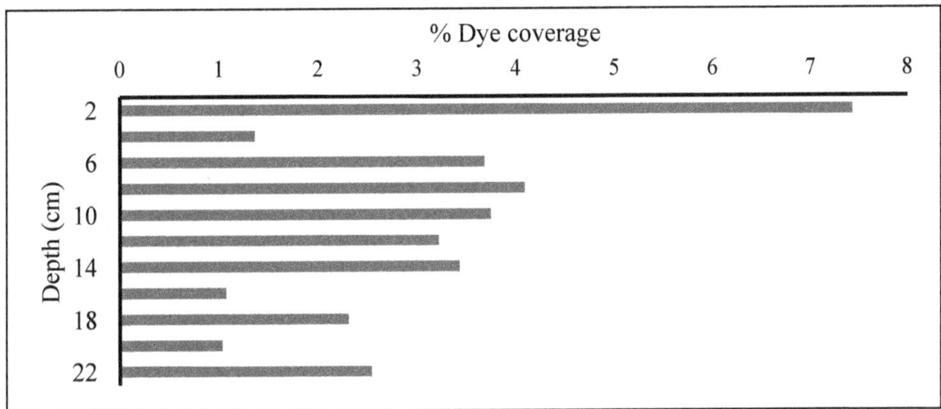

Figure 8. Depth-wise dye coverage distribution in the soil column.

4.1.3. Subsurface Flow Observation

The channel constructed at the downslope of the hillslope plot for collecting runoff was provided with holes on the side of the plot to collect the subsurface flow. The experiment for subsurface flow observation was carried out in a vegetated condition. From the experiment it was evident that in vegetated conditions due to presence of macropores, subsurface flow played an active part in the channel flow. Although the subsurface flow started late, it lasted for a longer time even after the rainfall experiment has stopped. Figure 9 shows the observed subsurface flow recorded during and after the rainfall simulation for a vegetated condition.

4.1.4. Overland Flow Observation

The water collected in the collecting channel flows to the stilling basin constructed at downslope of the hillslope plot. The water in the tank is measured at every 5-min interval using the digital water level recorder. Figure 10 shows the temporal variation of overland flow depth during and after the rainfall simulation experiment. It is observed from the figure that the peak overland flow occurs after around 8 min from the end of rainfall event.

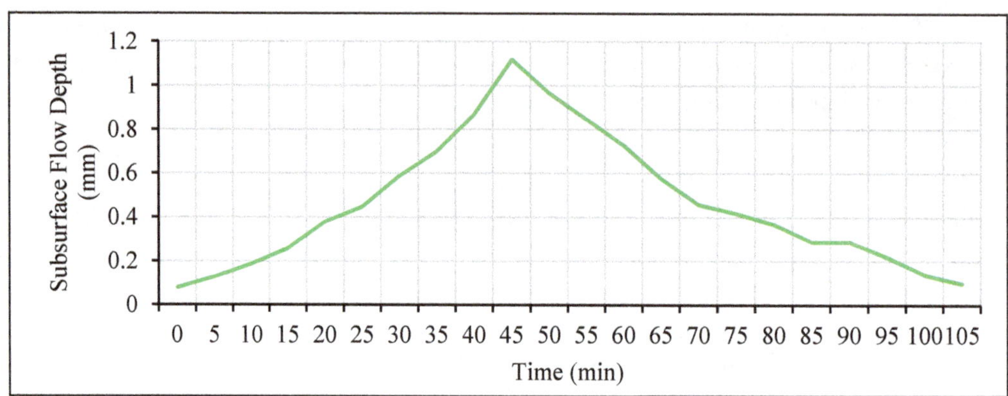

Figure 9. Subsurface flow for rainfall intensity of 100 mm/h for 30 min.

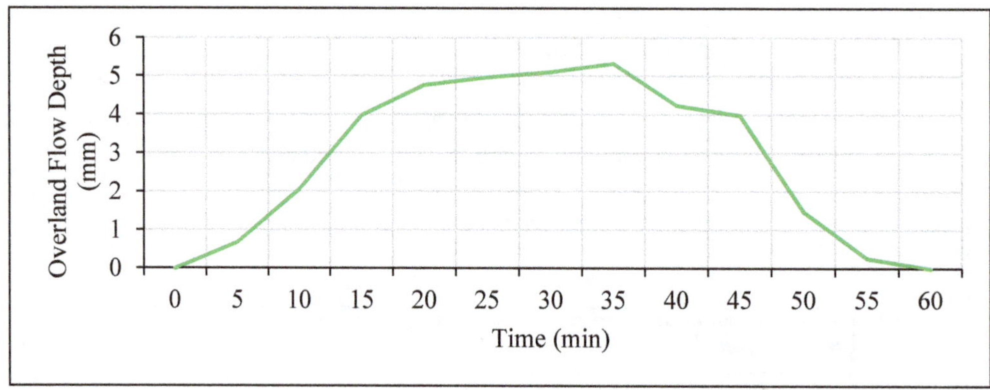

Figure 10. Overland flow hydrograph for rainfall intensity of 100 mm/h for 30 min.

4.2. Hydrological Modeling Results

This section shows the results of the rainfall simulator hillslope experiment. A comparison between the observed data of the hillslope rainfall simulator experiment and the results of the hydrological model was performed. The comparison is based on the obtained data of the rainfall–runoff relationship and soil moisture content.

Rainfall simulation experiments were performed for 100 mm/h rainfall intensity and field observations were taken for overland flow discharge and subsurface flow discharge. The hillslope hydrological model was run for the rainfall intensity 100 mm/h. The comparison of the observed and simulated overland flow hydrograph was done using the statistical parameter known as Nash–Sutcliffe model efficiency coefficient (E) [39]. The formula for the computation of Nash-Sutcliffe model efficiency is:

$$E = 1 - \left[\frac{\sum (Q_o - Q_s)^2}{\sum (Q_o - \overline{Q}_o)^2}\right] \tag{18}$$

where, $\overline{Q}_o$ is the mean of observed discharges, and Q_s is modeled discharge. Q_o is observed discharge at time t.

Nash–Sutcliffe efficiency can range from $-\infty$ to 1. An efficiency of 1 ($E = 1$) corresponds to a perfect match of modeled discharge to the observed data. An efficiency of 0 ($E = 0$) indicates that the model predictions are as accurate as the mean of the observed data, whereas an efficiency less than zero ($E < 0$) occurs when the observed mean is a better predictor than the model or, in other words, when the residual variance (described by the numerator in the expression above), is larger than the data variance (described by the denominator). Essentially, the closer the model efficiency is to 1, the more accurate the model is.

For comparison of observed and simulated overland flow hydrograph (Figures 11 and 12), correlation coefficient (R^2) and Nash–Sutcliffe efficiency coefficient were computed. The coefficients were found to be $R^2 = 0.95$ and $E = 0.91$, which is quite satisfactory.

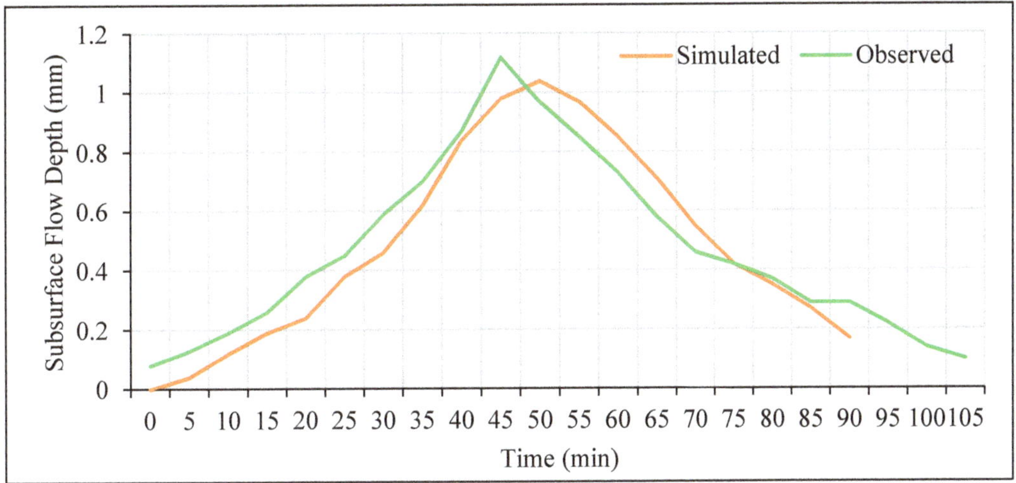

Figure 11. Subsurface flow hydrograph for rainfall intensity of 100 mm/h for 30 min.

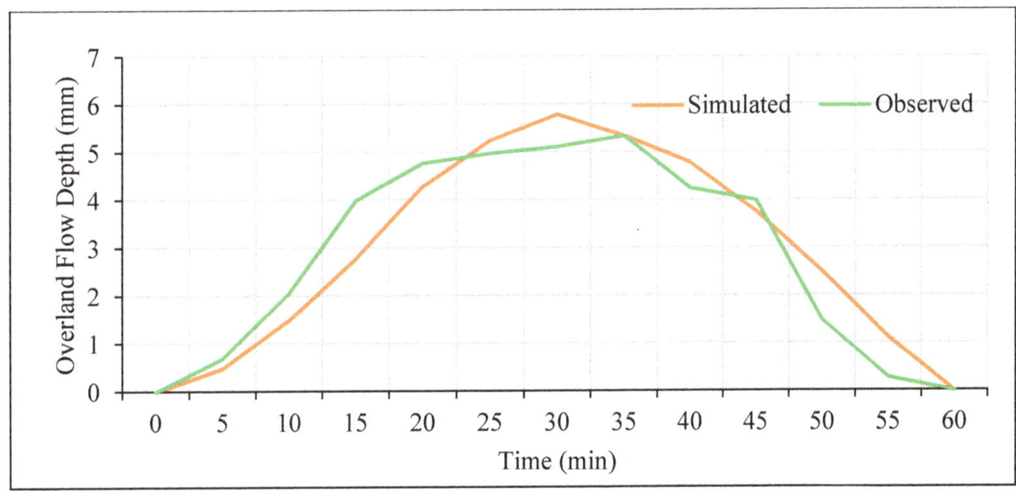

Figure 12. Overland flow hydrograph for rainfall intensity of 100 mm/h for 30 min.

The developed conceptual hillslope hydrological model results also include components of macropore storage i.e., lateral infiltration into unsaturated matrix (Q_{lu}), lateral exfiltration out of saturated soil matrix (q_{ls}), and lateral exfiltration out of soil matrix as interflow (Q_{li}). As the rainfall starts, water enters into the macropore by directly falling into it or through overland flow (runoff). The infiltrated water initially enters the main bypass (MB) domain and the internal catchment (IC) domain. The macropores in the MB domain are well connected throughout the depth of the soil while IC domain ends at different depths with no connection between them. After rainfall initiation, the soil gets saturated and swells. As the rainfall continues, the water pressure starts building up on the saturated macropores because of which water fraction already present in the macropores starts moving laterally into the groundwater (Q_{ls}). After the rainfall, the Q_{ls} component starts decreasing with a decrease in water pressure on the macropores. In contrast, Q_{lu} increases with time as water moves laterally to the unsaturated zone from the macropore domain. This occurs where the stored macropore water is in contact with unsaturated soil matrix. The lateral infiltration occurs due to absorption of macropore water because of capillary force. As the soil gets saturated, the water starts flowing out of it into the macropores, which is termed as Q_{li}. This is the special case of exfiltration. The hillslope hydrological model results for macropore storage components for rainfall intensities of 100 mm/h are shown in Figure 13.

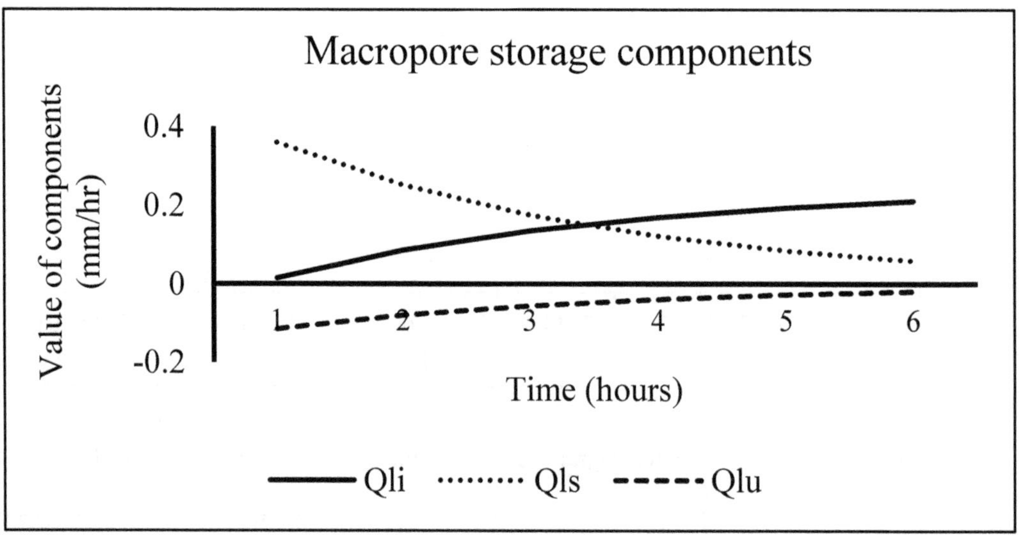

Figure 13. Change in macropore storage with respect to time for 100 mm/h rainfall intensity. Soil matrix; Q_{li}: lateral exfiltration out of the saturated soil matrix by interflow out of a zone with perched groundwater; Q_{ls}: lateral exfiltration out of the saturated soil matrix; Q_{lu}: lateral infiltration into the unsaturated soil matrix.

4.3. Water Balance Analysis

The water stored in the soil column (soil moisture) is very dynamic in space and time, however, in an event-based rainfall-runoff experiment, quantification of water absorbed/stored by the soil is essential. Traditionally, the accuracy of hydrological models is only tested by comparing model predicted runoff/discharge against the observed runoff/discharge. However, in the present case the developed hillslope hydrological model not only estimates surface and subsurface runoff but also quantifies the macropore storage in the study area. As such, an attempt has been made in the present study to validate the macropore storage results of developed model by using water balance approach. The water balance has been solved using observed components of the short term (temporal) water balance equation. The residual in the observed water balance excursive was assumed as water stored in the macropore/soil column. Thus, estimated soil water storage is used to validate the macropore storage predicted by the model. The results of water balance are shown in Figure 14. The model overpredicted the soil/macropore storage component by 0.38 mm, which is 13% of the observed residual water content. It was observed from the experimental results that around 20% of the total runoff comes in the form of subsurface flow. This higher fraction of subsurface contribution in total runoff is mainly due to the macropore-dominated flow pathways present in the soil column of the study plot.

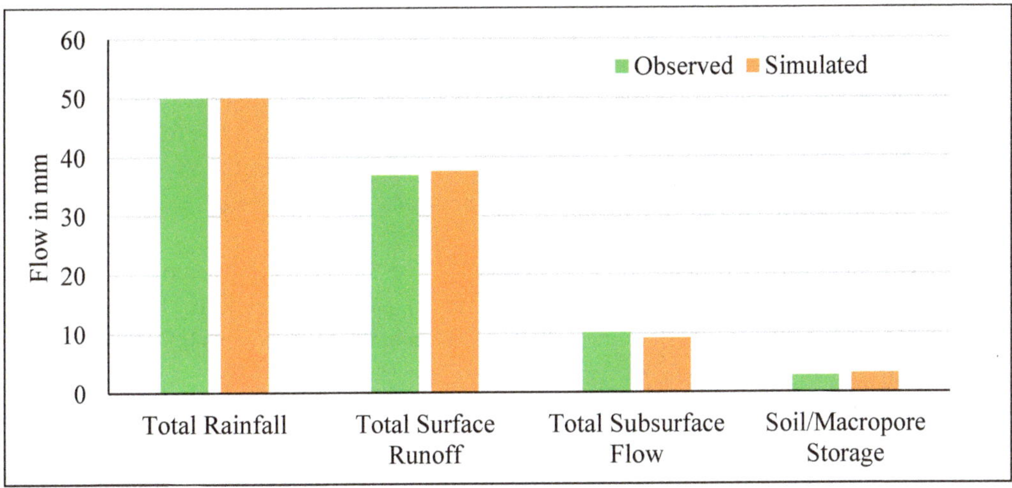

Figure 14. Water balance for rainfall intensity of 100 mm/h for 30 min.

5. Summary and Conclusions

This study delivers a useful description of a hillslope rainfall simulator experiment to estimate the hydrological response to high intensity rainfall over a hillslope plot size of 50 m^2. The rainfall simulator was designed to reflect the physical characteristics of natural rainfall event in terms of raindrop size, intensity, uniformity, continuity and kinetic energy. The rainfall simulator allowed us to vary rainfall intensity and duration with the help of nozzle pressure and control valves. The rainfall event was simulated for 100 mm/h intensity for 30 min. The surface runoff, subsurface runoff response and soil moisture were monitored throughout the experiment. The soil moisture profile observed during and after the rainfall simulation experiment revealed that soil moisture plays an important role in the runoff generation process only during inception phase. After some time, soil moisture remained constant mostly in bottom soil layer at a depth of 600–1000 mm. The runoff and macropore storage components were also estimated through a hillslope hydrological model. To calibrate the model, several field investigations were carried out to estimate the vertical profile of soil macropore fraction, moisture variability and soil characteristics. The observed and simulated water balance components were then compared by Nash–Sutcliffe model efficiency coefficient and found a satisfactory match between them. The correlation coefficient and Nash model efficiency were computed as 0.95 and 0.91, respectively, for surface runoff. The model accuracies in predicting surface runoff, subsurface runoff and soil/macropore storage were 98%, 90% and 86%, respectively.

This study showed that the influence of soil and macropore characteristics on drainage for the hillslope experimental conditions depends largely on the soil characteristics. It was also found that subsurface flow plays important role in generation of surface runoff in hilly watersheds. Therefore, based on our results, advances can be made for future experiments to quantify subsurface storm flow, saturation excess flow, overland flow, return flow and pipe storage processes under varying topography, vegetative cover, and rainfall intensities. In conclusion, this study is good food for thought for hydrologists and soil scientists with respect to future experiments.

Acknowledgments: Authors would like to thank the Director, Indian Institute of Remote Sensing, Dehradun, India for providing the necessary facilities. The study was funded under Technology Development Program of the Indian Space Research Organisation. Authors acknowledge the financial support received for the project. Authors are also thankful to Vaibhav Garg for his valuable contribution in the preparation of the manuscript.

Author Contributions: The idea of this study was conceived by Arpit Chouksey and Bhaskar R. Nikam. Experimental data collection and analysis was carried out by Vinit Lambey. Development of hillslope hydrological model and manuscript preparation was jointly done by Arpit Chouksey, Bhaskar R. Nikam, Shiv Prasad Aggarwal and Subashisa Dutta.

Conflicts of Interest: Authors declare no conflict of interests.

References

1. Band, L.E.; Patterson, P.; Nemani, R.; Running, S.W. Forest ecosystem processes at the watershed scale: Incorporating hillslope hydrology. *Agric. For. Meteorol.* **1993**, *63*, 93–126. [CrossRef]
2. Nolan, S.C.; van Vliet, L.J.P.; Goddard, T.W.; Flesch, T.K. Estimating storm erosion with a rainfall simulator. *Can. J. Soil Sci.* **1997**, *77*, 669–676. [CrossRef]
3. Beven, K.J.; Germann, P.J. Macropores and water flow in soils. *Water Resour. Res.* **1982**, *18*, 1311–1325. [CrossRef]
4. Shougrakpam, S.; Sarkar, R.; Dutta, S. An experimental investigation to characterize soil macroporosity under different land use and land covers of northeast India. *J. Earth Syst. Sci.* **2010**, *119*, 655–674. [CrossRef]
5. Yvonne, S.; Martine, J.; Ploeg, V.D.; Teuling, A.J. Rainfall Simulator Experiments to Investigate Macropore Impacts on Hillslope Hydrological Response. *Hydrology* **2016**, *3*, 39.
6. Adams, R.; Parkin, G.; Rutherford, J.C.; Ibbitt, R.P.; Elliott, A.H. Using a rainfall simulator and a physically based hydrological model to investigate runoff processes in a hillslope. *Hydrol. Process.* **2005**, *19*, 2209–2223. [CrossRef]
7. Sheridan, G.J.; Noske, P.; Lane, P.; Sherwin, C. Using rainfall simulation and site measurements to predict annual interrill erodibility and phosphorus generation rates from unsealed forest roads: Validation against in-situ erosion measurements. *CATENA* **2008**, *73*, 49–62. [CrossRef]
8. Arnaez, J.; Lasanta, T.; Ruiz-Flaño, P.; Ortigosa, L. Factors affecting runoff and erosion under simulated rainfall in Mediterranean vineyards. *Soil Tillage Res.* **2007**, *93*, 324–334. [CrossRef]
9. Verbist, K.; Cornelis, W.M.; Gabriels, D.; Alaerts, K.; Soto, G. Using an inverse modelling approach to evaluate the water retention in a simple water harvesting technique. *Hydrol. Earth Syst. Sci.* **2009**, *13*, 1979–1992. [CrossRef]
10. Humphry, J.B.; Daniel, T.C.; Edwards, R.D.; Sharpley, A.N. A portable rainfall simulator for plot scale runoff studies. *Appl. Eng. Agric.* **2002**, *18*, 199–204. [CrossRef]
11. Sousa Júnior, S.F.; Siqueira, E.Q. Development and Calibration of a Rainfall Simulator for Urban Hydrology Research. In Proceedings of the 12th Intenational Conference on Urban Drainage, Porto/Alegre, Brazil, 11–16 September 2011.
12. Pe' rez-Latorre, F.J.; Castro, L.; Delgado, A. A comparison of two variable intensity rainfall simulators for runoff studies. *Soil Tillage Res.* **2010**, *107*, 11–16. [CrossRef]
13. Abudi, I.; Carmi, G.; Berliner, P. Rainfall simulator for field runoff studies. *J. Hydrol.* **2012**, *454–455*, 76–81. [CrossRef]
14. Bubenzer, G.D. Inventory of rainfall simulators. In *Proceedings of the Workshop on Rainfall Simulators*; Agricultural Research, Science and Education Agency, USDA: Washington, DC, USA, 1979; pp. 120–130.
15. Meyer, L.D.; McCune, D.L. Rainfall simulator for runoff plots. *Agric. Eng.* **1958**, *39*, 644–648.
16. Swanson, N.P. Rotating–boom rainfall simulator. *Trans. ASAE* **1965**, *8*, 71–72. [CrossRef]
17. Foster, G.R.; Neibling, W.H.; Natterman, R.A. *A Programmable Rainfall Simulator*; American Society Agricultural Engineers: St. Joseph, MI, USA, 1982.
18. Moore, I.D.; Hirschi, M.C.; Barfield, B.J. Kentucky rainfall simulator. *Trans. ASAE* **1983**, *26*, 1085–1089. [CrossRef]
19. Shelton, C.H.; von Bernuth, R.D.; Rajbhandari, S.P. A continuous–application rainfall simulator. *Trans. ASAE* **1985**, *28*, 1115–1119. [CrossRef]
20. Miller, W.P. A solenoid–operated, variable intensity rainfall simulator. *Soil Sci. Soc. Am. J.* **1987**, *51*, 832–834. [CrossRef]
21. Meyer, L.D. Simulator of rainfall for soil erosion research. *Trans. ASAE* **1965**, *8*, 63–65. [CrossRef]
22. Moore, I.D. Effect of surface sealing on infiltration. *Trans. ASAE* **1981**, *24*, 1547–1552. [CrossRef]
23. Kathiravelu, G.; Lucke, T.; Nichols, P. Rain drop measurement techniques: A review. *Water* **2016**, *8*. [CrossRef]
24. Hudson, N.W. The Influence of Rainfall on the Mechanics of Soil Erosion with Particular Reference to Southern Rhodesia, Unpub. Master's Thesis, University of Cape Town, Cape Town, South Africa, 1965.
25. Kara, T.; Ekmekci, E.; Apan, M. Determining the Uniformity Coefficient and Water Distribution Characteristics of Some Sprinklers. *Pak. J. Biol. Sci.* **2008**, *11*, 214–219. [CrossRef] [PubMed]
26. Grierson, I.T.; Oades, J.M. A rainfall simulator for field studies of rainfall and runoff. *J. Agric. Res.* **1977**, *22*, 37–44.

27. Wischmeier, W.H.; Smith, D.D. Rainfall energy and its relation to soil loss. *Trans. Am. Geophys. Union* **1958**, *39*, 285–291. [CrossRef]
28. Lows, J.O.; Parson, D.A. The relationship of raindrop size to intensity. *Trans. Am. Geophys. Union* **1943**, *24*, 452–460. [CrossRef]
29. Weiler, M.; Flühler, H. Inferring flow types from dye patterns in macroporous soils. *Geoderma* **2004**, *120*, 137–153. [CrossRef]
30. Dutta, S.; Zade, M. RISE-A Distributed Hydrologic Model for Rice Agriculture: Concept and Evaluation. In *Watershed Hydrology*; Singh, V.P., Yadava, R.N., Eds.; Allied Publisher: New Delhi, India, 2003; pp. 240–251.
31. Agnese, A.; Baiamonte, G.; Corrao, C. A simple model of hillslope response for overland flow generation. *Hydrol. Process.* **2001**, *15*, 3225–3238. [CrossRef]
32. Anderson, S.P.; Dietrich, W.E.; Montgomery, D.R.; Torres, R.; Conrad, M.E.; Loague, K. Subsurface flow paths in a steep unchanneled catchment. *Water Resour. Res.* **1997**, *33*, 2637–2653. [CrossRef]
33. Horton, R.E. The role of infiltration in the hydrological cycle. *Trans. Am. Geophys. Union* **1933**, *14*, 446–460. [CrossRef]
34. Freeze, R.A. Role of subsurface flow in generating surface runoff. 2. Upstream source areas. *Water Resour. Res.* **1972**, *8*, 1273–1283. [CrossRef]
35. Dunne, T. Field studies of hillslope processes. In *Hillslope Hydrology*; Kirkby, M.J., Ed.; McGraw-Hill: New York, NY, USA, 1978; pp. 227–293.
36. Kroes, J.G.; van Dam, J.C.; Groenendijk, P.; Hendriks, R.F.A.; Jacobs, C.M.J. *SWAP Version 3.2. Theory Description and User Manual*; Alterra Report 1649; Alterra: Wageningen, The Netherlands, 1964.
37. Shakya, N.M.; Chander, S. Modelling of hillslope runoff processes. *Environ. Geol.* **1998**, *35*, 115. [CrossRef]
38. Knasiak, K.; Schick, R.J.; Kalata, W. Multiscale Design of Rain Simulator. In Proceedings of the 20th Annual Conference on Liquid Atomization and Spray Systems, Chicago, IL, USA, 15–18 May 2007.
39. Nash, J.E.; Sutcliffe, J.V. River flow forecasting through conceptual models part I—A discussion of principles. *J. Hydrol.* **1970**, *10*, 282–290. [CrossRef]

6

Climate Change and Its Impacts on Water Resources in the Bandama Basin, Côte D'ivoire

Gneneyougo Emile Soro [1,*], Affoué Berthe Yao [2], Yao Morton Kouame [1] and Tié Albert Goula Bi [1]

1 Unit Training and Research in Science and Environment Management, University Nangui Abrogoua, 02 BP 801 Abidjan 02, Abidjan, Ivory Coast; mortonkouame@ymail.com (Y.M.K.); goulaba2002@yahoo.fr (T.A.G.B.)

2 Unit Training and Research in Environment, Université Lorougnon Guédé, BP150 Daloa, Daloa, Ivory Coast; y_berth@yahoo.fr

* Correspondence: ge_soro@yahoo.fr

Abstract: This study aims to assess future trends in monthly rainfall and temperature and its impacts on surface and groundwater resources in the Bandama basin. The Bandama river is one of the four major rivers of Côte d'Ivoire. Historical data from 14 meteorological and three hydrological stations were used. Simulation results for future climate from HadGEM2-ES model under representative concentration pathway (RCP) 4.5 and RCP 8.5 scenarios indicate that the annual temperature may increase from 1.2 °C to 3 °C. These increases will be greater in the north than in the south of the basin. The monthly rainfall may decrease from December to April in the future. During this period, it is projected to decrease by 3% to 42% at all horizons under RCP 4.5 and by 5% to 47% under RCP 8.5. These variations will have cause an increase in surface and groundwater resources during the three periods (2006–2035; 2041–2060; 2066–2085) under the RCP 4.5 scenario. On the other side, these water resources may decrease for all horizons under RCP 8.5 in the Bandama basin.

Keywords: rainfall; temperature; runoff; groundwater recharge; Bandama river

---

1. Introduction

Climate change is inevitably resulting in changes in climate variability and in the frequency, intensity, spatial extent, duration, and timing of extreme weather and climate events [1]. The works of [2,3] showed that 46% of cultivated areas in the world are not suitable for rained agriculture because of climate changes and other meteorological conditions. Many studies have concluded that the impacts of climate change will not be equally shared among the population of the world [4–7]. The distribution of impacts will vary as both the ability to respond to impacts and the availability of resources with which to do so vary across nations [8]. There is high confidence that developing countries will be more vulnerable to climate change than developed countries, and there is medium confidence that climate change would exacerbate income inequalities between and within countries [7]. Sub-Saharan Africa is considered the most vulnerable to the impacts of climate change because of its high dependence on agriculture and natural resources, warmer baseline climates, low precipitation, and limited ability to adapt [9]. Several studies have shown that surface water and groundwater evolutions over the past decades in Sub-Saharan Africa have been strongly affected by rainfall variations. Climate models project important climate changes for the 21st century in West Africa as well as in the rest of the world, with potential impacts on the hydrological cycle [10]. This vulnerability is also due to the fact that the current climate is already severe, present information is not sufficient, and technological change has been slowest in Sub-Saharan Africa [11].

Côte d'Ivoire, located in West Africa is no exception to this situation because its economy is based on rain-fed agriculture and it has a strong dependence on river flow for the power generation and fisheries. In this country, the impacts of climate change have led to recurrent droughts, changes in rain amount distribution, reduction of arable land, coastal erosion, and flooding [12]. Many studies on the impacts of climate variability were carried out in Côte d'Ivoire [13–18]. Few studies have been conducted to investigate the impact of climate change on the water resources in the Bandama basin, despite the increase in water requirements for water supply of populations, agriculture, livestock, mining, and hydroelectric dams. Moreover, the few previous studies have used the Special Report on Emissions Scenarios (SRES) that explicitly consider the effects of prescribed levels of emissions into the atmosphere. However, there was enormous uncertainty regarding contributing factors such as population growth, economic development, and technological advances, hence, the move towards representative concentration pathways (RCPs) in this study. The better understanding of potential future changes on water resources is fundamental to inform populations in order to increase awareness and to support the development of adaptation strategies on the Bandama basin. This study examined how projected changes in temperature and precipitation regimes impact simulated runoff and groundwater recharge in the Bandama catchment using statistical downscaled climatological data.

2. Data and Methods

2.1. Study Area

Figure 1 shows the Bandama basin located in Côte d'Ivoire. The Bandama river is one of the four major rivers of the country. Its source is located north of Côte d'Ivoire between Korhogo and Boundiali. There are three major tributaries: White Bandama (22,293 km^2), N'zi (3500 km^2), and Marahoue (19,800 km^2). The Bandama basin spreads over three different climatic and hydrographic regions because its regime follows the rainy season. Its northern part is characterized by dry sub-tropical climate (between 1000 mm and 1700 mm). This area has a unimodal rainfall distribution or pattern with distinct wet (rainy) and dry seasons. The central (equatorial climate) and southern (humid equatorial climate) parts of the basin are characterized by two rainy seasons. In the equatorial climate, the annual rainfall is greater than 1500 mm. The amount of rainfall is higher in the humid equatorial climate, with a yearly mean of 1800 mm. As described in previous studies [19,20], the lithology is characterized by the Birimian formations (volcanic, volcanogenic, and sedimentary formations) and granitoid Eburnean comprised of granitic solid masses in which several generations of granites are distinguished. Alterites and fracture aquifers provide a year groundwater supply linked to the underground grid of fractures [21]. Previous studies [22,23] have shown that groundwater recharge is highly dependent on rainfall and varies between 50 mm and 354 mm in the Bandama basin. Vegetation cover in the Bandama catchment varies from north (savannah) to south (forest). Topography of the Bandama catchment is gentle, with a maximum elevation of 809 m above sea level.

2.2. Dataset

2.2.1. Historical Time Series Data

Historical meteorological and hydrological data were collected from the Department of meteorology and hydraulic infrastructures division, Government of Côte d'Ivoire. In this study, the climate data include details of rainfall, temperature, and the potential evapotranspiration. The hydrological data includes mean monthly flow of the Bandama river and its tributaries (Marahoue and N'zi). The stations were selected to provide a good spatial coverage of the different climatic areas across in the Bandama basin (Figure 2). The meteorological stations outside the watershed were used because the climatic parameters measured at these stations influence the flows on the basin. The list of historical data used is presented in Table 1.

Table 1. Characteristics of the selected stations.

Code	Station	Data	Longitude (Decimal Degree)	Latitude (Decimal Degree)	Data Availability
1090012000	Korhogo	Meteorological	−5.61	9.41	1971–2005
1090006400	Boundiali		−6.46	9.51	1922–2005
1090005600	Bouaké		−5.06	7.73	1966−2005
1090018700	Tafiré		−5.15	9.06	1950−2005
1090007300	Dabakala		−4.43	8.38	1945−2005
1090015100	M'Bahiakro		−4.33	7.45	1944−2005
1090022200	Zuénoula		−6.05	7.41	1953−2005
1090009100	Dimbokro		−4.70	6.65	1921−2005
1090016900	Oumé		−5.41	6.36	1944−2005
1090005200	Bouaflé		−5.75	6.98	1924−2005
1090010300	Gagnoa		−5.95	6.13	1922−2005
1090019600	Tiassalé		−4.83	5.88	1959−2005
1090000100	Abidjan		−3.93	5.25	1961−2005
1090010000	Ferkessédougou		−5.20	9.60	1961−2005
1090101006	Marahoué	Hydrological	−5.75	6.97	1954−2005
1090102515	N'Zi		−4.81	6.00	1953−2005
1090100154	Tiassalé		−4.82	5.89	1954−2005

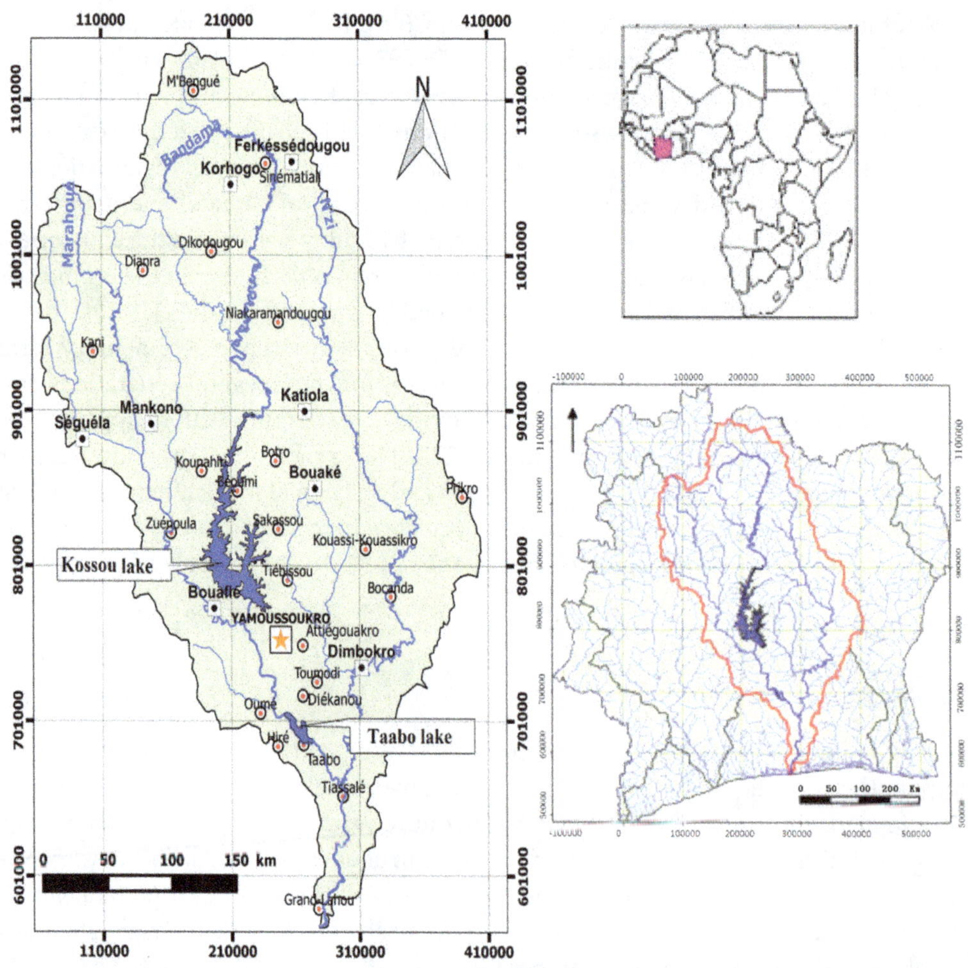

Figure 1. Geographical location of the study area.

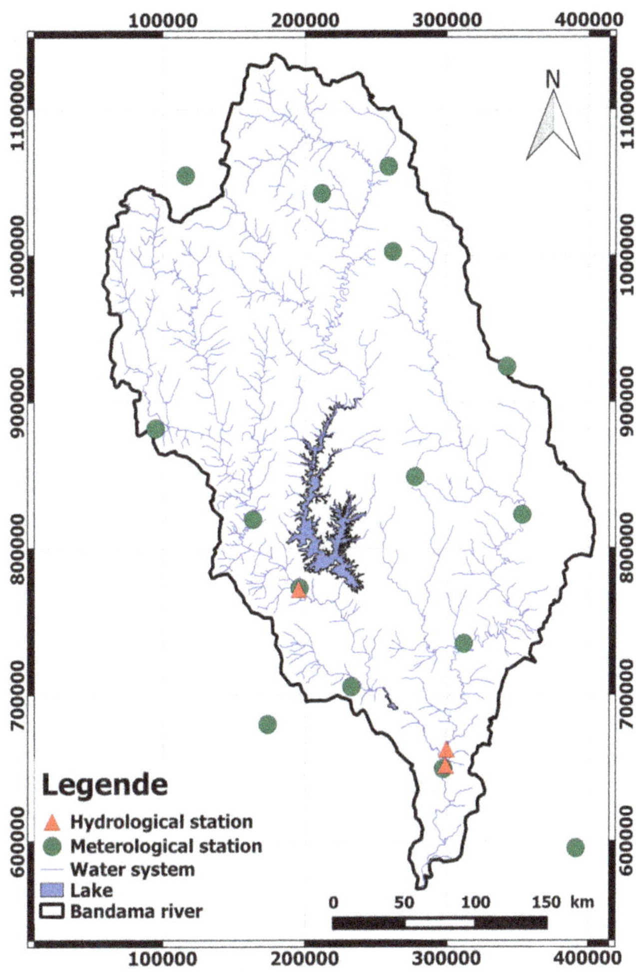

Figure 2. Network of meteorological and hydrological stations used in this study.

2.2.2. Projected Future Climate Data

Simulation results (monthly rainfall and temperature) for future climate from the HadGEM2-ES model were used. It is an Earth system model based on the HadGEM2 atmosphere-ocean general circulation model with additional representation of global-scale processes of biology and chemistry [24]. The simulations described here were driven by prescribed CO_2 concentrations from the representative concentration pathways (RCPs) as part of the 5th Coupled Model Intercomparison Project [25]. The model used the RCP scenarios of changes in other anthropogenic greenhouse gases such as methane, nitrous oxide, and halocarbons, and anthropogenic aerosols such as sulfate and black carbon [26]. For this study, the RCP 4.5 and RCP 8.5 were used. RCP 4.5 is a scenario that stabilizes radiative forcing at 4.5 Watts per meter squared in the year 2100 without ever exceeding that value. This scenario is consistent with a future with relatively ambitious emission reductions. It is similar to SRES B1 [27]. Contrariwise, RCP 8.5 corresponds to a high greenhouse gas emissions pathway compared to the scenario literature. This RCP is consistent with a future with no policy changes to reduce emissions. It is comparable to the SRES A1F1 Emission Scenario [28].

The data used in this work are available from 1900–2100. The simulations were performed to monitor the climate in the short, medium, and long term on the following periods: 2006–2035, 2041–2060, and 2066–2085. The baseline period is 1986–2005. The choice of baseline period is due to the availability of the historical meteorological and hydrological data. Indeed, since 2006, the data transmission has been interrupted in a few stations. The data transmission was re-established in 2015.

2.3. Methodology

2.3.1. Climate Change Simulations

The differences between the climate (rainfall and temperature) of the baseline period and the future climate are calculated for each cell of the general circulation model (GCM) and for each time step (month i, year j over the baseline period 1986–2005, and year k over the 2006–2085 period). They are then expressed in change ratios (Horizons scenario) based on a mean climatology drawn from the HadGEM2-ES model simulations over the same baseline period. The change ratios of monthly rainfall and temperature between the baseline period and the three horizons, 2025 (2006–2035), 2050 (2036–2065), and 2075 (2066–2085), are calculated as:

$$\Delta_{horiz,i} = 100 \times \left(\overline{X}_{horiz,i} - \overline{X}_{ref,i}\right) / \overline{X}_{ref,i} \tag{1}$$

where $\overline{X}_{horiz}$ is the mean value of the simulated series calculated over a given time horizon and $\overline{X}_{ref}$ is the mean value of the simulated series calculated over the baseline period.

2.3.2. Evaluation of Potential Climate Change Impacts on Water Resources

Climate Change Scenarios

General circulation model (GCMs) often show bias in their simulation of the present climate. Various studies have shown that these models overestimate or underestimate the climatic parameters [29–33]. To reduce the estimation errors, it is recommended to build climate change scenarios. In this study, the deltas method proposed in [34] was used. The delta method used allows the implementation of two types of disturbances:

(1) Disturbances "additive" to the temperature:

$$X_{add(t,m,y)} = X_{act(t,m,y)} + M_{cc(m)} - M_{ref(m)} \tag{2}$$

(2) Disturbances "multiplicative" to the rainfall:

$$X_{mult(t,m,y)} = X_{act(t,m,y)} \times \frac{M_{cc(m)}}{M_{ref(m)}} \tag{3}$$

where $X_{add(t,m,y)}$ and $X_{mult(t,m,y)}$ are disturbed variables, $X_{act(t,m,y)}$ is a baseline variable, and $M_{cc(m)}$ and $M_{ref(m)}$ are the monthly averages of the variable calculated from HadGEM2-ES simulations on the time horizons and baseline period. The disturbance of month m is imputed to the current variable (i.e., observed) X_{act}, at all-time steps t of the month, for all years.

Hydrological Modeling

The climate change scenarios are evaluated by means of the GR2M model to observe the impacts of rainfall and temperature changes on water resources in the Bandama river basin. The GR2M model is widely used for hydrological modeling of river basins in Sub-Saharan Africa [29–33,35]. The GR2M model is a spatially lumped and a monthly time-step model developed by IRSTEA [36]. The hydrological functioning of the model is based on two reservoirs: the production reservoir with capacity $X1$ and the routing reservoir with a capacity of 60 mm controlled by parameter $X2$. The hydrological balance of the model is defined by the five terms as:

$$P = R + I + ET + \Delta S \tag{4}$$

where P is the rainfall, R is the runoff at the outlet, I is the infiltration, ET is the evapotranspiration (transpiration and direct evaporation), and ΔS is variation in the water in the basin.

The model is run with monthly data of two climate parameters (the rainfall and the potential evapotranspiration) and the monthly discharge data of the watershed outlet. The implementation of the model consists of the determination of the two parameters $X1$ and $X2$ during a calibration period and then a validation where the representativeness of the two parameters is evaluated (different with the calibration period) [35]). GR2M is calibrated and validated through the Nash efficiency criterion [37]. For both periods (calibration and validation), the Nash efficiency should be higher than 60% for simulations with the optimum parameters. The efficiency criterion evaluates the representativeness of the simulated monthly discharges by the model from:

$$Nash = 100 \times \left[1 - \frac{\sum_{i=1}^{n} (Q_{obs,i} - Q_{cal,i})^2}{\sum_{i=1}^{n} (Q_{obs,i} - \overline{Q_{cal,i}})^2}\right] \quad (5)$$

where $Q_{i\ obs\ \ ios}$ is the observed discharge for month i, $Q_{i\ cal}$ is the simulated discharge for month i, and $\overline{Q}_{cal,i}$ is the mean observed discharge over the given period. A full description of the GR2M model is presented in [30].

3. Results and Discussion

3.1. Assessment of the Changes in Climate Parameters

3.1.1. Changes in Temperatures

The results showed that the temperatures from the Bandama basin will increase in the future (Figure 3). The findings are in accordance with the latest Intergovernmental Panel on Climate Change (IPCC) report. In this study, the temperature in all the months might increase in both the scenarios, but it will be higher in the RCP 8.5 than that in the RCP 4.5. The works of [38] identify tropical West Africa as a hotspot of climate change for both RCP 4.5 and RCP 8.5 pathways, and unprecedented climates are projected to occur earlier (late 2030s to early 2040s) in these regions. It shows that in RCP 8.5, the Bandama basin will experience a temperature rise of 1.5 °C with the minimum temperature rise of about 1.2 °C and a maximum of about 1.7 °C by 2025. The monthly temperatures may vary from 2.2 °C to 3 °C by 2050. A more pronounced increase in temperature is expected in 2066–2085, with annual temperature predictions approximately 20% higher than the baseline temperature. Under RCP 4.5, changes in monthly temperatures may vary from by 0.9 °C in July to 2 °C in January with an average annual of 2.5 °C by 2025. The temperature increases will be greater in the north (Ferkessedougou station) than in the south (Abidjan station) by the 2050's and 2075. Several works in West Africa show a warming range of 3 and 6 °C above the late 20th Century baseline [39–42]. As in the Bandama watershed, the work [43] has shown that the magnitude of temperature is higher for the higher emission scenarios of RCP 8.5 than for the medium–low emission scenarios of RCP 4.5.

3.1.2. Changes in Rainfall

The per cent changes in monthly rainfall based on ground-based observations and projected simulation from the HadGEM2-ES model are presented in the Figure 4. The rates of increase or decrease in rainfall are relatively more in RCP 8.5 than RCP 4.5 over the basin. Under RCP 8.5 and RCP 4.5, the rainfall may decrease from December to April in future periods. This period corresponds to the long dry season in the basin. It is projected to decrease by 3% to 42% at all horizons under RCP 4.5 and by 5% to 47% under RCP 8.5.

During the wet months (June–July and September–November) in all future periods, the rainfall may increase with respect to the reference period (1986–2005). The works of [44] show that the dry-season decrease ranges from 4% to 25% and the wet-season increase ranges from 5% to 23%. However, the increase in the wet months' rainfall may be higher in the far future (2041–2060 or

2066–2085) than in the near future (2006–2035). In West Africa, the rainfall season is predicted to be wetter and delayed by the end of the 21st Century [45]. Otherwise, the work in West Africa [33] showed that although the GCMs manage to reproduce these seasonal dynamics (except for HadCM3), they have real difficulty in accurately simulating the volume of rainfall.

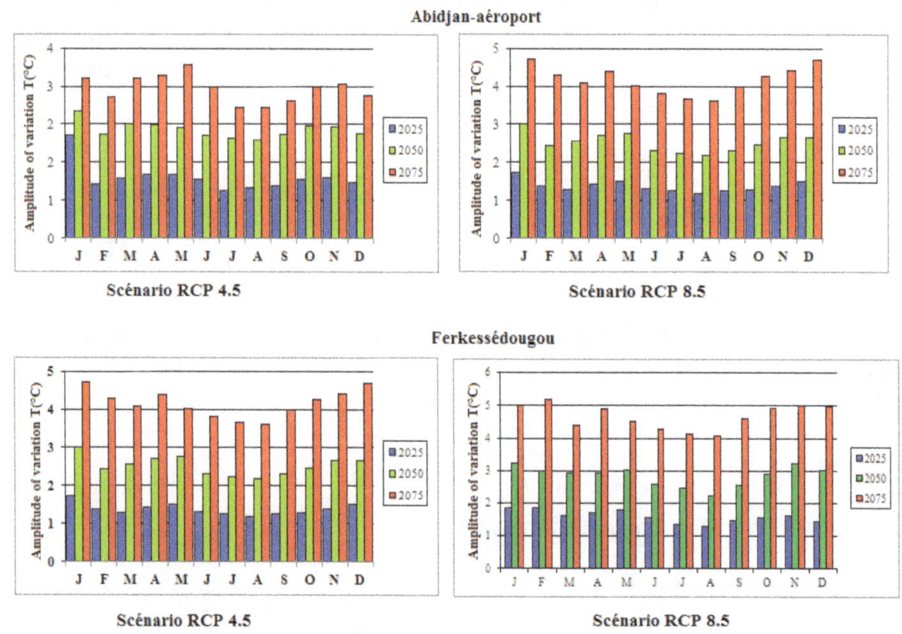

Figure 3. Amplitude of variations in monthly temperatures at different horizons according to the HadGEM2-ES model in the Bandama basin.

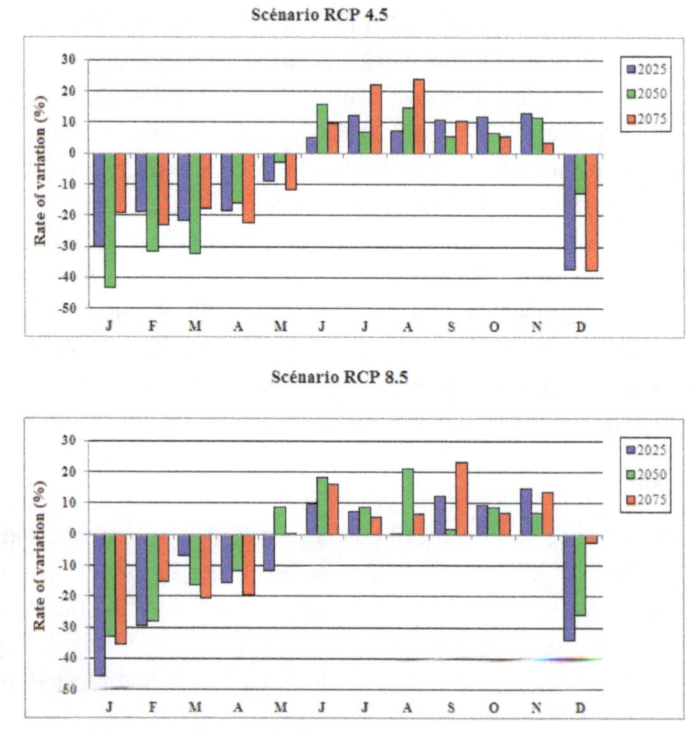

Figure 4. Percentage change in the monthly rainfall for the period 1978–2004 versus 1951–1977 in the Climate Research Unit (CRU) and University of Delaware (UD) observations (top left panels) and in each selected Coupled Model Intercomparison Project Phase 5 (CMIP5)model simulation in the Bandama basin.

3.2. Hydrological Model Calibration

Impacts of climate on water resources were assessed using a hydrological model. The hydrological model GR2M calibration was performed for the periods 1980–1989 (N'zi river), 1980–1987 (Marahoué river), and 1969–1975 (Bandama river). Calibration is the process of choosing the best sets of model parameters, by automatically adjusting their numerical values to better mimic the response observed at the outlet. For these different periods, the Nash performance coefficients are 72.3%, 77.3%, and 76.7% respectively (Table 2). Validating a model means to the check the reproducibility of the results by the calibrated parameters. The model validation was done using one period by selected river: 1990–1998 (N'Zi river), 1967–1977 (Marahoué river), and 1969–1975 (Bandama river). The results shows that the simulated and observed monthly runoffs are very similar with the Nash coefficients greater than 70% (Figure 5).

Table 2. Calibration and vadiation periods with Nash criteria value from the GR2M model.

River (Station)	Calibration		Validation	
	Period	Nash Criteria Value (%)	Period	Nash Criteria Value (%)
Marahoué (Bouaflé)	1980–1987	72.3	1967–1977	74.5
N'Zi (Zianouan)	1980–1989	77.3	1990–1998	72.5
Bandama (Tiassalé)	1969–1975	76.7	1981–1989	78.4

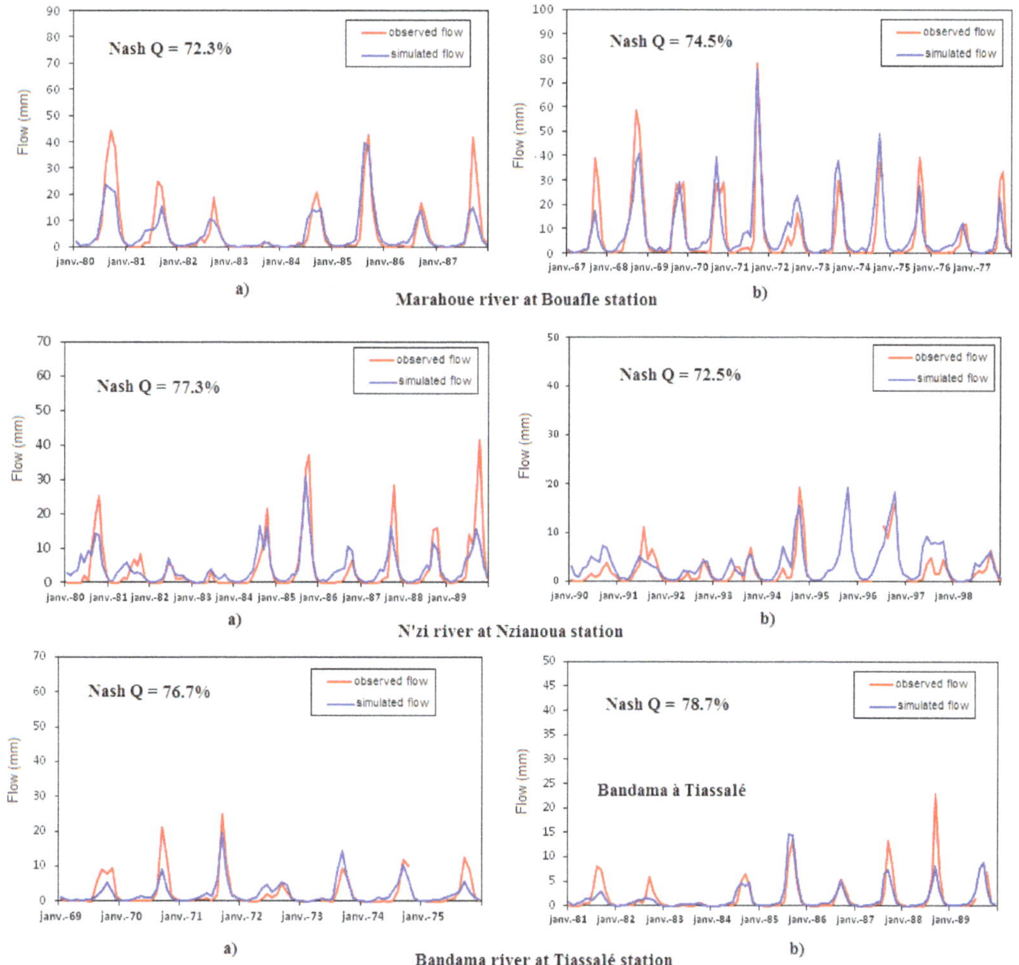

Figure 5. Monthly flow from the observations and the simulations with GR2M with (**a**) representing the period of calibration and (**b**) the validation period.

3.3. Impacts of Climate Change on Surface Water

Comparison of the mean monthly hydrographs for the three time horizons with that of the baseline period shows that the annual hydrological pattern for the Bandama, Nzi, and the Marahoué catchments remains unchanged (Figure 6a,b). The changes caused by climate change affect runoff volume in the watershed. However, the trends in river discharge are different for each scenario. Under RCP 4.5, mean monthly runoff increases for all horizons. For the Marahoué and Bandama river, the changes are important from July to October. For the N'zi river, mean monthly runoff increases gradually from March to November. The findings are in accordance with the works of [31] on the Sassandra basin in Côte d'Ivoire under several scenario (A1, A2, B1, B2, A1B). Under RCP 8.5, runoff is projected to slightly decrease up to the horizon 2025, then to decrease at the horizons 2050 and 2075. Runoff is projected to vary mainly in tributary rivers. The works in the Comoé basin of [33] with the ReGcm model and A1 scenario, revealed a decrease in runoff of 18.8% to 34% in 2031–2040 and 40% to 73% in the 2091–2100 horizon.

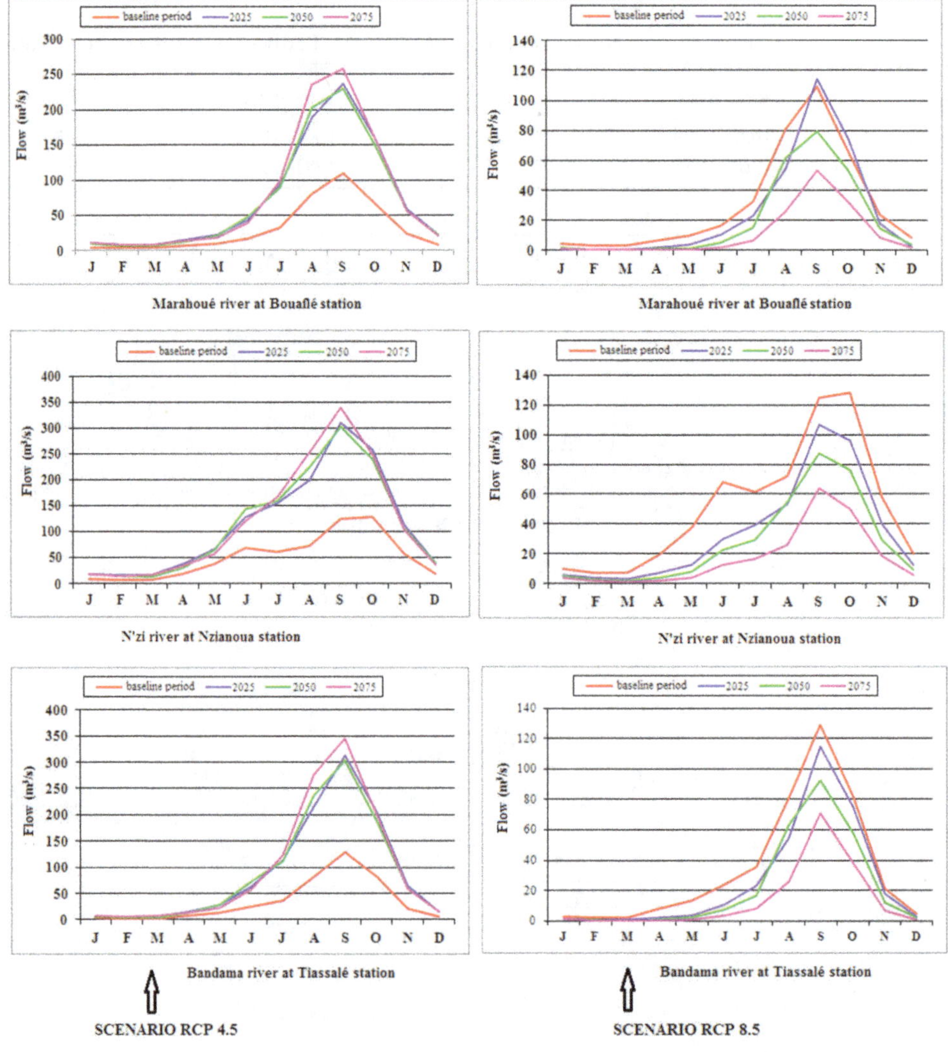

Figure 6. Monthly flows from the periods 1986–2005 (basline period), 2016–2035, 2041–2060, and 2066–2085 of the hydrological stations in the Bandama basin.

These results show that there is considerable uncertainty about the impacts of climate change on runoff. According to [45], overall projections of impacts of climate change on water resources in Sub-Saharan Africa are associated with large uncertainties. Apart from addressing the lack of observational data, key challenges for assessing climatic risks to water availability relate to their responses to heat waves, seasonal rainfall variability, as well as the relationship between land use changes, evapotranspiration, and soil moisture at different levels of global warming [43].

3.4. Impacts of Climate Change on Groundwater

Under RCP 8.4 and RCP 4.5, the climate change is likely to affect groundwater due to changes in precipitation and temperature. The sensitivity of groundwater recharge to climate change is shown in Figure 7. The scenarios under RCP 8.5 show that a trend towards increasing greenhouse gases may significantly decrease in groundwater recharge. The groundwater recharge may decrease from 136.6 mm to 73.8 mm by 2025. By 2075, groundwater may decrease from 60.2% to 55.4% compared to the baseline period. In the far future (2066–2095), the groundwater recharge may decrease in the Bandama aquifers. In the Comoé basin, infiltration could decrease by 7% to 13% in the 2031–2040 horizon and 49.3% to 70% from 2091 to 2100 [26]. Contrary to RCP 8.5, RCP 4.5 indicates climate change should induce an increase in groundwater recharge of the coming decades (Figure 7).

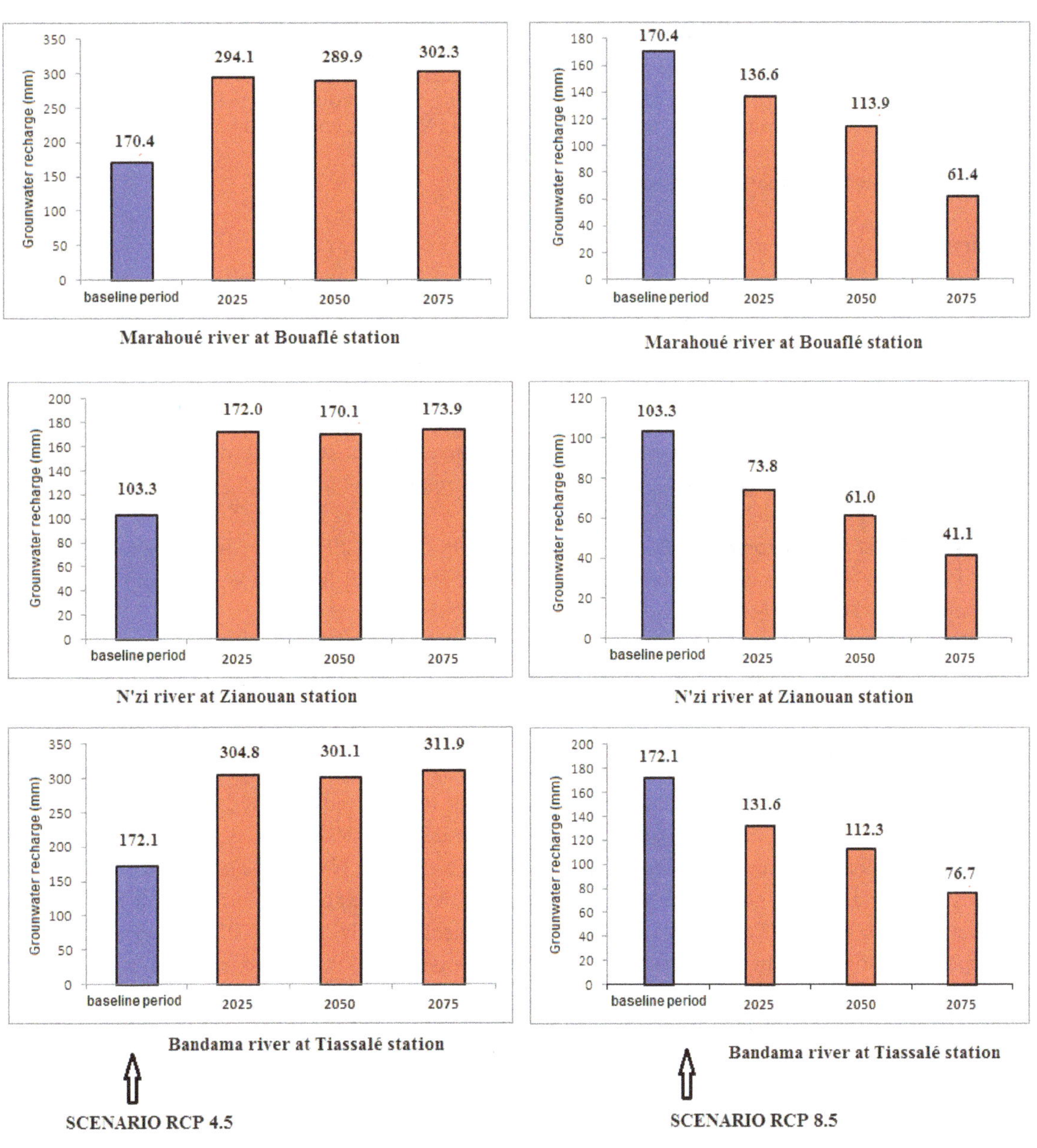

Figure 7. Evolution of annual average groundwater recharge in the Bandama basin.

4. Conclusions

This paper showed that the Bandama basin is highly vulnerable to climate change. Indeed, analyses from this work showed that the temperature may increase and monthly rainfall may decrease from December to April in the future. In addition, the climate change analysis shows that the impacts are very different from RCP 4.5 to RCP 8.5. Under RCP 4.5, mean monthly runoff and groundwater recharge may increase for all horizons. Changes of runoffs and groundwater recharge are mainly dominated by the variations in projected precipitation. Especially in the long-term, increasing precipitation in the wet-season would make it wetter, resulting in higher runoff and aquifer recharge in the watershed. This phenomenon is due to the strong aquifer-river relationship on the basin. Indeed, the increase in aquifer levels would lead to an increase in the flow of watercourses supplied by the aquifers. On the other side, these two parameters may decrease for all horizons. These results highlight the large uncertainties associated with the impacts of climate change on water resources through global models (climatic and hydrological).

Acknowledgments: The authors thank the Department of Meteorology and Hydraulic Infrastructures division, Government of Côte d'Ivoire for data acquisition.

Author Contributions: Gneneyougo Emile Soro developed the ideas; Affoué Berthe Yao and Yao Morton Kouame contributed to the realization of the map and the data processing. Gneneyougo Emile Soro analyzed data and wrote the paper with inputs from Affoué Berthe Yao, Yao Morton Kouame, and Tié Albert Goula Bi.

Conflicts of Interest: The authors declare no conflict of interest.

References

1. Intergovernmental Panel on Climate Change. *Special Report: Managing the Risks of Extreme Events and Disasters to Advance Climate Change Adaptation*; Field, C.B., Barros, V., Stocker, T.F., Dahe, Q., Eds.; Cambridge University Press: Cambridge, UK, 2012.
2. Valipour, M. Necessity of irrigated and rainfed agriculture in the world. *Irrig. Drain. Syst. Eng.* **2013**, *S9*, e001. [CrossRef]
3. Valipour, M. Need to update of irrigation and water resources information according to the progresses of agricultural knowledge. *Agrotechnology* **2013**, *S10*, e001. [CrossRef]
4. Kurukulasuriya, P.; Mendelsohn, R. *Modeling Endogenous Irrigation: The Impact of Climate Change on Farmers in Africa*; Center for Environmental Economics and Policy in Africa (CEEPA) Discussion Paper N°8; Special Series on Climate Change and Agriculture in Africa; World Bank Policy Research Working and University of Pretoria: Pretoria, South Africa, 2006.
5. Burlando, P.; Rosso, R. Effects of transient climate change on basin hydrology. 1. Precipitation scenarios for the Arno River, central Italy. *Hydrol. Proc.* **2002**, *16*, 1151–1175. [CrossRef]
6. Guo, S.; Wang, J.; Xiong, L.; Ying, A.; Li, D. A macro-scale and semi-distributed monthly water balance model to predict climate change impacts in China. *J. Hydrol.* **2002**, *268*, 1–15. [CrossRef]
7. Smith, J.B.; Schellnhuber, H.J.; Mirza, M.Q.; Fankhauser, S.; Leemans, R.; Erda, L.; Ogallo, L.; Pittock, B.; Richels, R.; Rosenzweig, C.; et al. Vulnerability to climate change and reasons for concern: A synthesis. In *Climate Change 2001: Impacts, Adaptation, and Vulnerability*; McCarthy, J., Canziana, O., Leary, N., Dokken, D., White, K., Eds.; Cambridge University Press: New York, NY, USA, 2001; pp. 913–967.
8. Kurukulasuriya, P.; Rosenthal, R. *A Ricardian Analysis of the Impact of Climate Change on African Cropland*; Policy Research Working Paper, No. 4305. World Bank: Washington, DC, USA, 2007. Avaliable online https: //openknowledge.worldbank.org/handle/10986/7508License:CCBY3.0Unported (accessed on 20 December 2016).
9. Hassan, R.; Nhemachena, C. Determinants of African farmers' strategies for adapting to climate change: Multinomial choice analysis. *Afr. J. Agric. Resour. Econ.* **2008**, *2*, 83–104.
10. Roudier, P.; Ducharne, A.; Feyen, L. Climate change impacts on runoff in West Africa: A review. *Hydrol. Earth Syst. Sci.* **2014**, *18*, 2789–2801. [CrossRef]
11. Brown, M.E.; McCusker, B. Climate Change and Agriculture in Africa: Impact Assessment and Adaptation Strategies. *Eos Trans. AGU* **2008**, *89*, 474–474. [CrossRef]

12. United Nations Development Programme. *Programme d'Appui à la Réduction de la Pauvreté PNUD 2009–2013, Sous Programme Protection de l'Environnement et Gestion Durable des Ressources Naturelles (PGDRN)*; UNDP: Abidjan, Côte d'Ivoire, 2009; p. 19. (In French)
13. Soro, N.; Lasm, T.; Kouadio, B.H.; Soro, G.; Ahoussi, K.E. Variabilité du régime pluviométrique du sud de la Côte d'Ivoire et son impact sur l'alimentation de la nappe d'Abidjan. *Sud-Sci. Technol.* **2004**, *14*, 12–19.
14. Goula, B.T.A.; Savane, I.; Konan, B.; Fadika, V.; Kouadio, G.B. Impact de la variabilité climatique sur les ressources hydriques des bassins de N'Zo et N'Zi en Côte d'Ivoire (Afrique tropicale humide). *Vertigo* **2006**, *1*, 1–12. [CrossRef]
15. Kouassi, A.M.; Kouamé, K.F.; Koffi, Y.B.; Djé, K.B.; Paturel, J.E.; Oularé, S. Analyse de la variabilité climatique et de ses influences sur les régimes pluviométriques saisonniers en Afrique de l'Ouest: Cas du bassin versant du N'Zi (Bandama) en Côte d'Ivoire. *Cybergéo* **2010**, *513*, 29–In. (In French) [CrossRef]
16. Yao, A.B. Evaluation des Potentialités en eau du Bassin Versant de la Lobo en vue D'une Gestion Rationnelle (Centre-Ouest de la Côte D'ivoire). Ph.D. Thèse, Université Nangui Abrogoua, Abidjan, Côte d'Ivoire, 2015.
17. Otchoumou, K.F.; Saley, M.B.; Aké, G.E.; Savane, I.; Djê, K.B. Variabilité climatique et production de café et cacao dans la zone tropicale humide: Cas de la région de Daoukro (Centre-Est de la Côte d'Ivoire). *Int. J. Innov. Appl. Stud.* **2012**, *1*, 194–215.
18. Sorokoby, V.M.; Saley, M.B.; Kouamé, K.F.; Djagoua, E.M.V.; Affian, K.; Biemi, J. Variabilité spatio-temporelle des paramètres climatiques et son incidence sur le tarissement dans les basins versants de Bô et Debo (département de Soubré au Sud-Ouest de la Côte d'Ivoire). *Int. J. Innov. Appl. Stud.* **2013**, *2*, 287–299.
19. Kouamelan, A.N. Géochronologie et Géochimie de Formations Archéennes et Protérozoïques de la Dorsales de Man en Côte D'ivoire: Implication Pour la Transition Archéen-Protérozoïque. Ph.D. Thèse, Université de Renne, Rennes, France, 1996.
20. Lemoine, S. Evolution Géologique de la Région de Dabakala (NE de la Côte D'Ivoire) au Protérozoïque Inférieur. Possibilité D'extension au Reste de la Côte D'ivoire et au Burkina Faso: Similitudes et Différences; les Linéaments Greenville-Ferkessédougou et Grand Cess-Niankaramadougou. Ph.D. Thèse, Université Blaise Pascal (Clermont Ferrand II), Aubière, France, 1998.
21. Biémi, J. Contribution à L'étude Géologique, Hydrogéologique et par Télédétection des Bassins Versants Subsaheliens du Socle Précambrien d'Afrique de L'ouest: Hydrostructurale, Hydrodynamique, Hydrochimie et Isotopie des Aquifères Discontinus de Sillons et Aires Granitiques de la Haute Marahoué (Côte D'Ivoire) D'Abidjan. Ph.D. Thèse, Naturelles Université Nationale, Abidjan, Côte d'Ivoire, 1992.
22. Kouakou, E.; Koné, B.; N'Go, A.; Cissé, G.; Ifejika Speranza, C.; Savané, I. Ground water sensitivity to climate variability in the white Bandama basin, Ivory Coast. *SpringerPlus* **2014**, *3*, 226. [CrossRef] [PubMed]
23. Amani, K.M. Caractérisation D'une Modification Éventuelle de la relation pluie-débit et Ses Impacts Sur les Ressources en eau en Afrique de L'Ouest: Cas du Bassin Versant du N'zi (Bandama) en Côte D'Ivoire. Ph.D. Thèse, Université de Cocody, Abidjan, Côte d'Ivoire, 2007.
24. Martin, G.M.; Bellouin, N.; Collins, W.J.; Culverwell, I.D.; Halloran, P.R.; Hardiman, S.C.; Hinton, T.J.; Jones, C.D.; McDonald, R.E.; McLaren, A.J.; et al. The hadGEM2 family of met office unified model climate configurations. *Geosci. Model Dev.* **2011**, *4*, 723–757.
25. Taylor, K.; Stouffer, R.; Meeh, G. An overview of cmip5 and the experiment design. *Bull. Am. Meteorol. Soc.* **2012**, *93*, 485–498. [CrossRef]
26. Betts, R.A.; Golding, N.; Gonzalez, P.; Gornall, J.; Kahana, R.; Kay, G.; Mitchell, L.; Wiltshire, A. Climate and land use change impacts on global terrestrial ecosystems and river flows in the HadGEM2-ES Earth system model using the representative concentration pathways. *Biogeosciences* **2015**, *12*, 1317–1338. [CrossRef]
27. Van Vuuren, D.; Edmonds, J.; Kainuma, M.; Riahi, K.; Thomson, A.; Hibbard, K.; Hurtt, G.; Kram, T.; Krey, V.; Lamarque, J.F.; et al. The representative concentration pathways: An overview. *Clim. Chang.* **2011**, *109*, 5–31. [CrossRef]
28. Riahi, K.; Rao, S.; Krey, V.; Cho, C.; Chirkov, V.; Guenther, F.; Georg, K.; Nakicenovic, N.; Rafaj, P. RCP 8.5-A scenario of comparatively high greenhouse gas emissions. *Clim. Chang.* **2011**, *109*, 33. [CrossRef]
29. Ardoin, B.S. Variabilité Hydroclimatique et Impacts sur les Ressources en eau de Grands Bassins Hydrographiques en Zone Soudano-Sahélienne. Ph.D. Thèse, Université de Montpellier II, Montpellier, France, 2004.
30. Sighomnou, D. Analyse et Redéfinition des Régimes Climatiques et Hydrologiques du Cameroun:

Perspectives D'évolution des Ressources en eau. Ph.D. Thèse, Université de Yaoundé I, Yaounde, Cameroun, 2004.

31. Ardoin-Bardin, S.; Dezetter, A.; Servat, E.; Paturel, J.E.; Mahé, G.; Niel, H.; Dieulin, C. Using general circulation model outputs to assess impacts of climate change on runoff for large hydrological catchments in West Africa. *Hydrol. Sci. J.* **2009**, *54*, 77–89. [CrossRef]
32. Kouakou, K.E. Impacts de la Variabilité Climatique et du Changement Climatique sur les Ressources en eau en Afrique de L'ouest: Cas du Bassin Versant de la Comoé. Ph.D. Thèse, Université Abobo-Adjamé, Abidjan, Côte d'Ivoire, 2011.
33. Kouakou, K.E.; Goula, B.T. A.; Kouassi, A.M. Analyze of climate variability and change impacts on hydro-climate parameters: Case study of Côte d'Ivoire. *Int. J. Sci. Eng. Res.* **2012**, *3*, 1–8.
34. Ducharne, A.; Théry, S.; Viennot, P.; Ledoux, E.; Gomez, E.; Déqué, M. Influence du changement climatique sur l'hydrologie du bassin de la Seine. *VertigO* **2003**, *4*. [CrossRef]
35. Ibrahim, B.; Karambiri, H.; Polcher, J. Hydrological Impacts of the Changes in Simulated Rainfall Fields on Nakanbe Basin in Burkina Faso. *Climate* **2015**, *3*, 442–458. [CrossRef]
36. Institut National de Recherche en Sciences et Technologies Pour L'environnement et L'agriculture. Available online: http://www.irstea.fr/ (accessed on 20 March 2016).
37. Nash, J.; Sutclie, J. River flow forecasting through conceptual models. Part I: A discussion of principle. *J. Hydrol.* **1970**, *10*, 282–290. [CrossRef]
38. Diffenbaugh, N.S.; Giorgi, F. Climate change hotspots in the CMIP5 global climate model ensemble. *Clim. Chang.* **2012**, *114*, 813–822. [CrossRef] [PubMed]
39. Meehl, G.A.; Stocker, T.F.; Collins, W.D.; Friedlingstein, P.; Gaye, A.T.; Gregory, J.M.; Kitoh, A.; Knutti, R.; Murphy, J.M.; Noda, A.; et al. Global Climate Projections. In *Climate Change*; Solomon, S., Qin, D., Manning, M., Chen, Z., Marquis, M., Averyt, K.B., Tignor, M., Miller, H.L., Eds.; The Physical Science Basis; Contribution of Working Group I to the Fourth Assessment Report of the Intergovernmental Panel on Climate Change; Cambridge University Press: Cambridge, UK; New York, NY, USA, 2007.
40. Diallo, I.; Sylla, M.B.; Giorgi, F.; Gaye, A.T.; Camara, M. Multimodel GCM-RCM ensemble-based projections of temperature and precipitation over West Africa for the early 21st century. Hindawi Publishing Corporation. *Int. J. Geophys.* **2012**. [CrossRef]
41. Monerie, P.A.; Fontaine, B.; Roucou, P. Expected future changes in the African monsoon between 2030 and 2070 using some CMIP3 and CMIP5 models under a medium-low RCP scenario. *J. Geophys. Res. Atmos.* **2012**, *117*, 1–12. [CrossRef]
42. Niang, I.; Ruppel, O.C.; Abdrabo, M.A.; Essel, A.; Lennard, C.; Padgham, J.; Urquhart, P. Africa. In *Climate Change 2014: Impacts, Adaptation, and Vulnerability*; Part B: Regional Aspects; Contribution of Working Group II to the Fifth Assessment Report of the Intergovernmental Panel on Climate Change; Cambridge University Press: Cambridge, UK, 2014.
43. Ayele, H.S.; Li, M.-H.; Tung, C.-P.; Liu, T.-M. Impact of Climate Change on Runoff in the Gilgel Abbay Watershed, the Upper Blue Nile Basin, Ethiopia. *Water* **2016**, *8*, 380. [CrossRef]
44. Christensen, J.H.; Hewitson, B.; Busuioc, A.; Chen, A.; Gao, X.; Held, I.; Jones, R.; Koli, R.K.; Kwon, W.-T.; Laprise, R.; et al. Regional climate projections. In *Climate Change 2007: The Physical Science Basis. Contribution of Working Group I to the Fourth Assessment Report of the Intergovernmental Panel on Climate Change*; Solomon, S., Qin, D., Manning, M., Chen, Z., Marquis, M., Averyt, K.B., Tignor, M., Miller, H.L., Eds.; Cambridge University Press: Cambridge, UK, 2007; pp. 847–940.
45. Serdeczny, O.; Adams, S.; Baarsch, F.; Coumou, D.; Robinson, A.; Hare, W.; Schaeffer, M.; Perrette, M.; Reinhardt, J. Climate change impacts in Sub-Saharan Africa: From physical changes to their social repercussions. *Reg. Environ. Chang.* **2015**, 1–16. [CrossRef]

7

Highlighting the Role of Groundwater in Lake–Aquifer Interaction to Reduce Vulnerability and Enhance Resilience to Climate Change

Yohannes Yihdego [1,2,*], John A Webb [2] and Babak Vaheddoost [3]

1 Snowy Mountains Engineering Corporation (SMEC), Sydney, New South Wales 2060, Australia
2 Environmental Geoscience, La Trobe University, Melbourne, Victoria 3086, Australia; john.webb@latrobe.edu.au
3 Hydraulic Lab., Istanbul Technical University, Istanbul 34467, Turkey; babakwa@gmail.com
* Correspondence: yohannesyihdego@gmail.com

Abstract: A method is presented to analyze the interaction between groundwater and Lake Linlithgow (Australia) as a case study. A simplistic approach based on a "node" representing the groundwater component is employed in a spreadsheet of water balance modeling to analyze and highlight the effect of groundwater on the lake level over time. A comparison is made between the simulated and observed lake levels over a period of time by switching the groundwater "node "on and off. A bucket model is assumed to represent the lake behavior. Although this study demonstrates the understanding of Lake Linlithgow's groundwater system, the current model reflects the contemporary understanding of the local groundwater system, illustrates how to go about modeling in data-scarce environments, and provides a means to assess focal areas for future data collection and model improvements. Results show that this approach is convenient for getting first-hand information on the effect of groundwater on wetland or lake levels through lake water budget computation via a node representing the groundwater component. The method can be used anywhere and the applicability of such a method is useful to put in place relevant adaptation mechanisms for future water resources management, reducing vulnerability and enhancing resilience to climate change within the lake basin.

Keywords: lake–groundwater interaction; water balance; wetland; ecosystem; hydrology; climate change; adaptation

---

1. Introduction

When water demand exceeds water availability, water scarcity is inevitable. Climate change, population growth, and economic development add to water scarcity risks mainly in arid regions [1–6]. Concerted data collection efforts are usually lacking in developing countries, especially when it comes to groundwater systems. Many lake studies encountered difficulties in estimating groundwater or defining a plausible, appropriate conceptual model of the aquifer system at hand [7–11]. Prudent understanding of the (ground) water system forms a major inhibiting factor for effective water management. Water management based on such models may have unintended or even detrimental consequences.

Wetland's importance has been recognized by the Ramsar convention due to the possible impacts it may have on people in the next decades and how its conservation can ameliorate poverty conditions. Recent research has been done to assess the relationship between groundwater and surface water, due to the dependability of ecosystems on groundwater contributions [12–16]. The pressure on groundwater resources by these activities has alerted the interest of environmental authorities, who need to assess the hydrodynamic between wetlands, adjacent groundwater, and surface water [17–64]. Many activities have been carried out to validate the hydrogeological conceptual model in the wetland vicinity. A group of observation wells installed around the wetland has enabled the assessment through non-linear Darcy's expression to approximate volumes of recharge and discharge from the aquifer to the wetland.

Groundwater is important for understanding lake systems due to its influence on a lake's water budget, nutrient budget, and acid buffering capacity [21,22]. As a result, groundwater flows to and from lakes have often been estimated using simple flow grids and one-dimensional Darcian calculations. Groundwater interaction with lakes can be spatially and temporally variable, however. Other water balance approaches have also been used that employed a representative groundwater head underneath the lake for calculating the flux over the entire lake area. The most sophisticated way of investigating lake–groundwater interactions is by explicitly including lakes in groundwater flow models [10,28–30]. Water resources managers rely on tools that assist with streamlining supply and demand [25–27,31,49].

Modeling lake–aquifer interaction is an essential milestone in limnological studies [32]. Groundwater is one of the most important hydrological variables of the water budget in lakes; however, this variable, due to its nature, cannot be addressed without uncertainties [32,33]. Many scientists, however, have tried to model the interaction between groundwater and surface water using various statistical, conceptual, or empirical models (e.g., Lohman [34]; Edelman [2]; Lewis et al. [35]; Post et al. [36]; Jakovovic et al. [37]; Yihdego et al. [38]).

The role of groundwater in wetland water budget is of great concern to ecologists, water managers, and environmental scientists [39]. Groundwater is critical for understanding most lake systems because it influences a lake's water budget and nutrient budget [40]. Several studies have reported on the use of a mass balance approach to simulate lake levels from hydrological and meteorological data [41,42]. There have been many different empirical, analytical, and numerical approaches for simulating lake–groundwater interactions (e.g., fixed lake stages, High-K nodes, and LAK3 package through numerical modeling). The advantages and disadvantages of these approaches have been documented by many researchers. Many highly sophisticated models like LAK Package are not widely used due to data and resource constraints [17,35,43]. The limited use of LAK Package is attributable to the lack of standardization and associated graphical user interfaces, complex three-dimensional discretization and data needs, and spatial and temporal complexity inherent in including surface water features in a groundwater model. While sophisticated approaches offer more detail, their advantages may be offset by associated complexity and even instability of the solution procedure [44]. Consequently, a full-featured LAK package is not an automatic choice for practitioners, regulators, and the wider community; rather, the chosen method should depend on both the hydrogeological conditions and the modeling objectives [39,45–48].

Within the Glenelg Hopkins Catchment Management Authority area, Lake Linlithgow and the nearby shallow lakes are considered to be of environmental importance due to the drainage and/or degradation of many other wetlands throughout this region [32,52,53]. However, in recent years Lake Linlithgow has been affected by algal blooms and high lake salinities, and has become dry over the summer during most years since 2000. The normal seasonal fluctuation for Lake Linlithgow is 10,000 μS/cm in winter and spring, rising to 16,000 μS/cm in late autumn [60]. However, during 1999, the lowest salinity levels recorded were 27,000 μS/cm and they peaked at 58,000 μS/cm (seawater) in the autumn, before sharply increasing to 63,000 μS/cm prior to drying out in February 2000. Extensive research was carried out to characterize future changes in groundwater salinization within the basalt aquifers in the Hamilton area, including Lake Linlithgow, using hydrogeological, chemical, and isotopic techniques [14,32,50]. The aim of this study is to improve our understanding of lake–aquifer interaction through analyzing the groundwater component of Lake Linlithgow as a case study.

2. Study Area

Lake Linlithgow covers an area of 9.65 km^2 and is the largest in a series of highly to moderately saline wetlands, located approximately 16 km east of Hamilton, western Victoria (Figure 1). The lake is fed by Boonawah Creek and there is no surface outlet. The catchment area for Lake Linlithgow is 85 km^2. The volcanic plains surrounding Lake Linlithgow are topographically subdued, comprising a flat to undulating plain with average elevation of ~200 m Australian Height Datum (AHD), gently increasing in elevation from south to north (Figure 1). The plain is dotted with several prominent eruption points and a number of smaller low-relief volcanic cones. The volcanic plain is deeply dissected (up to 80 m) by streams on its western and southwestern margins.

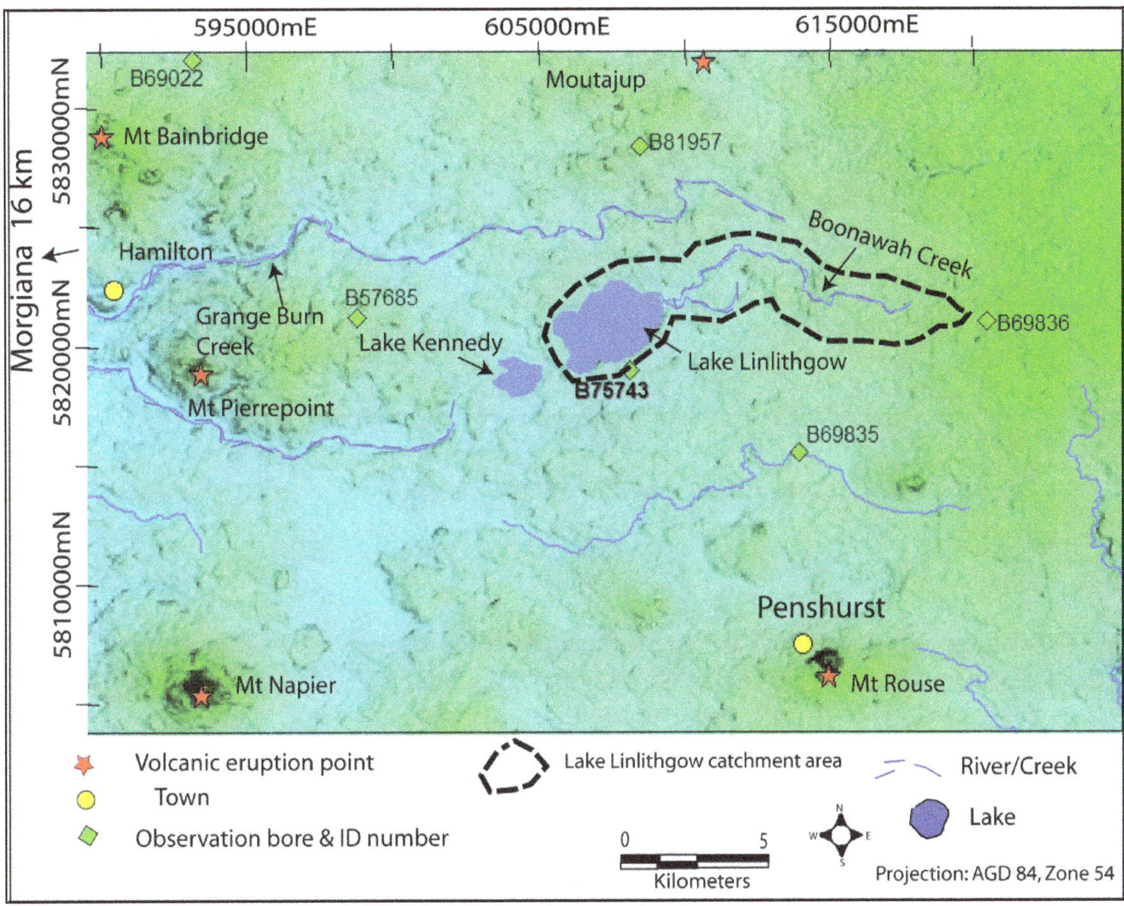

Figure 1. Digital elevation model showing the relative elevation of the terrain surrounding Lake Linlithgow.

The pre-European vegetation comprised River Red Gum, Swamp Gum, Manna Gum, Blackwood, and Lightwood along moderately incised drainage lines, and Tea-tree, Silver Banksia, and Lightwood on the black self-mulching clays associated with the margins of swamps and lakes (i.e., Buckley Swamp) [55]. The poorly developed drainage lines associated with the headwaters of Grange Burn, Violet Creek, and Muddy Creek were often dominated by grasslands. Lake Linlithgow and Lake Kennedy (Figure 2) were presumably vegetated with salt-tolerant species such as plantain, Australian salt grass, and streaked arrow grass, and are relatively unaltered since European settlement.

The most significant land use change occurred between 1900 and 1920, with the conversion to introduced pasture [55].

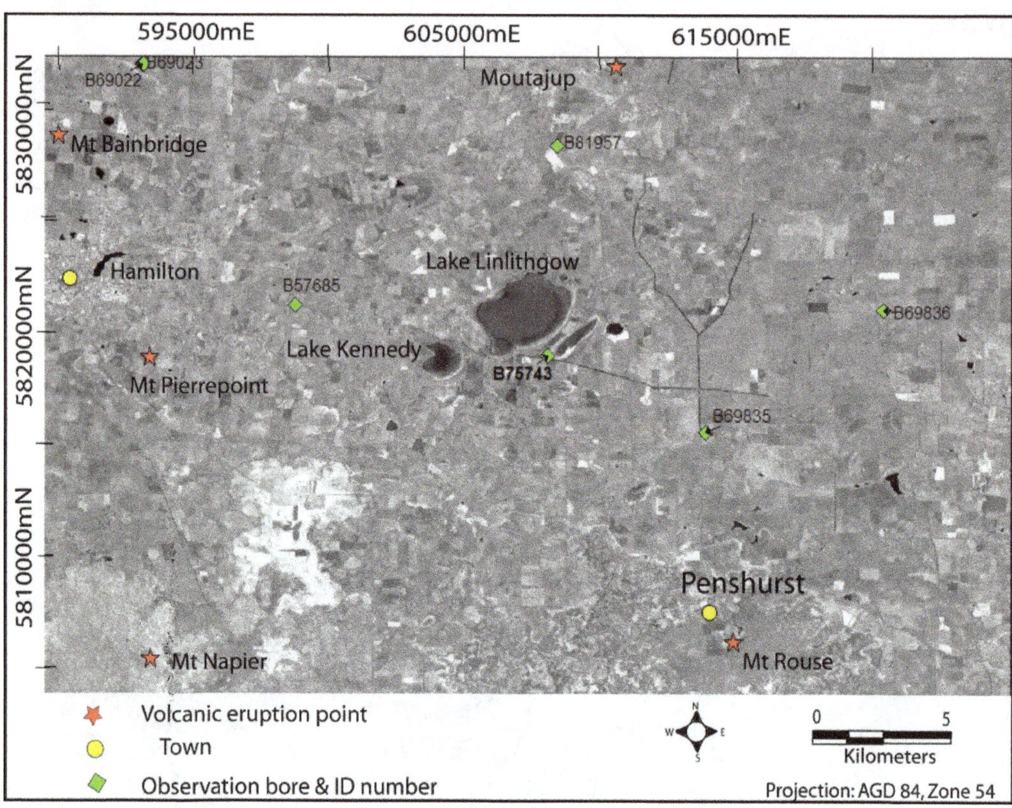

Figure 2. Landsat imagery (taken in February 2004) showing the Lake Linlithgow area.

2.1. Hydrology

The average rainfall over the catchment is about 689 mm, recorded at Hamilton Research Centre. Rainfall records over the last 45 years clearly show that rainfall is highly variable, but over the last decade there has been a substantial drop, with annual rainfall below the long-term average. Maximum rainfall is received over winter (the wettest months are July and August) and exceeds or equals evaporation for May to September, when groundwater recharge is most likely to occur. June to September rainfall contributes 45% to the mean annual precipitation in the catchment; pan evaporation is highest from October to April and totals 1053 mm.

2.2. Geology

The basement geology of the area consists of Cambrian volcanics, Early Paleozoic turbidites, and Silurian sandstone, intruded by Devonian granites (Figure 3). These basement rocks do not outcrop around Lake Linlithgow.

The disruption of the drainage system by the basalt flows formed the lakes in the center of the catchment. Lake Linlithgow sits on first phase (~4 Ma) basalts, and is encircled by second phase (~2 Ma) basalt flows, forming a relatively flat lakebed with steep banks. Lunettes occur on the eastern and northeastern margin of the lake, ~4 m above the current maximum lake level [50]. There is a local groundwater flow system around Lake Linlithgow [32,55], and a regional to intermediate groundwater flow system in the surrounding volcanic plain (Figure 4).
56]. Potentiometric surface contours for the Newer Volcanics aquifer indicate that Lake Linlithgow strong correlation with the lake level. The declining trend of 20 cm/year from 1997 to 2001 is clearly a response to below-average rainfall [14]. The palaeosols lying between successive basalt phases hinder outflow from the lake and act as barriers to groundwater flow (Figure 6).

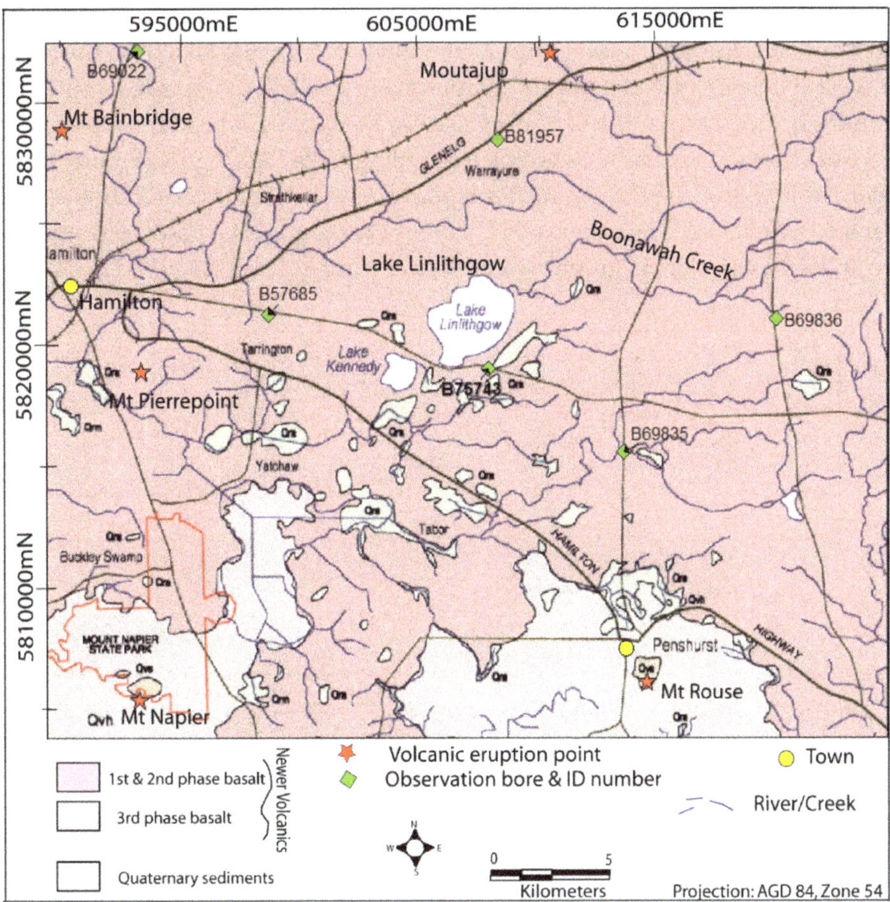

Figure 3. Geological map of the area surrounding Lake Linlithgow.

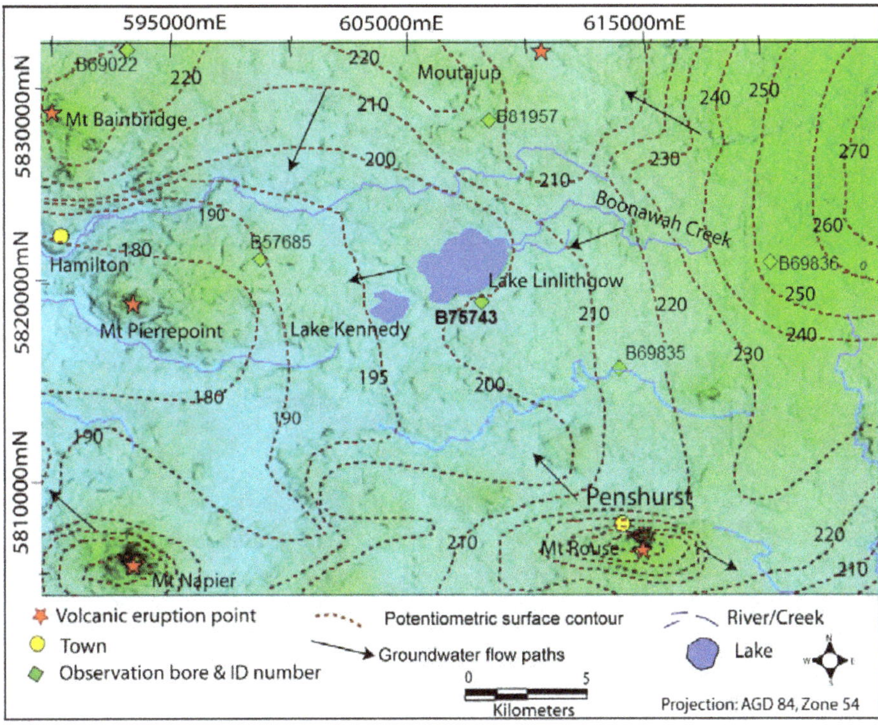

Figure 4. Lake Linlithgow and associated wetlands (gray) and basalt aquifer potentiometric surface contours (m AHD).

Potentiometric surface contours for the Newer Volcanics aquifer indicate that Lake Linlithgow is a groundwater through-flow lake, with groundwater flow entering the lake from the east and leaving towards the west (Figure 4). Groundwater enters through a series of springs and seeps along the eastern lake margin, mostly from the second phase basalt aquifer, which terminates here [50,54–56]. Hydrographs from bores in the basalt aquifer surrounding Lake Linlithgow (Figure 5) show a strong correlation with the lake level. The declining trend of 20 cm/year from 1997 to 2001 is clearly a response to below-average rainfall [14]. The palaeosols lying between successive basalt phases hinder outflow from the lake and act as barriers to groundwater flow (Figure 6).

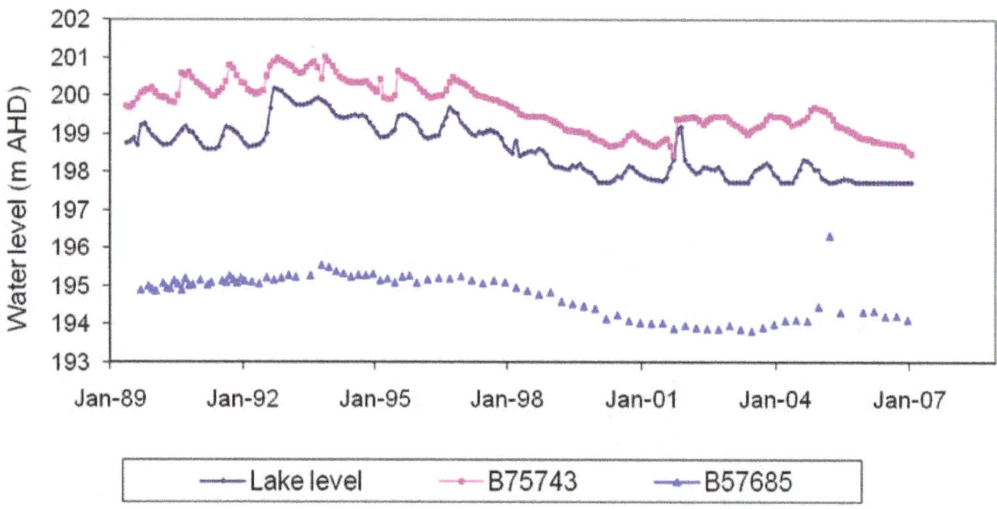

Figure 5. Comparison of bore hydrographs with the lake level. The positions of the bores are shown in Figure 4; screened depth of B57685 is from 82.5 to 88.5 m.

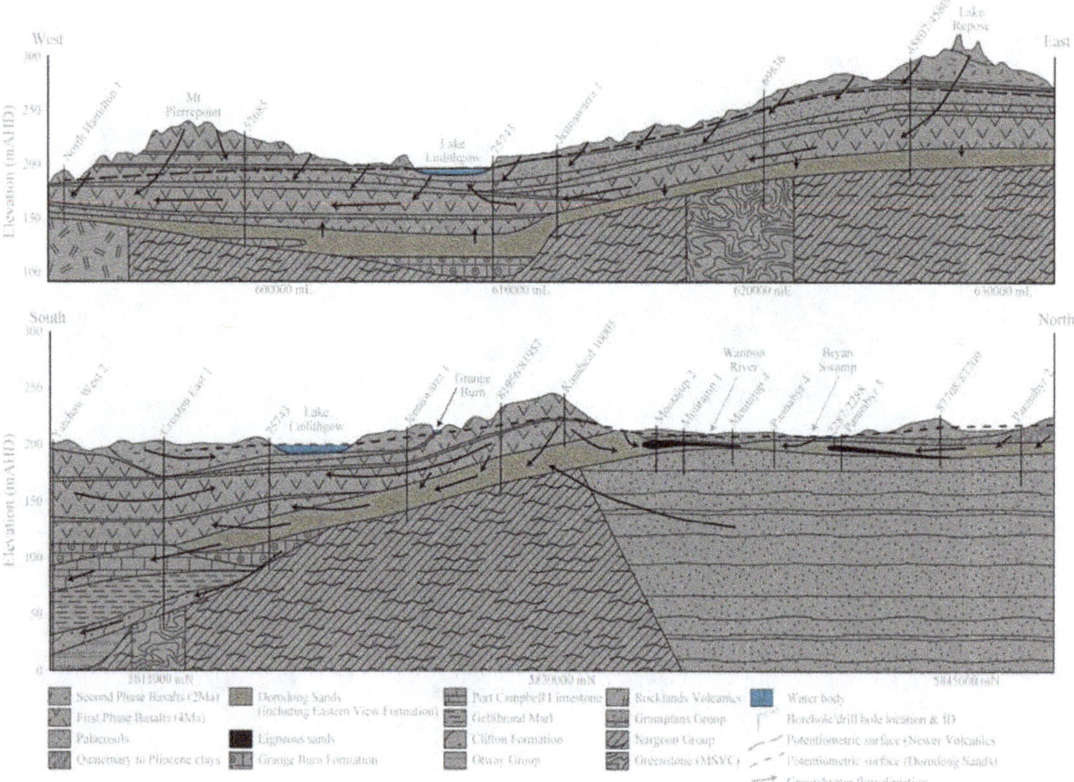

Figure 6. West–East and South–North hydrogeological cross sections through Lake Linlithgow [50,55].

2.3. Lake Hydrology

Monthly lake level and salinity data are available for Lake Linlithgow from 1964 to 2007, when the lake dried out, although the salinity measurements are available only sporadically from 1964 to 1974 (Figure 7). Lake Linlithgow has a median depth of 1.45 m, but typically varies seasonally by up to 1 m (Figure 7). A graph of cumulative deviation of rainfall from the mean for Hamilton Research Station shows a strong correlation with lake levels; peaks in the lake level generally correspond to periods of above-average rainfall (Figure 7). In years of below-average rainfall, the lake dries out during summer, as occurred in January–April of 1983, 2000, and 2001, but after a succession of high rainfall years, lake levels rise substantially (Figure 7).

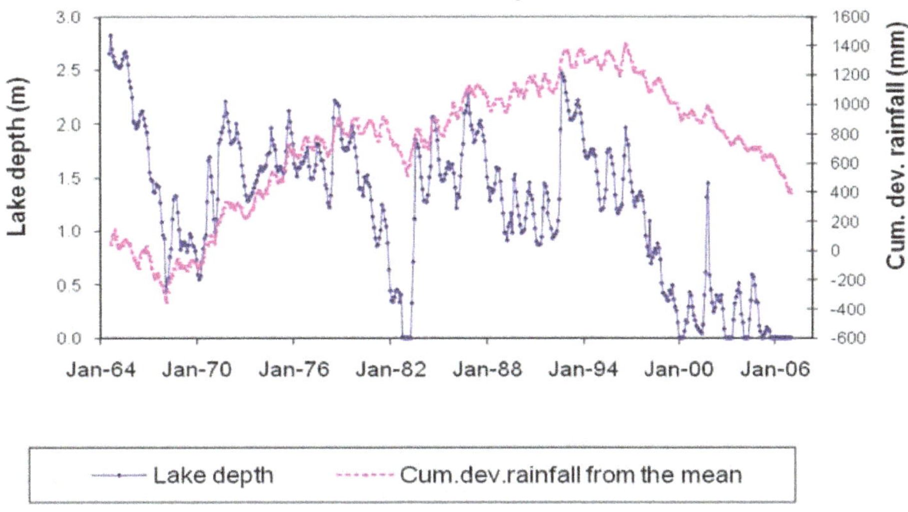

Figure 7. Lake level with variation in rainfall.

Lake Linlithgow is typically saline (median 12400 μS/cm). The large seasonal variation in lake level is correlated with a substantial range in lake salinity (3300–98,900 μS/cm; Figure 8). There is a general tendency for Lake Linlithgow to become more saline until the lake dries out (Figure 8).

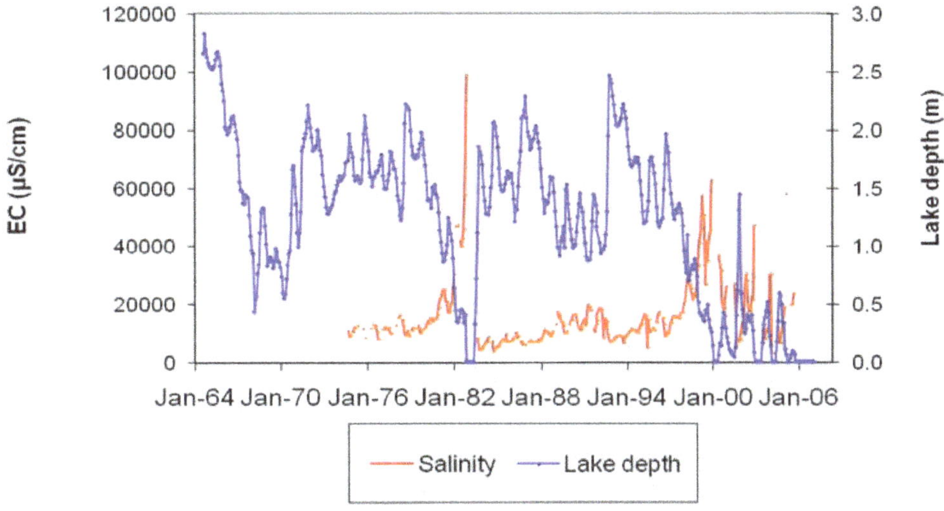

Figure 8. Relationship between measured lake depth and salinity for the period 1964–2007.

3. Methodology

A spreadsheet model was employed to analyze for the current study [32]. An explanation of the model is given below.

3.1. Water Balance Model

Monthly time-step modeling of the lake water level was carried out using Excel spreadsheets, and the resulting water budget was used as input for the salt budget. The lake water balance is calculated by estimating all the lake's water gains and losses, and the corresponding change in volume is expressed as:

$$\text{Volume change} = \text{Surface water inflow} + \text{Rainfall} + Q_{in} - \text{Evaporation} - \text{Surface water outflow} - Q_{out}, \quad (1)$$

where Q_{in} is the groundwater inflow and Q_{out} the groundwater outflow.

The net flux of the groundwater flow ($Q_{in} - Q_{out}$) can be calculated as:

$$Q = C\,(H_{lake} - H_{aquifer}) \text{ in } m^3{\cdot}month^{-1}, \quad (2)$$

where C is the conductance of the lakebed sediments ($m^2{\cdot}month^{-1}$) and H is the water level in the lake and surrounding aquifer (m). Hydraulic conductivity can be expressed as

$$C = K \times A/L \text{ in } m^2{\cdot}month^{-1}, \quad (3)$$

where K is the hydraulic conductivity of the lakebed sediments ($m{\cdot}month^{-1}$) and A and L are the lake area (m^2) and lakebed sediment thickness (m), respectively. The formulation is similar to that used in the river (RIV) and lake packages, which specifies the flux through the riverbed or lakebed as a function of stage, potentiometric head in the connected cells, and the riverbed or lakebed conductance in which the lakebed conductance, COND, at each cell is either specified by the user in the lake package input file, or calculated from the lakebed geometry and hydraulic conductivity. As with the river package, flow from the lake to the groundwater in the LAK package is limited when the head in a cell falls below the lakebed bottom. Also, if the stage of the lake is below the top of the lakebed, the lake cell is dry and seepage into the groundwater is cut off for that cell. In this spreadsheet model, a similar procedure was applied to formulate the boundary conditions that control the solution of potentiometric head.

The temporal area is estimated from the lake stage–area–volume relationship built in [42], while the lakebed sediment is estimated using the soil erosion model of the catchment. The water level in the surrounding aquifer is updated ($H_{aquifer\text{-}new}$) using the inflow and outflow calculated for the previous month ($H_{aquifer\text{-}pre}$) is

$$H_{aquifer\text{-}pre} = Q/A \times S_y \ (m) \quad (4)$$

$$H_{aquifer\text{-}new} = H_{in\text{-}old} + H_{in} \ (m), \quad (5)$$

where A is the surface area of the interacting aquifer and S_y is the specific yield of the aquifer. The model requires known hydrometeorological data (inflow from the rivers, rainfall on the lake surface, aquifer area, and evaporation from the lake) and estimates the unknown net groundwater flux due to interaction of the lake with the surrounding aquifer by comparing the simulated and recorded lake levels and calculating a residual. The model was calibrated using solver and iteration. The net groundwater component is represented as a node in the equation to switch on and off and assess the significance of the groundwater in the overall lake water budget, as explained through the fluctuation and lake level trend over time.

3.2. *Lake Water Budget and Model Parameters*

3.2.1. Lake Storage

Bathymetry data are not available for the lake; however, the area–depth relationship was estimated. The lake area was measured from 12 Landsat images taken between 1972 and 2004 and correlated with the measured lake depth in the month when the images were taken (Figure 9). The lake area was estimated using ENVI software (with threshold method and grow button); this is more accurate than the GIS (vector) method used by [50] because it finds similar pixels selected for the water body and thereby better identifies the natural boundary. The line of best fit (with $r^2 = 0.99$) to the area/depth data gives the relationship for Lake Linlithgow as:

$$A(t) = 1.545(D(t))^5 - 23.09(D(t))^4 + 135.9(D(t))^3 - 393.7(D(t)^2 + 561.1(D(t) - 305.8. \quad (6)$$

The polynomial relationship between lake area (A(t)) and lake depth (D(t)) is due to the lake's relatively flat lakebed and steep banks, and gives a better fit for the depth–area relationship than the logarithm function used by [50], which had a smaller r^2 value.

The lake volume at the beginning of a given month can therefore be calculated from the depth and area at the end of the preceding month. The lake depth must be first adjusted, because the minimum reading on the base of the lake level gauge is ~1.3 m. Thus, this value is deducted from the recorded lake level to get the true lake depth. For the modeling, the initial volume in September 1964 was calculated from the measured water depth in that month (2.79 m).

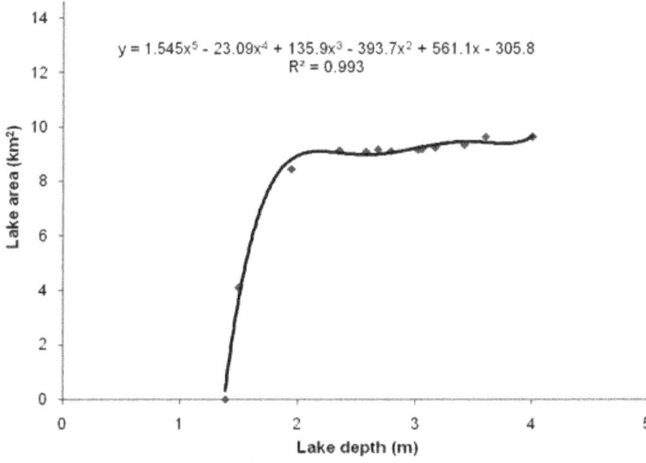

Figure 9. Relationship between lake area derived from Landsat imagery and measured lake depth.

3.2.2. Precipitation on Lake

Monthly precipitation is taken from the nearest rainfall station (Hamilton Research Centre), which lies approximately 13 km southwest of Lake Linlithgow, and is multiplied by the lake area to give the volume of direct rainfall into the lake. The precipitation data are only available up until 2001; values for the subsequent time period have been extrapolated from the rainfall at Carinya station (~65 km from Hamilton Research Centre) using the correlation between rainfall at Hamilton Research Centre and Carinya.

3.2.3. Surface Flow to the Lake

Surface inflow to Lake Linlithgow is received through Boonawah Creek, for which there are no gauging data available. Surface inflows can, however, be estimated by the *tanh* cumulative surplus rainfall approach [42] using the Grange Burn flow, which is gauged at Morgiana (gauge no.

238219; Figure 1), because the Grange Burn and Lake Linlithgow catchments are adjacent and have similar topography, soil type, vegetation, land use, and rainfall. The tanh cumulative surplus rainfall approach provides a tool with which runoff in Grange Burn can be predicted for any given monthly precipitation/evaporation (Figure 10). Using area scaling, this can be converted to a flow in Boonawah Creek; the catchment areas for Boonawah Creek (at Lake Linlithgow) and Grange Burn at Morgiana are 85 km^2 and 997 km^2, respectively.

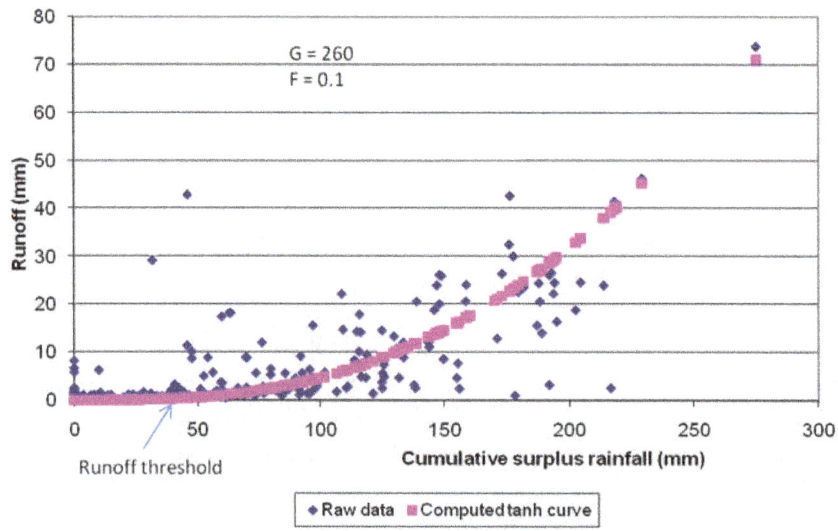

Figure 10. Stream flow modeling for the Grange Burn at Morgiana (gauge no. 238219). The location of the gauge station is shown in Figure 1.

3.2.4. Outflow from the Lake

The evaporation from Lake Linlithgow is estimated using monthly evaporation data at Hamilton Research Station, where pan evaporation was measured from 1968 to June 2000. The evaporation data from July 2000 has been extrapolated from White Swan Reservoir station by establishing a correlation between pan evaporation at Hamilton Research Centre and White Swan Reservoir. Evaporation data from 1964 to 1968 are lacking at White Swan Reservoir, so evaporation for this period was estimated from the correlation of evaporation at the Hamilton Research Centre with that at the Melbourne regional office. A local calibration coefficient was used to adjust the seasonal pan evaporation data from Hamilton Research Centre for the best fit of the model. The optimized local calibration coefficient is 1.18; this takes into account the spatial variation in position, elevation, and storage effect between Hamilton Research Centre and Lake Linlithgow.

3.2.5. Groundwater Inflow and Outflow Estimation

Groundwater inflow/outflow was initially estimated using Darcy's Law. The widths of the groundwater inflow and outflow zones along the lake perimeter are ~6.1 km and 3.6 km respectively. The cross-sectional area is calculated by multiplying the width of the groundwater inflow/outflow zone by the saturated thickness (8 m) of the first phase basalt aquifer in hydraulic contact with the lake. The average hydraulic conductivity of the basalt aquifer (0.09 m/day) is derived from groundwater flow rates calculated using groundwater radiocarbon ages by Bennetts [50]. The hydraulic gradient either side of the lake, derived directly from the potentiometric contours, is 3.7×10^{-3} (Figure 4). Similar to Lake Burrumbeet, an average value of 0.000864 m/d has been chosen for the permeability of the lake floor in order to estimate the groundwater outflow through the lakebed. Using these figures, the monthly groundwater outflow and inflow from/to the lake were estimated as ~0.29 ML and ~0.48 ML, respectively.

Model calibration, carried out by adjusting input parameters so that the simulated lake levels fit the observed lake level, gives the net groundwater flow estimation. In this case the optimized lake–aquifer conductance (C) value is 2.71×10^3 m^2/day, which is equivalent to a hydraulic conductivity value of lake shore sediments of about 7 m/day; this falls within the range of previous hydraulic conductivity estimates for these sand beaches. The optimized specific yield and aquifer area are 0.1 and 45 km^2, respectively [32].

The optimized specific yield compares reasonably with previous specific yield estimates of the basalt aquifer, which range up to 0.18 [51]. The aquifer area in hydraulic contact with the lake (45 km^2) is significant compared with the catchment area of Lake Linlithgow (85 km^2), indicating that lake level fluctuations directly impact about half of the basalt aquifer in the catchment area.

The best possible fit of the water budget model was attained at ~0.37×10^6 L and 3.1×10^6 L monthly average groundwater outflow and inflow, with the exception of dry periods. The value of optimized groundwater outflow compares closely with the initial estimate.

4. Results and Discussion

4.1. Lake Water Levels

The predicted lake levels show good agreement with the measured lake level data (Figure 11), with an r^2 value of ~0.85 (Figure 12). The sum of the squared differences did not exceed 0.4 m, except for a few outliers (Figure 13).

The water balance shows that the major influence on lake levels is evaporation (Table 1), accounting for an average of 54% of the total water budget. It has the greatest influence in summer, reaching up to 99% of the total budget, but decreased during dry periods, because it is proportional to the lake area [32]. Groundwater inflow contributes about 1% of the lake water budget. Even though groundwater outflow is minor (0.1%), it dominates the lake water losses during early winter (June), a time when evaporation is very small (Figure 14).

Thus, Lake Linlithgow is a groundwater through-flow lake, and the lake–groundwater interaction is important since it affects the environmental health of this major wetland.

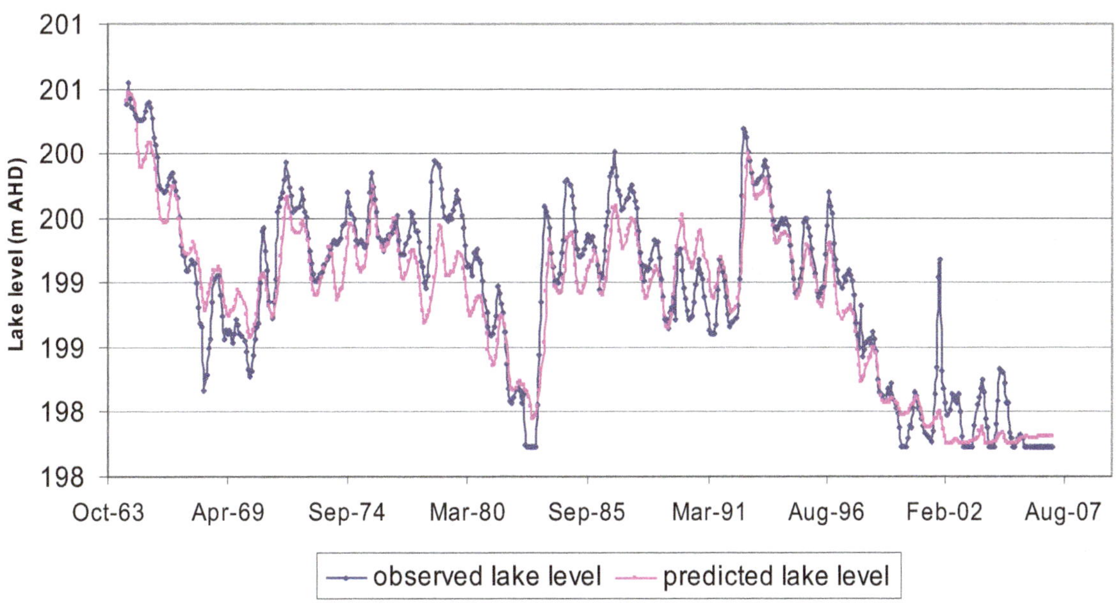

Figure 11. Lake Linlithgow measured water levels and a comparison to modeled results for the period 1964–2007.

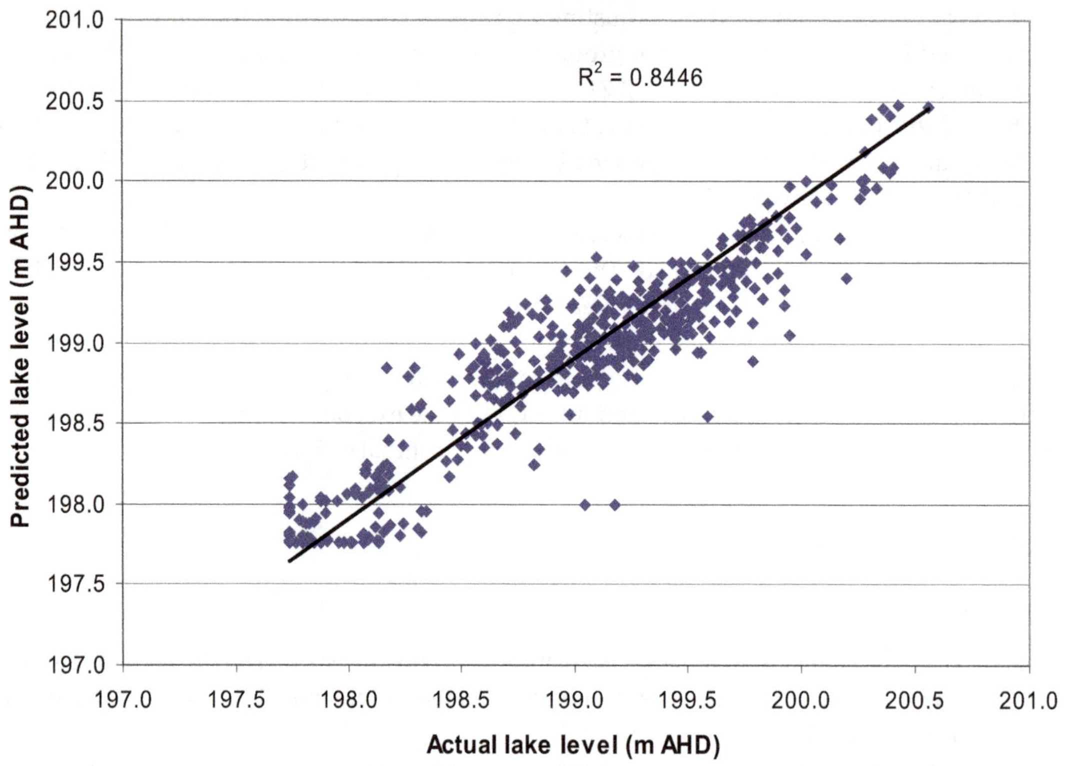

Figure 12. Correlation between observed and calculated lake elevation in AHD m.

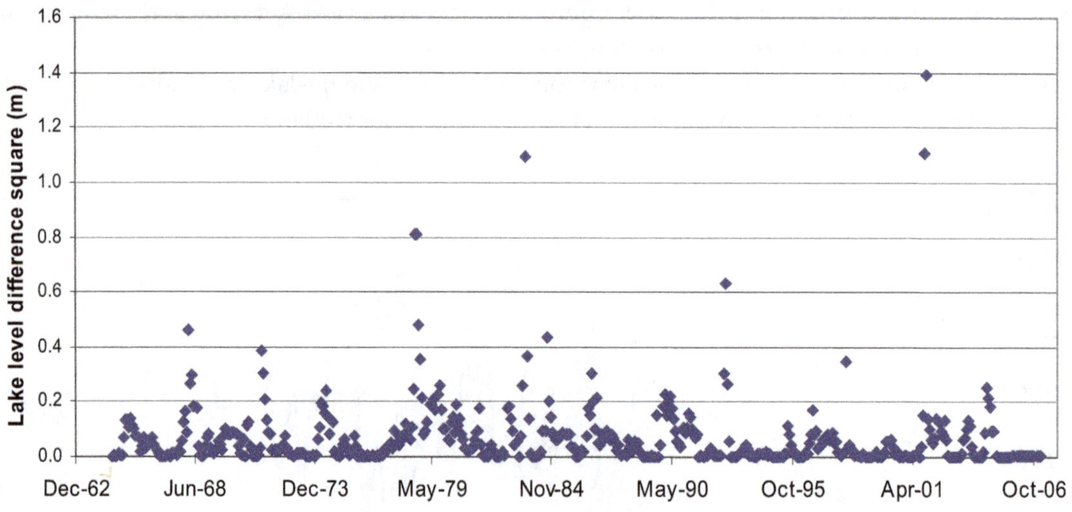

Figure 13. Temporal distribution of square difference between observed and calculated lake levels.

Table 1. Average long-term (1964–2007) monthly contribution (in percent) of each Lake Linlithgow water budget component to the overall lake water budget.

Evaporation (%)	Groundwater Outflow (%)	Precipitation (%)	Surface Inflow (%)	Groundwater Inflow (%)
54	0.1	37	8	1

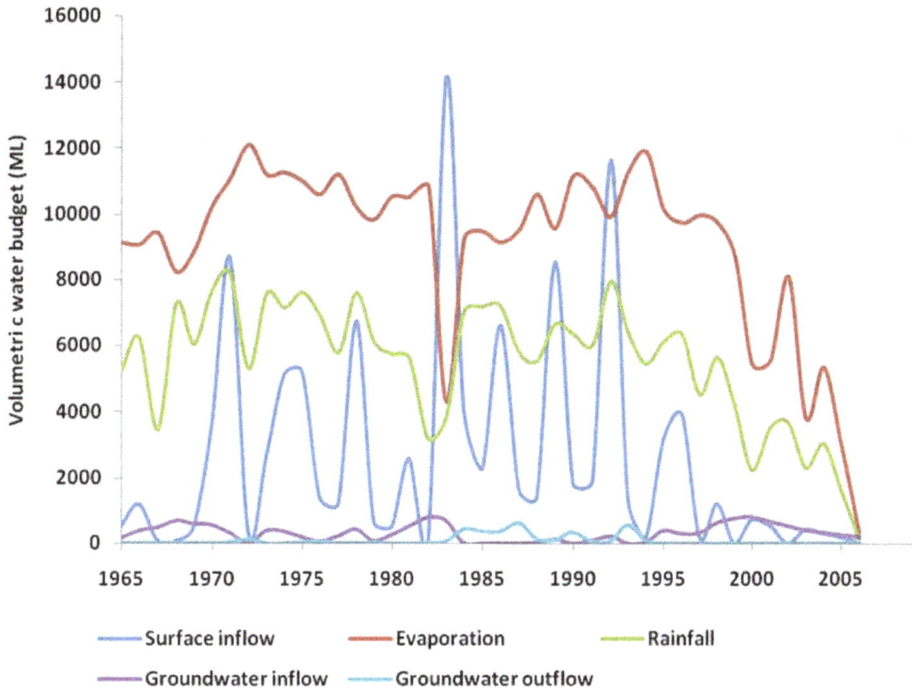

Figure 14. Annual water balance of Lake Linlithgow (1965–2006).

4.2. Water Budget Errors and Sensitivity Analysis

A sensitivity analysis shows that the model is most sensitive to evaporation and precipitation (Figure 15), and surface inflow to a lesser extent. However, the most likely source of error within the model is estimating the ungauged surface inflow (i.e., Boonawah Creek) by the tanh relationship from Grange Burn; this could be in error if the threshold value at which cumulative surplus rainfall becomes runoff in Grange Burn at Morgiana is different to that of the smaller catchment of Boonawah Creek.

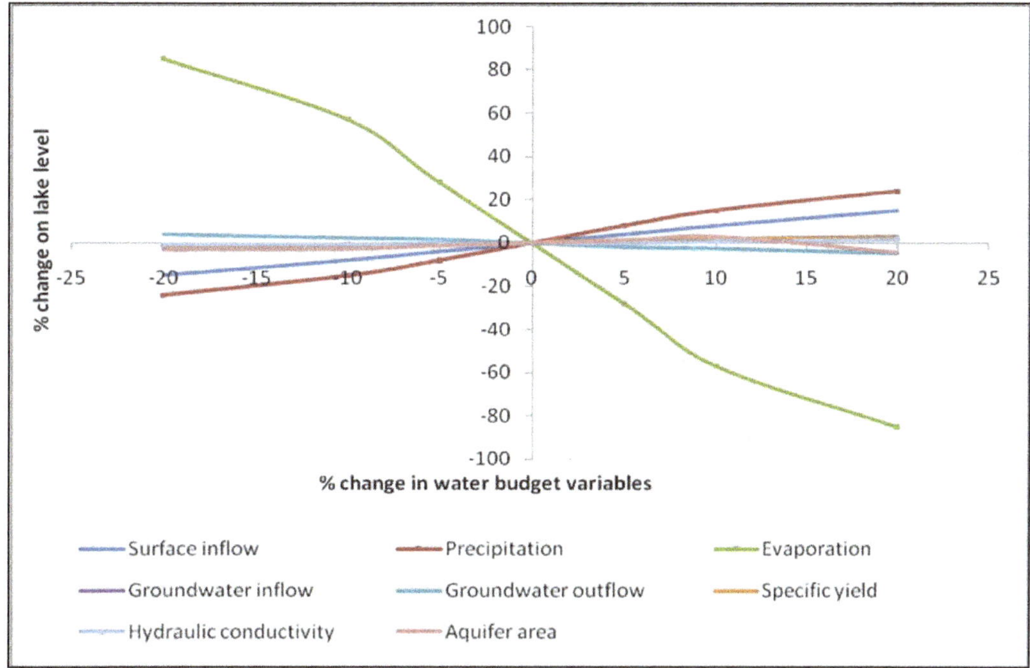

Figure 15. Relative sensitivity of Lake Linlithgow water balance model to changes in the water budget components.

4.3. Interpretation

To assess the effect of groundwater component, which is the main objective of this paper, further analysis was carried out through a second model run without a groundwater component. The calculated water levels followed the same trend as observed lake levels, but were on average lower than the observed values. The lower calculated lake level implies that the total observed lake storage is lower than would be expected if the lake did not have groundwater inflow (Figure 16).

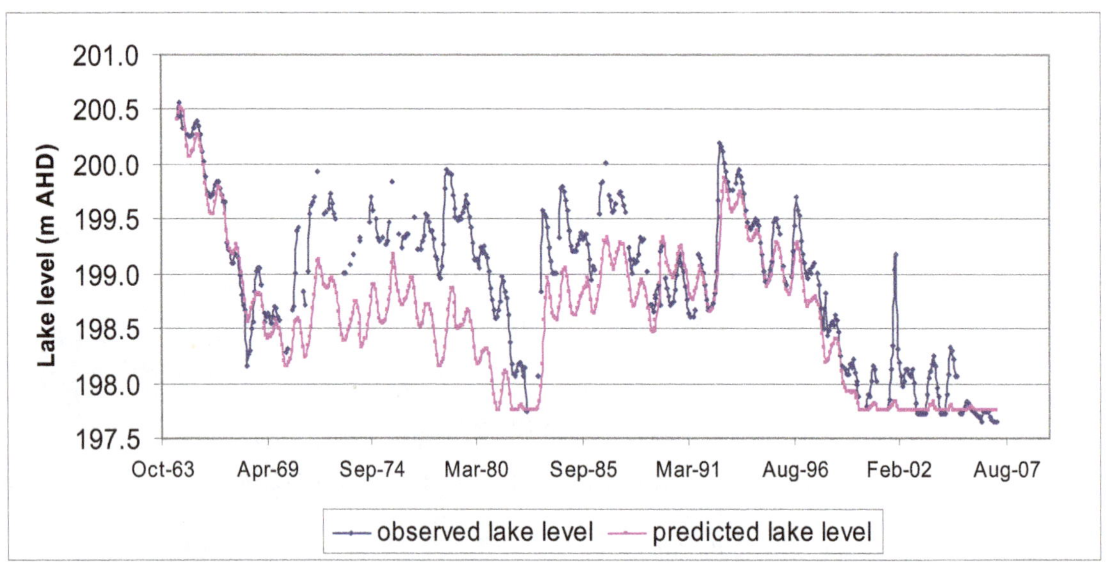

Figure 16. Observed and simulated lake level (deactivating the groundwater node from the model).

The effect of groundwater is evident from the graph. The simulated volumetric component of the groundwater of the lake water budget is shown in Figure 17. The base flow is estimated (G = 260) (Figure 10) using the tanh cumulative surplus rainfall approach. This simplistic model seems a reliable and modest tool to see the lake–groundwater relationship at a glance and will be useful if constrained by the mass balance and reliable estimate of parameters [9,32,51]. This approach could be adopted for much wetland management. This model will give insight into the groundwater–surface water relationship and guide our future data gathering from sensitive parameters. The parameters used are small and convenient for management practice.

The need for study of the interaction of lakes and groundwater stems from the fact that groundwater is commonly ignored or is the residual term in lake water. Because all variables in a lake water budget are rarely measured, it is impossible to adequately evaluate the errors or the residual.

Even if groundwater is included as part of a lake water balance study, improper placement of wells can lead to a misunderstanding of the interaction between lakes and groundwater. No matter how many wells are used to define the water table in the settings, the maps shows a gradient towards the lake, and there would be no way to detect the out-seepage that occurs. If only one or a few wells are placed near a lake, the groundwater flow system would not be adequately defined and would be subject to misinterpretation [16,55,58].

Over a long period of falling lake levels from 1968 up to the drought period 1982/3, the simulated lake levels are more than one meter lower than the observed levels due to the lack of groundwater inflow. However, there is a good match between the simulated and actual lake levels during a period of rising lake levels, indicating that the groundwater outflow from Lake Linlithgow recharging the aquifer is very small. When the groundwater component is added to the model, it accurately follows the observed levels (Figure 11).

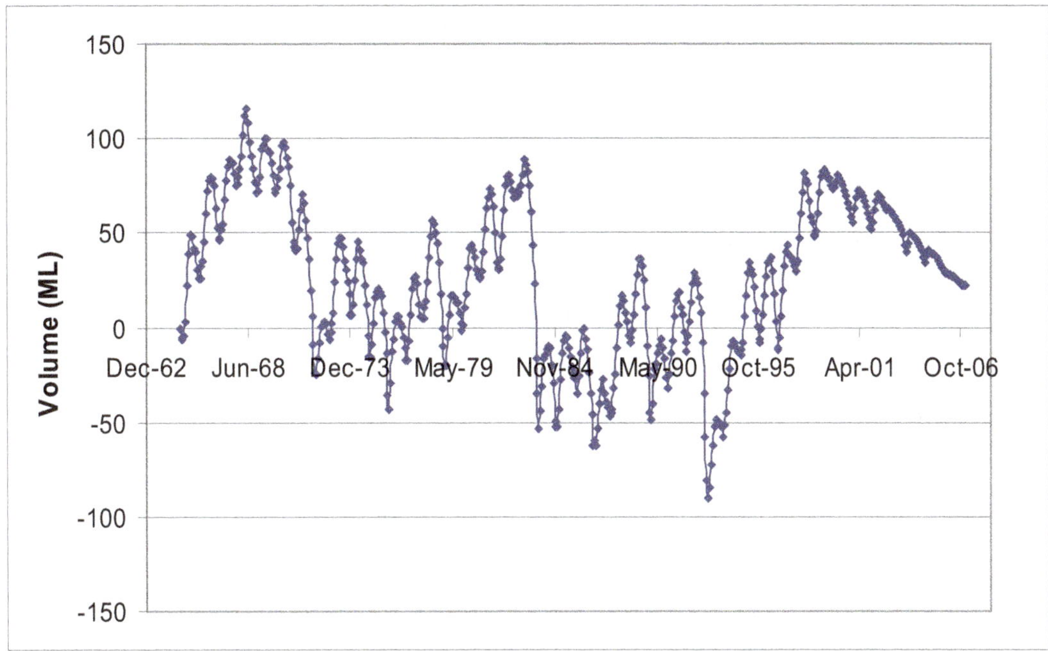

Figure 17. Volumetric groundwater component in the Lake Linlithgow water budget.

Groundwater is an important component of the water balance of Lake Linlithgow, and its flow is influenced by the lake level. If the lake level rises, groundwater outflow and therefore lake recharge into the surrounding aquifer will increase. If the lake level decreases, groundwater discharge from the aquifer into the lake will increase. This interaction causes inertia in the lake–groundwater system, delaying reactions to external (meteorological) stresses (Figure 17). This phenomenon can be shown using the lake water balance model. If the model is run with no groundwater components, it overshoots after periods of rise or recession (Figure 16).

5. Conclusions

This paper aims at an understanding of the groundwater system in wetlands, via a case study at Lake Linlithgow, Australia.

When the model was run based on surface water balance components (via switching the "node" off, i.e., in the absence of groundwater), there was a progressive separation between observed and calculated lake levels. The calculated levels imply that the lake should accumulate more storage than is actually observed. This separation could not be attributed to systematic errors in surface runoff, precipitation, and evaporation measurements. Rather, it is an indication of subterranean water fluxes (i.e., groundwater).

The results of such analysis lead to a better understanding of the groundwater system within a lake/wetland ecosystem.

Uncertainties associated with the groundwater component of the lake water budget increase unpredictability and unreliability and further complicate the future management of wetlands. No matter how many wells are used to define the water table in the settings, the groundwater flow system would not be adequately defined and would be subject to misinterpretation. The present method gives a better way of studying surface water–groundwater interaction, which is in great demand in the current strategy for better wetland and integrated water resources management. The result underscores the need to put in place relevant adaptation mechanisms to reduce vulnerability and enhance resilience to climate change within the lake basin.

Acknowledgments: We would like to thank the anonymous reviewers. The manuscript has benefitted from the reviewers' and editors' comments.

Author Contributions: Y.Y. and J.W. conceived and designed the research; Y.Y. performed the research; Y.Y. and J.W. analyzed the data; B.V. contributed analysis tools; Y.Y. wrote the paper.

Conflicts of Interest: The authors declare no conflict of interest.

References

1. Dausman, A.M.; Doherty, J.; Langevin, C.D.; Dixon, J. Hypothesis testing of buoyant plume migration using a highly parameterized variable-density groundwater model. *Hydrogeol. J.* **2010**, *18*, 147–160. [CrossRef]
2. Edelman, J.H. *Groundwater Hydraulics of Extensive Aquifers (No. 13)*; International Livestock Research Institute, Wageningen University: Wageningen, The Netherlands, 1983.
3. Yihdego, Y.; Drury, L. Mine water supply assessment and evaluation of the system response to the designed demand in a desert region, central Saudi Arabia. *Environ. Monit. Assess. J.* **2016**, *188*, 619. [CrossRef] [PubMed]
4. Yihdego, Y.; Drury, L. Mine dewatering and impact assessment: Case of Gulf region. *Environ. Monit. Assess. J.* **2016**, *188*, 634. [CrossRef] [PubMed]
5. Yihdego, Y.; Paffard, A. Predicting open mine inflow and recovery depth in the Durvuljin soum, Zavkhan Province, Mongolia. In *Mine Water and the Environment*; Springer: Berlin/Heidelberg, Germany, 2016; pp. 1–10.
6. Yihdego, Y.; Paffard, A. Hydro-engineering solution for a sustainable groundwater management at a cross border region: Case of Lake Nyasa/Malawi basin, Tanzania. *Int. J. Geo-Eng.* **2016**, *7*, 23. [CrossRef]
7. Christensen, S.; Doherty, J. Predictive error dependencies when using pilot points and singular value decomposition in groundwater model calibration. *Adv. Water Resour.* **2008**, *31*, 674–700. [CrossRef]
8. Tonkin, M.; Doherty, J. Calibration-constrained Monte Carlo analysis of highly-parameterized models using subspace techniques. *Water Resour. Res.* **2008**, *45*, 12. [CrossRef]
9. Yihdego, Y.; Webb, J.A. Hydrogeological constraints on the hydrology of Lake Burrumbeet, southwestern Victoria, Australia. In *21st VUEESC Conference, Victorian Universities Earth and Environmental Sciences Conference*; Hagerty, S.H., McKenzie, D.S., Yihdego, Y., Eds.; Abstracts No 88, 12; Geological Society of Australia: Sydney, Australia, 2007.
10. Yihdego, Y. Three Dimensional Groundwater Model of the Aquifer around Lake Naivasha Area, Kenya. Master's Thesis, Department of Water Resources, ITC, University of Twente, Enschede, The Netherlands, March 2005.
11. Yihdego, Y.; Webb, J.A. Characterizing groundwater dynamics in Western Victoria, Australia using Menyanthes software. In Proceedings of the 10th Australasian Environmental Isotope Conference and 3rd Australasian Hydrogeology Research Conference, Perth, Australia, 1–3 December 2009.
12. Rosenberry, D.O.; Lewandowski, J.; Meinikmann, K.; Nützmann, G. Groundwater-the disregarded component in lake water and nutrient budgets. Part 1: Effects of groundwater on hydrology. *Hydrol. Process.* **2015**, *29*, 2895–2921. [CrossRef]
13. Scibek, J.; Allen, D.M.; Cannon, A.J.; Whitfield, P.H. Groundwater–surface water interaction under scenarios of climate change using a high-resolution transient groundwater model. *J. Hydrol.* **2007**, *333*, 165–181. [CrossRef]
14. Yihdego, Y.; Webb, J.A. Modeling of bore hydrographs to determine the impact of climate and land use change in a temperate subhumid region of southeastern Australia. *Hydrogeol. J.* **2011**, *19*, 877–887. [CrossRef]
15. Yihdego, Y.; Webb, J.A.; Leahy, P. Response to Parker: Rebuttal: ENGE-D-13-00994R2 "Modelling of lake level under climate change conditions: Lake Purrumbete in south eastern Australia". *Environ. Earth Sci. J.* **2016**, *75*, 1–4. [CrossRef]
16. Yihdego, Y.; Webb, J.A. Assessment of wetland hydrological dynamics in a modified catchment basin: Case of Lake Buninjon, Victoria, Australia. *Water Environ. Res. J.* **2017**, *89*, 144–154. [CrossRef] [PubMed]
17. Doherty, J.; Hunt, R.J. Response to comment on "Two statistics for evaluating parameter identifiability and error reduction". *J. Hydrol.* **2010**, *380*, 489–496. [CrossRef]
18. Doherty, J.; Johnston, J.M. Methodologies for calibration and predictive analysis of a watershed model. *J. Am. Water Resour. Assoc.* **2003**, *39*, 251–265. [CrossRef]
19. Yihdego, Y. Drought and Pest Management Initiatives (Book chapter 11). In *Handbook of Drought and Water Scarcity (HDWS): Vol. 3: Management of Drought and Water Scarcity*; Eslamian, S., Eslamian, F.A., Eds.; Francis and Taylor, CRC Group: Burlington, MA, USA, 2016.

20. Yihdego, Y. Drought and Groundwater Quality in Coastal Area (Book chapter 15). In *Handbook of Drought and Water Scarcity (HDWS): Vol. 2: Environmental Impacts and Analysis of Drought and Water Scarcity*; Eslamian, S., Eslamian, F.A., Eds.; Francis and Taylor, CRC Group: Burlington, MA, USA, 2016.
21. Lewandowski, J.; Meinikmann, K.; Nützmann, G.; Rosenberry, D.O. Groundwater–the disregarded component in lake water and nutrient budgets. Part 2: Effects of groundwater on nutrients. *Hydrol. Process.* **2015**, *29*, 2922–2955. [CrossRef]
22. Shaw, G.D.; White, E.S.; Gammons, C.H. Characterizing groundwater–lake interactions and its impact on lake water quality. *J. Hydrol.* **2013**, *492*, 69–78. [CrossRef]
23. Yihdego, Y.; Al-Weshah, R. Hydrocarbon assessment and prediction due to the Gulf War oil disaster, North Kuwait. *J. Water Environ. Res.* **2016**. [CrossRef]
24. Yihdego, Y.; Al-Weshah, R. Gulf war contamination assessment for optimal monitoring and remediation cost –benefit analysis, Kuwait. *Environ. Earth Sci.* **2016**, *75*, 1–11. [CrossRef]
25. Gallagher, M.R.; Doherty, J. Predictive error analysis for a water resource management model. *J. Hydrol.* **2007**, *34*, 513–533. [CrossRef]
26. Hunt, R.J.; Doherty, J.; Tonkin, M.J. Are models too simple? Arguments for increased parameterization. *Ground Water* **2007**, *45*, 254–262. [CrossRef] [PubMed]
27. Yihdego, Y.; Eslamian, S. Drought Management Initiatives and Objectives (Book chapter 1). In *Handbook of Drought and Water Scarcity (HDWS): Vol. 3: Management of Drought and Water Scarcity*; Eslamian, S., Eslamian, F.A., Eds.; Francis and Taylor, CRC Group: Burlington, MA, USA, 2016; p. 41.
28. Shaw, G.D.; Mitchell, K.L.; Gammons, C.H. Estimating groundwater inflow and leakage outflow for an intermontane lake with a structurally complex geology: Georgetown Lake in Montana, USA. *Hydrogeol. J.* **2016**, 1–15. [CrossRef]
29. Skahill, B.; Doherty, J. Efficient accommodation of local minima in watershed model calibration. *J. Hydrol.* **2006**, *329*, 122–139. [CrossRef]
30. Tonkin, M.; Doherty, J. A hybrid regularised inversion methodology for highly parameterized models. *Water Resour. Res.* **2005**, *41*, W10412. [CrossRef]
31. Yihdego, Y. Data visualization tool as a framework for groundwater flow and transport models. In *Proceedings of the International Conference on "MODFLOW and more 2013", Translating Science into Practice*; Integrated Groundwater Modelling Centre (IGWMC), Colorado School of Mines, University: Golden, CO, USA, 2013.
32. Yihdego, Y.; Webb, J.A. Modelling of seasonal and long-term trends in lake salinity in south-western Victoria, Australia. *J. Environ. Manag.* **2012**, *112*, 149–159. [CrossRef] [PubMed]
33. Gallagher, M.R.; Doherty, J. Parameter estimation and uncertainty analysis for a watershed model. *Environ. Model. Softw.* **2006**, *22*, 1000–1020. [CrossRef]
34. Lohman, S.W. *Ground-Water Hydraulics*; US Government Printing Office: Washington, DC, USA, 1972; p. 70.
35. Lewis, M.A.; Cheney, C.S.; ÓDochartaigh, B.E. *Guide to Permeability Indices*; CR/06/160N; British Geological Survey, Natural Environmental Council: London, UK, 2006.
36. Post, V.; Kooi, H.; Simmons, C. Using hydraulic head measurements in variable-density ground water flow analyses. *Ground Water* **2007**, *45*, 664–671. [CrossRef] [PubMed]
37. Jakovovic, D.; Werner, A.D.; de Louw, P.G.; Post, V.E.; Morgan, L.K. Saltwater upcoming zone of influence. *Adv. Water Resour.* **2016**, *94*, 75–86. [CrossRef]
38. Yihdego, Y.; Reta, G.; Becht, R. Hydrological analysis as a technical tool to support strategic and economic development: Case of Lake Navaisha, Kenya. *Water Environ. J.* **2016**, *30*, 40–48. [CrossRef]
39. Yihdego, Y.; Reta, G.; Becht, R. Human impact assessment through a transient numerical modelling on The UNESCO World Heritage Site, Lake Navaisha, Kenya. *Environ. Earth Sci.* **2017**, *76*, 9. [CrossRef]
40. Yihdego, Y.; Webb, J.A. Validation of a model with climatic and flow scenario analysis: Case of Lake Burrumbeet in Southeastern Australia. *Environ. Monit. Assess.* **2016**, *188*, 1–14. [CrossRef] [PubMed]
41. Jones, R.N.; McMahon, T.A.; Bowler, J.M. Modelling historical lake levels and recent climate change at three closed lakes, Western Victoria, Australia (c.1840–1990). *J. Hydrol.* **2001**, *246*, 159–180. [CrossRef]
42. Yihdego, Y.; Webb, J.A. An empirical water budget model as a tool to identify the impact of land-use change on stream flow in southeastern Australia. *Water Resour. Manag.* **2012**, *27*, 4941–4958. [CrossRef]
43. James, S.C.; Doherty, J.; Eddebarh, A.A. Post-calibration uncertainty analysis: Yucca Mountain, Nevada, USA. *Ground Water* **2009**, *47*, 851–869. [CrossRef] [PubMed]

44. Tonkin, M.; Doherty, J.; Moore, C. Efficient nonlinear predictive error variance analysis for highly parameterized models. *Water Resour. Res.* **2007**, *43*, 7. [CrossRef]
45. Doherty, J.; Skahill, B. An advanced regularization methodology for use in watershed model calibration. *J. Hydrol.* **2006**, *327*, 564–577. [CrossRef]
46. Doherty, J. Groundwater model calibration using pilot points and regularization. *Ground Water* **2003**, *41*, 170–177. [CrossRef]
47. Moore, C.; Doherty, J. The cost of uniqueness in groundwater model calibration. *Adv. Water Resour.* **2006**, *29*, 605–623. [CrossRef]
48. Yihdego, Y.; Becht, R. Simulation of lake–aquifer interaction at Lake Navaisha, Kenya using a three-dimensional flow model with the high conductivity technique and a DEM with bathymetry. *J. Hydrol.* **2013**, *503*, 111–122. [CrossRef]
49. Yihdego, Y.; Al-Weshah, R. Engineering and environmental remediation scenarios due to leakage from the Gulf War oil spill using 3-D numerical contaminant modellings. *J. Appl. Water Sci.* **2016**. [CrossRef]
50. Bennetts, D. Hydrology, Hydrogeology and Hydro-Geochemistry of Groundwater Flow Systems within the Hamilton Basalt Plains, Western Victoria, and Their Role in Dry Land Salinisation. Ph.D. Thesis, Department of Earth Sciences, La Trobe University, Melbourne, Australia, September 2005.
51. Yihdego, Y.; Webb, J.A. Use of a conceptual hydrogeological model and a time variant water budget analysis to determine controls on salinity in Lake Burrumbeet in southeast Australia. *Environ. Earth Sci.* **2015**, *73*, 1587–1600. [CrossRef]
52. Yihdego, Y.; Webb, J.A.; Leahy, P. Modelling of lake level under climate change conditions: Lake Purrumbete in southeastern Australia. *Environ. Earth Sci.* **2015**, *73*, 3855–3872. [CrossRef]
53. Yihdego, Y.; Webb, J.A.; Leahy, P. Modelling water and salt balances in a deep, groundwater-throughflow lake—Lake Purrumbete, southeastern Australia. *Hydrol. Sci. J.* **2016**, *61*, 186–199. [CrossRef]
54. Yihdego, Y. *Modelling of Lake Level and Salinity for Lake Purrumbete in Western Victoria, Australia*; A Co-Operative Research Project between La Trobe University and EPA Victoria; EPA Victoria: Melbourne, Australia, 2010.
55. Yihdego, Y. Modelling Bore and Stream Hydrograph and Lake Level in Relation to Climate and Land Use Change in Southwestern Victoria, Australia. Ph.D. Thesis, Faculty of Science, Technology and Engineering, Melbourne, La Trobe University, Melbourne, Australia, May 2010.
56. Yihdego, Y.; Webb, J.A. Characterizing groundwater dynamics using Transfer Function-Noise and auto-regressive modelling in Western Victoria, Australia. In Proceedings of the 5th IASME/WSEAS International Conference on Water Resources, Hydraulics and Hydrology (WHH '10), Cambridge, UK, 23–25 February 2010.
57. Yihdego, Y.; Al-Weshah, R. Assessment and prediction of saline sea water transport in groundwater using using 3-D numerical modelling. *Environ. Process. J.* **2016**. [CrossRef]
58. Winter, T.C. Effects of water table configuration on seepage through lakebeds. *Limnol. Oceanogr.* **1981**, *26*, 925–934. [CrossRef]
59. Yihdego, Y. Evaluation of Flow Reduction due to Hydraulic Barrier Engineering Structure: Case of Urban Area Flood, Contamination and Pollution Risk Assessment. *J. Geotech. Geol. Eng.* **2016**, *34*, 1643–1654. [CrossRef]
60. Al-Weshah, R.; Yihdego, Y. Modelling of Strategically Vital Fresh Water Aquifers, Kuwait. *Environ. Earth Sci.* **2016**, *75*, 1315. [CrossRef]
61. Yihdego, Y.; Danis, C.; Paffard, A. 3-D numerical groundwater flow simulation for geological discontinuities in the Unkheltseg Basin, Mongolia. *Environ. Earth Sci. J.* **2015**, *73*, 4119–4133. [CrossRef]
62. Yihdego, Y.; Webb, J.A. Characterizing groundwater dynamics in Western Victoria, Australia using Menyanthes software. In Proceedings of the 10th Australasian Environmental Isotope Conference and 3rd Australasian Hydrogeology Research Conference, Perth, Australia, 1–3 December 2009.
63. Yihdego, Y.; Webb, J.A. Modelling of Seasonal and Long-term Trends in Lake Salinity in Southwestern Victoria, Australia. *J. Environ. Manag.* **2012**, *112*, 149–159.
64. Yihdego, Y. Engineering and enviro-management value of radius of influence estimate from mining excavation. *J. Appl. Water Eng. Res.* **2017**. [CrossRef]

8

Application of HEC-HMS in a Cold Region Watershed and Use of RADARSAT-2 Soil Moisture in Initializing the Model

Hassan A. K. M. Bhuiyan [1,*], Heather McNairn [2], Jarrett Powers [1] and Amine Merzouki [2]

1 Science and Technology Branch, Agriculture and Agri-Food Canada, Winnipeg, MB R3C 3G7, Canada; Jarrett.Powers@AGR.GC.CA

2 Science and Technology Branch, Agriculture and Agri-Food Canada, Ottawa, ON K1A 0C6, Canada; Heather.McNairn@AGR.GC.CA (H.M.); Amine.Merzouki@AGR.GC.CA (A.M.)

* Correspondence: akmh.bhuiyan@gmail.com

Abstract: This paper presents an assessment of the applicability of using RADARSAT-2-derived soil moisture data in the Hydrologic Modelling System developed by the Hydrologic Engineering Center (HEC-HMS) for flood forecasting with a case study in the Sturgeon Creek watershed in Manitoba, Canada. Spring flooding in Manitoba is generally influenced by both winter precipitation and soil moisture conditions in the fall of the previous year. As a result, the soil moisture accounting (SMA) and the temperature index algorithms are employed in the simulation. Results from event and continuous simulations of HEC-HMS show that the model is suitable for flood forecasting in Manitoba. Soil moisture data from the Manitoba Agriculture field survey and RADARSAT-2 satellite were used to set the initial soil moisture for the event simulations. The results confirm the benefit of using satellite data in capturing peak flows in a snowmelt event. A sensitivity analysis of SMA parameters, such as soil storage, maximum infiltration, soil percolation, maximum canopy storage and tension storage, was performed and ranked to determine which parameters have a significant impact on the performance of the model. The results show that the soil moisture storage was the most sensitive parameter. The sensitivity analysis of initial soil moisture in a snowmelt event shows that cumulative flow and peak flow are highly influenced by the initial soil moisture setting of the model. Therefore, there is a potential to utilize RADARSAT-2-derived soil moisture for hydrological modelling in other snow-dominated Manitoba watersheds.

Keywords: RADARSAT-2; flood forecasting; soil moisture accounting (SMA); HEC-HMS

---

1. Introduction

Flooding is a common occurrence in the Red and Assiniboine River sub-basins; part of the larger Lake Winnipeg basin in Southern Manitoba, Canada. Of the ten highest recorded floods on the Red River dating back to the 1800s, four have occurred in the last twenty years. This includes the 1997 flood, which stands as the third largest flood ever recorded on the Red River. In 2011, the Assiniboine River experienced a one in 145 year flood and the largest in recorded history lasting over 120 days [1]. Floods of this magnitude have a devastating impact, resulting in damage to homes, infrastructure and lost agricultural production. Costs for recovery programs and investments in flood infrastructure are shared by all levels of government and cost billions of dollars [2].

The majority of Manitoba floods are caused by spring snowmelt (freshet) events in late April and May [3]. Spring flooding in Manitoba watersheds is greatly influenced by the soil moisture condition of the previous fall along with the snow received in the watershed [2,4]. The freeze-thaw cycle in

Manitoba is connected to flooding and the timing of the peak. The ground often stays frozen while the surface snow begins to melt. This can create dramatically large volumes of surface runoff, as there is low to null soil infiltration. The freeze-thaw cycle can act as the trigger of flooding [5]. These physical processes need to be considered in order to provide accurate flood forecasting.

Watershed models are tools to incorporate all relevant surface processes to provide runoff volumes for flood forecasting. Presently, the Hydrologic Forecast Center (HFC) of Manitoba is using the Manitoba Antecedent Precipitation Index (MANAPI) model for flood forecasting in Manitoba. MANAPI is an event-based model, computing a single runoff value for a watershed or sub-watershed from rain or snowmelt events. The model computes snowmelt from historical events representing either average, rapid or gradual melt. MANAPI uses a relationship that relates runoff to the 'total winter precipitation' and the 'antecedent precipitation index (API)'. The relationship is based on historical events and is unique to the watershed. Therefore, it cannot simulate a unique runoff response that has not been experienced in the past and which is significantly different from the average of the historical events. MANAPI is not capable of computing runoff from events that involve a significant variation in input, such as freeze-melt cycles. There are other known limitations of MANAPI in addressing complex watershed processes, such as precipitation and depression storage [6]. The MANAPI was last reviewed in 1985. Since then, many developments in hydrologic modelling procedures have occurred. The 2013 Flood Task Force recommended that the Province of Manitoba should examine other hydrological models to assess which model may best meet its forecasting requirements [2]. As a result of the recommendation, the HFC has selected the Hydrologic Modelling System developed by the Hydrologic Engineering Center (HEC-HMS) as one of the models to be tested. HEC-HMS was selected due to its flexibility and applicability in other regions for flood forecasting.

The uncertainty in flood forecasting is largely associated with hydro-meteorological input and the selection of hydrological model parameters [7,8]. There are known sources of uncertainty in initialization of the model for soil moisture. Soil moisture estimates from satellite data are increasingly used for hydrological modelling, as measured data are sparse [9,10]. Tramblay et al. [11] stated that satellite data products are able to reproduce reasonably accurate daily soil moisture dynamics at the catchment scale. Li et al. [12] presented recent advancements on integrating remotely-sensed satellite soil moisture data using a rainfall-runoff model for rain fall-driven flood forecasting. Massari et al. [13] used the initial wetness condition from globally available soil moisture retrievals in a simplified rainfall-runoff model to simulate rainfall events in a Mediterranean catchment. Xu et al. [14] provided a review on the integration of remote sensing data and hydrological modelling. Soil moisture measurements derived from satellite data were reported to be an improvement over field measured data due to improved spatial scale. Furthermore, several studies confirmed the use of satellite estimates of antecedent wetness conditions for flood modelling [9,11]. Knowing the importance of antecedent soil moisture in flood event modelling, none of these studies attempted integrating remotely-sensed soil wetness to provide the initial setting of a cold region's hydrological model. McNairn et al. [15] tested the accuracy of RADARSAT-2 data to estimate surface soil moisture and were able to estimate volumetric soil moisture with a root mean square error (RMSE) of 5.37%. Given these advancements, this study will examine the applicability of using soil moisture data derived from the RADARSAT-2 satellite as initial setting values of HEC-HMS in simulating flood events in a cold region watershed. The specific objectives of the current research are as follows:

(a) to assess the usability/applicability and potential benefits of HEC-HMS using RADARSAT-2-derived soil moisture estimates in the snowmelt-dominated Sturgeon Creek watershed and;
(b) to test the sensitivity of initial soil moisture in setting the HEC-HMS model for flood forecasting.

2. Study Area and Data

The Sturgeon Creek watershed is located northwest of the City of Winnipeg, Manitoba, Canada. Figure 1 illustrates the geographic location of the watershed along with the location of a flow gauge (ID#05MJ004) and three weather stations. The watershed has an effective drainage area of 545 km^2. The watershed slopes towards the southeast and flows through the Rural Municipality of Rosser and through the City of Winnipeg before discharging into the Assiniboine River. The landscape is relatively flat with elevations ranging from a high of 279 m (upstream) and 231 m. The upper reaches of the watershed are higher sloped (up to 1.2%) compared to the middle and lower reaches (as low as 0.05%). Soils in the upstream portion of the watershed are composed of a thin layer of black to dark grey clay loams overlying a mixed parent material of lacustrine clay and extremely calcareous clay loam tills. The middle reach of the watershed is a nearly level landscape with stratified layers of loam, fine sand and deep lacustrine clay deposits. The lower reach is a level landscape with thick lacustrine clay deposits. As such, surface drainage is very slow, resulting in the development of surface drains and stream channelization to improve the flow of water off agricultural lands. Agriculture is the dominant land use in the watershed with 75% of the land base devoted to annual crop production. Forage and pasture grasslands account for 16% of the land cover. Wetlands are less than 1% [16,17]. LiDAR (light detection and ranging) elevation data, provided by the Province of Manitoba, are used for GIS analysis to derive topographic information for HEC-HMS modelling and to delineate the watershed. Different watershed data, such as precipitation, snow depth and outflow, were collected and examined to identify dominant hydrological processes.

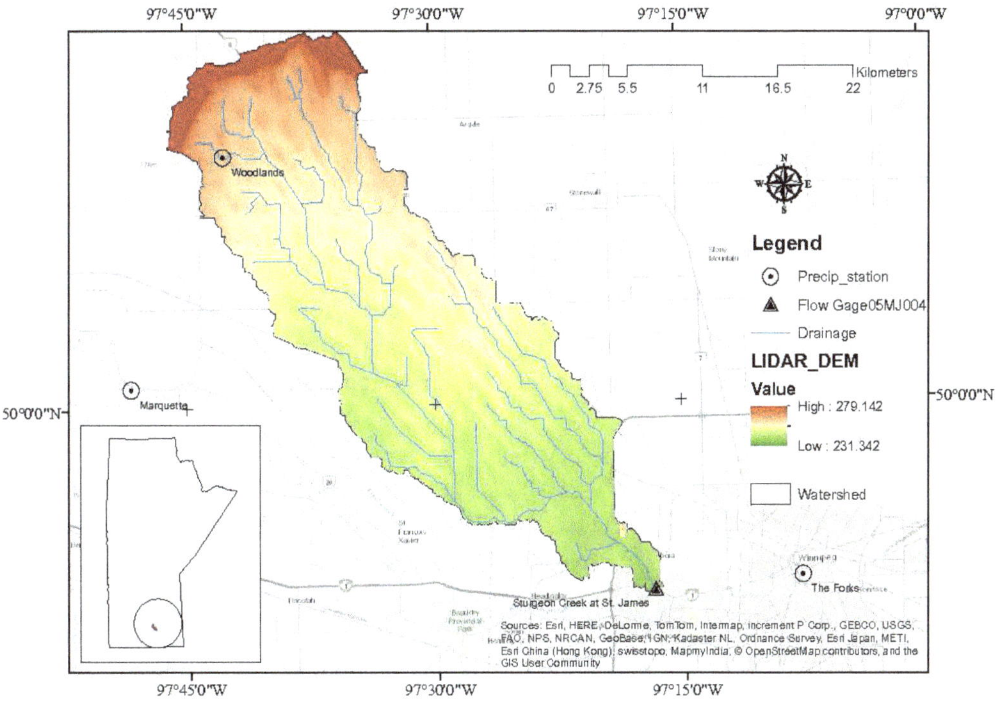

Figure 1. Study area: Sturgeon Creek watershed in Manitoba, Canada. The index map shows the provincial boundary and the location of the watershed in the circle.

2.1. Flow Data

Discharge data used in this study were collected from the Water Survey Canada (WSC) data portal. The Water office provides public access to real-time hydrometric data through the https://wateroffice.ec.gc.ca/ site, accessed on 4 July 2015 at the Sturgeon Creek (#05MJ004) gauging station located at St. James. The geographic location of the station is at latitude 49°52′ 54″ N and longitude

97° 16′ 47″ W. The gauge station measures seasonal flow data from 1 March–31 October. Annual peak flows (1965–2014) from the Sturgeon Creek gauge were analysed. During this time, the highest daily flow of 82.7 m^3/s was recorded in 1974. Two other high peak flows of 67.1 m^3/s and 63.2 m^3/s were recorded in 2009 and 1979, respectively. The time series of peak flows were segmented into five ten-year intervals in order to evaluate dry-wet hydrological cycles. Jacob and Lorenz [18] used ten-year segments to examine variability and trends of a hydrologic cycle. Figure 2 depicts the five ten-year intervals of daily average flow at the gauge station and shows that the peak of the creek appears during the spring snowmelt.

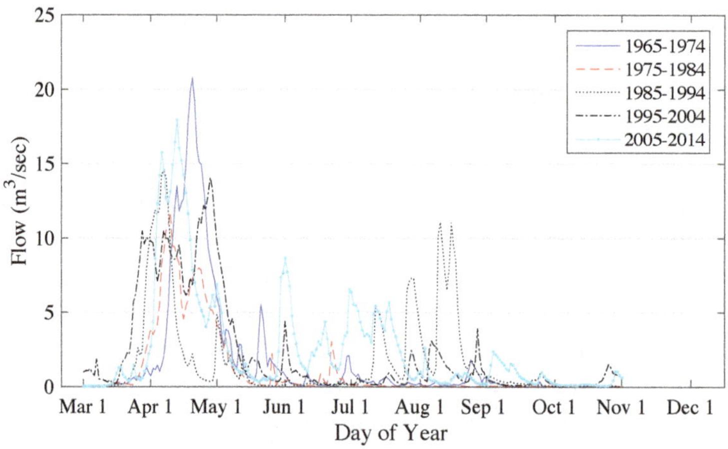

Figure 2. Flow data (1965–2014): average of ten-year segments at the Sturgeon Creek (#05MJ004) gauging station.

2.2. Weather Data

Precipitation and temperature data used in this study were obtained from two Environment Canada and Climate Change (ECC) weather stations at Winnipeg and Marquette, as well as one Manitoba Agriculture (MA) weather station at Woodlands. The locations of the weather stations are shown in Figure 1. The daily meteorological data used in the modelling were reviewed for consistency, and any missing records were replaced with data from nearest neighbouring stations. Sub-watershed temperatures were assigned from the closest neighbouring station, and a lapse rate of 5 °C/1000 m was used. Precipitation data were interpolated across the sub-watershed using the inverse-distance-squared weighting method.

The weather station at Marquette is selected as the representative station due to data availability and quality. The climate normal (1971–2000) of the station shows the average high and low temperature as 19.5 °C and −17.5 °C for the months of July and January, respectively. The mean annual average temperature is 2.9 °C. Precipitation statistics at the Marquette station shows that the watershed receives most summer rainfall starting from the middle of May to the end of July. Average annual precipitation (2005–2014) recorded at the Marquette station is 540 mm, whereas the highest and lowest annual precipitation received at the station is 790 mm and 328 mm, respectively.

Figure 3 provides a plot of snow depth over 10 years at the Marquette station. Depending on the snow year, maximum depth varies from 20 cm–60 cm. At the Marquette station, snow starts to accumulate in the middle of November. Snow melt typically begins the first week of March and is finished by the end of April. The snowmelt may act as the trigger of flooding [5]; therefore, evaluating the snow depth variable may provide more insight into the interaction of hydrological processes within the watershed.

Although during event modelling, water loss due to evaporation may be neglected, it must be included for continuous modelling. The monthly evaporation values were estimated for the entire watershed using Thornwaite's method [19]. Calculated monthly evapo-transpiration (ET) values were entered manually for each sub-watershed with a coefficient value of 0.7. The model calculates the potential ET as the product of the monthly value and the coefficient for all time periods of the month according to the model's setting [20]. Thornwaite's methodology is adopted to estimate ET, as only temperature data were available.

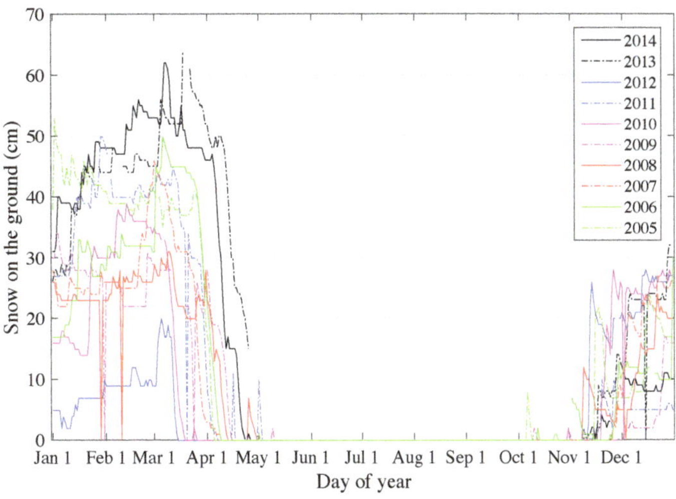

Figure 3. Snow depth (in cm) measured at the Marquette station.

2.3. Soil Moisture Estimates from RADARSAT-2 Satellite Data

RADARSAT-2 is a synthetic aperture radar (SAR) satellite operating at 5.4 GHz. The intensity of microwave energy scattered at this frequency is primarily driven by the dielectric constant (and hence, the amount of water) in a target illuminated by the satellite. The physically-based integral equation model (IEM) is used to estimate the volume of moisture in the top few centimetres of the soil, using backscatter intensity recorded by RADARSAT-2 [21,22]. The real dielectric constant is retrieved using backscatter at horizontal transmit-horizontal receive (HH) and vertical transmit-vertical receive (VV) polarizations and the local SAR incident angle. Volumetric soil moisture is then derived from the real dielectric constant using a dielectric mixing model. RADARSAT-2 was programmed to acquire an image over the Sturgeon Creek Watershed, on 15 October 2014 (Figure 4a). This was the last satellite acquisition date available prior to the soil freeze-up. As the soil temperature approaches zero, the dielectric properties of the water in the soil change. Under frozen conditions, backscatter is no longer sensitive to the soil dielectric, and thus, inversion of RADARSAT-2 data for soil moisture is not valid. Data were collected by the three Real-time In-situ Soil Monitoring for Agriculture (RISMA) stations. RISMA stations data operated by Agriculture and Agri-Food Canada (AAFC) can be obtained through the http://aafc.fieldvision.ca/ site, accessed on 14 November 2016. The stationslocated in the watershed confirmed that for all RADARSAT-2 acquisition after 15 October 2014, soil temperature was below freezing. The 15 October acquisition was the closest available image before the freeze-up and, thus, was used to establish the initial soil moisture state given that the soil moisture remains static once soils have frozen. The output image is a pixel-by-pixel estimate of percent volumetric soil moisture, θ (m^3m^{-3}), at a spatial resolution of 13.6 m. Pixel-based estimates of soil moisture were then binned into eight moisture intervals. In Figure 4a, pixels displayed in red shades represent soils at lower saturation, and green toned pixels represent higher saturation. White areas are regions outside of the satellite image or non-annual cropped areas (grasslands, trees, urban, open water) where soil moisture values cannot be retrieved. The estimated soil moisture ranges between 0.029 m^3m^{-3} and 0.550 m^3m^{-3} over the area. ArcGIS was used to overlay the pixel-based (rasterized) soil moisture product with the sub-watershed polygons to calculate average soil moisture values for each sub-watershed (Figure 4b) excluding no data pixels. The image did not cover the entire watershed (i.e., the missing part of Sub-watersheds W670 and W780). Average soil moisture for Sub-watersheds W670 and W780 were estimated assuming similar soil moisture retrievals for the missing part of the sub-watershed. The HEC-HMS model requires soil saturation (in percent), which is calculated using Equation (1).

$$S = \theta \times \phi^{-1} \times 100 \tag{1}$$

$$\phi = 1 - \rho_b / \rho_s \tag{2}$$

where S is the saturation percentage, θ is the volumetric water content from RADARSAT-2 data, ϕ is porosity, ρ_b is bulk density and ρ_s is specific density. Three measured bulk densities of 1.04, 1.39 and 1.32 gm/cm^3 are used in this study for the southern, middle and northern sub-watersheds of Sturgeon Creek, respectively [23]. Soil porosity (ϕ) is calculated following Equation (2) using specific density (ρ_s) as 2.65 gm/cm^3.

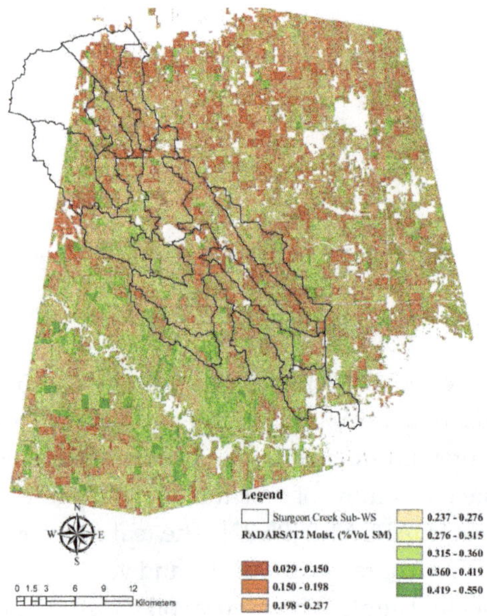

(**a**) Soil moisture binned into eight intervals.

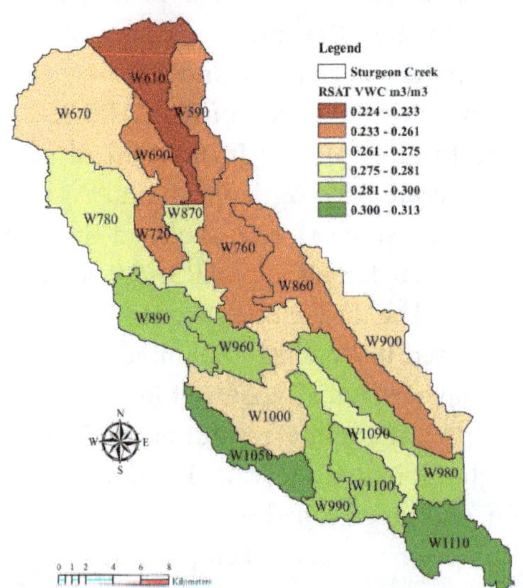

(**b**) Pixel-based soil moisture (m^{3}m^{-3}) averaged over the sub-watershed.

Figure 4. Soil Moisture retrieved using a RADARSAT-2 satellite image acquired on 15 October 2014.

2.4. Fall Soil Moisture Survey Data

Each year, Manitoba Agriculture prepares a fall soil moisture map using field survey data. Details of fall soil moisture maps can be seen from http://www.gov.mb.ca/agriculture/environment/soil-management/manitoba-fall-soil-moisture-survey.html. These maps are generated from core samples that are taken from approximately 100 fields across southern Manitoba. The fields are sampled in mid-late October, prior to freeze-up. The samples are weighed and oven-dried to determine their soil moisture level. The 2014 surface (0–30 cm) soil moisture map is used to establish the initial state of soil moisture for the event modelling. Due to the sparse measurements and highly generalized nature of the fall soil moisture survey, one soil moisture value of 40% is used to represent the initial saturation for the entire watershed.

3. HEC-HMS Model

The HEC-HMS model is designed to simulate the complete hydrological processes of a dendritic watershed system [24]. The model can be applied to analyse flooding, flood frequency, flood warning system planning, reservoir and spillway capacity studies, etc. [25]. The model can be used for both continuous and event-based modelling. Many researchers have successfully used HEC-HMS [26–28] for flood event modelling. The soil moisture accounting (SMA) algorithm has been successfully used in continuous simulation of the model [29–31]. A snow model is essential in order to capture flood peak and timing in the spring snow melt-dominated watersheds in Manitoba [2]. Gyawali and Watkins [32] tested the temperature index snow accumulation and melt algorithm of HEC-HMS in snow-affected watersheds in the Great Lakes basin. However, no studies have reported the application of HEC-HMS in a snow-dominated Manitoba, Canada, watershed.

The HEC-HMS (Version 4.0) model is grouped into four major input components, such as the watershed model, the meteorological model, the data manager and the control manager. The watershed model is the representation of real-world objects and describes the different elements of the hydrological system, such as sub-watershed, reaches, junctions, sources, sinks, reservoirs and diversions. Each of these elements needs some parameters to define their interaction in a hydrological system. These elements are inter-linked to facilitate the flow of water and to create a dendritic network [24]. Table 1 provides a list of different parameter methods selected in the watershed model. A simple canopy is selected, as no changes of canopy (i.e., dynamic canopy) are expected. A simple surface is selected to provide simple representation of the soil surface where rainfall on the soil surface is stored until the storage capacity of the surface is filled.

In order to set-up the HEC-HMS for the Sturgeon Creek Watershed, a hydrologically-corrected DEM was created from LiDAR data by re-sampling the DEM at a 15-m resolution. River network, road network and bridge/culvert data from the Manitoba Land Initiative (MLI) were used to create a hydrologically-conditioned DEM. Land cover and soil properties were also processed and re-classed using ArcGIS. Terrain pre-processing steps, such as filling sink, flow direction, flow accumulation, stream/drainage line processing and watershed delineation, were performed using ArcHydro Tools and ArcGIS. The HEC-GeoHMS was used to extract physical parameters necessary for the HEC-HMS model setup. The HEC-HMS can easily import the setup data from HEC-GeoHMS to construct a project and schematic for the model.

Figure 5 presents a schematic of the Sturgeon Creek Watershed prepared by HEC-HMS. The schematic shows sub-watersheds, reaches, junctions, sources and sinks of the watershed. The HEC-HMS was set up as a semi-distributed model by sub-dividing the catchment into 19 sub-watersheds. The sub-division of the catchment is performed by following the stream and road network, as well as underlying soil properties. The semi-distributed setup allowed us to examine governing hydrological processes in the sub-watersheds.

Table 1. Selected methods of HEC-HMS model.

Basin Model		Meteorological Model	
Parameter Method	**Selected Method**	**Parameter Method**	**Selected Method**
Canopy	Simple Canopy	Precipitation	Inverse Distance
Surface	Simple Surface	Evaporation	Monthly Average
Transform	SCS Unit Hydrograph	Snowmelt	Temperature Index
Base Flow	Recession	Shortwave	None
Routing	Muskingum		
Loss	Soil Moisture Accounting		

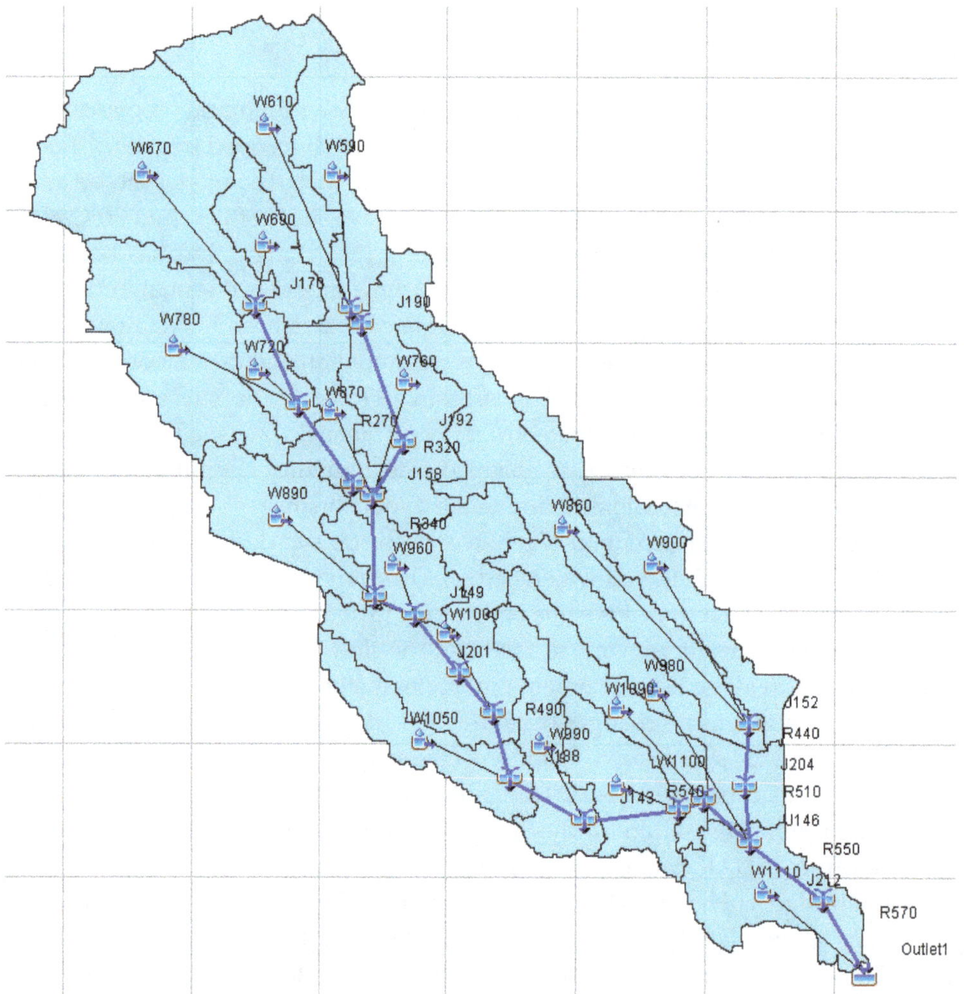

Figure 5. The schematic of the Sturgeon Creek Watershed created by HEC-HMS. The naming of sub-watersheds, reaches and junctions begins with W, R and J, respectively.

The HEC-HMS tracks snowmelt and accumulation using the temperature index method. Melt rates are calculated dynamically based on the current atmospheric condition and past conditions of the snow pack. The temperature index method was set up and calibrated in order to capture the spring snow melt peak. This method is governed by a threshold temperature, which separates snowfall from rainfall denoted by PXtemperature. There is a base temperature that distinguishesmelt from non-melt periods of snow. The temperature index method does not account for sublimation from and condensation to the snow pack. The final calibrated parameter values are shown in Table 2. The antecedent temperature index (ATI) melt-rate and cold-rate functions are specified separately in the model. The temperature index model includes parameter data for each sub-watershed in the meteorological model. Each sub-watershed must have one elevation band defined in the meteorological model.

In this study, the SMA method was used to account for vegetative canopy retention and to simulate the movement of water through the soil surface and the deeper soil profile to the groundwater layers [20]. These layers provide wetting and recovery cycles of soil moisture for long-term continuous hydrological simulations. SMA requires the initial soil moisture condition to be specified at the beginning of the simulation. The soil moisture map derived from RADARSAT-2 (Figure 4) was used as the basis of initial soil moisture. The HEC-GeoHMS was used to build a project setup for the SMA loss method. The parameters needed for SMA (maximum infiltration rate, soil storage, tension zone storage and soil zone percolation rate) were estimated using Manitoba land use, land cover and soil databases. Soil profile data were also used for the estimation of soil parameters of the SMA model. The soil percolation rate was based on the average hydraulic conductivity of soil profile data. The SCS unit hydrograph was used with lag time estimated by employing HEC-GeoHMS with an empirical relationship. The recession method is used for base flow calculation. The simple Muskingum routing is selected to route flow through the channel.

Table 2. Snow melt input parameters for the temperature index method.

Parameter	Unit	Value
PX Temperature [a]	°C	1.7847
Base Temperature	°C	0.6294
Wet Melt Rate	mm/°C-day	0.9876
Rain Rate Limit	mm/day	2
ATI [b]-Melt Rate Coefficient	-	0.9995
Cold Limit	mm/day	20
ATI-Cold Rate Coefficient	-	0.9995
Water Capacity	%	10
Ground Melt	mm/day	0

[a] The PXtemperature is used to differentiate between precipitation falling as rain or snow; [b] Antecedent Temperature Index.

3.1. Model Evaluation

Model calibration and validation were conducted based on simulated and observed daily flow data at the gauging station. The model parameters were first calibrated using automated calibration methods available in the HEC-HMS model. The automated calibration procedure uses an iterative method to minimize the objective function in order to obtain agreement between simulated and observed flow data [24]. The precise adjustments of parameters were obtained through manual calibration.

Many different test criteria have been developed to assess the efficiency of a hydrological model calibration [33–35]. For this study, the Nash–Sutcliffe (N_s) coefficient of model efficiency [36] and the deviation of runoff volumes (D_v) were used to measure the goodness-of-fit between the observed and simulated flow time series. Higher values of N_s (closer to one) indicate better agreement. Henriksen et al. [37], Table 4, suggested that values of N_s between 0.5 and 0.65 are good; 0.65–0.85 are very good; and >0.85 are excellent. For a perfect model the D_v is equal to zero. The D_v value emphasizes volume conservation and is not sensitive to errors in streamflow timing or seasonality.

A sensitivity analysis of the model was performed to understand the complex relationships amongst model parameters and variables. The sensitivity analysis determines which parameters significantly impact model performance and provides an estimate of the precise value of each parameter. This analysis of SMA parameters was conducted by varying each input parameter by $\pm$10% on each step without changing other parameters.

4. Results and Discussion

The HEC-HMS model simulations were performed for a single flood event, as well as continuous simulation over the Sturgeon Creek Watershed.

4.1. Event Modelling

The HEC-HMS event model was set up for a 28 March 2015 (12:00) to 6 April 2015 (12:00) flood event. The event hydrological modelling was performed using hourly time steps to understand fine-scale hydrological processes and to respond to the quantity of surface runoff, peak and timing of peak. Two event simulations are presented in this study to demonstrate the benefits of using the initial state of soil moisture measurement from satellite data. Simulation 1 utilizes one soil moisture value from the Fall Soil Moisture Survey. Simulation 2 is performed using soil moisture values for each sub-watershed estimated from RADARSAT-2. Results from the event simulation are presented in Figure 6.

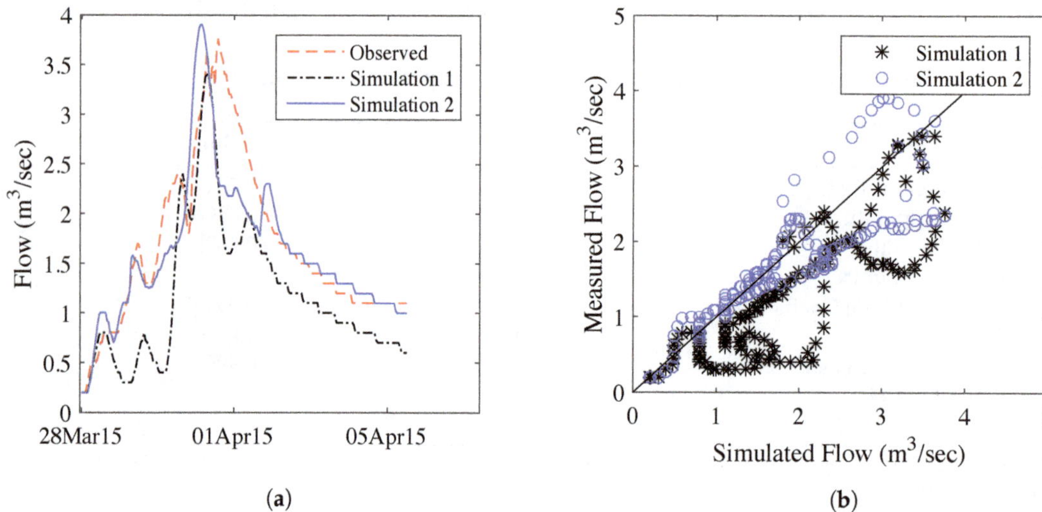

Figure 6. Event modelling of hourly flow series on an event of 28 March 2015 (12:00) to 6 April 2015 (12:00) for the Sturgeon Creek Watershed. Simulation 1 utilizes one soil moisture value from the Fall Soil Moisture Survey, while Simulation 2 utilizes initial soil moisture values from RADARSAT-2. (**a**) Comparison of the observed and simulated hydrograph; (**b**) the distribution of points is shown from the y = x line.

The comparison of observed and simulated hourly flow series of the event using RADARSAT-2 shows that the simulated peak matches well (within 3%) with the peak values of measured flow. However, the timing of the simulated peak was earlier than that observed. Furthermore, the small peaks of the simulated flow series did not match with the observed. These small peaks may be the result of localized melt events, which could not be captured by the model. The Nash–Sutcliffe coefficient of efficiency (N_s) and the deviation of runoff volume (D_v) are found to be 0.74 and −4.83%, respectively. These values are within acceptable ranges.

The simulated hourly flow series of the event using data from the Fall Soil Moisture Survey does not agree well with the observed flow event. The event peak was underestimated by −10%; and other small peaks were also underestimated. The model performance measures N_s and D_v were 0.31 and −34.2%, respectively. The two simulations differ only in the state of initial soil moisture. The difference in the generation of peaks in the event model simulations was clear. Given the small number of sampling points and the generalization of the data of the Fall Soil Moisture Survey, only one surface soil moisture value can be used to represent the initial saturation value of the watershed.

With the differences in soil, landscapes and precipitation, this value may not be representative of soil moisture within the individual sub-watersheds. The initial melt at the surface may have been retained at the sub-watershed and added to the soil water content to reach a threshold before contributing to runoff at the outlet. This could be the most probable reason of underestimating the peak flow and cumulative outflow. A sensitivity test of initial soil moisture is provided in Section 4.3.2.

4.2. Continuous Modelling

The parameters obtained from the calibrated HEC-HMS for event modelling were used to set up a continuous simulation. The continuous simulation was performed with the SMA model using a daily time step and compared with flow series of 1 March 2014–1 June 2014. A continuous multi-year simulation was not done in this watershed, as measured flow data are only available from 1 March–31 October each year.

Figure 7a presents a comparison of observed and simulated output from the continuous modelling of daily flow series for the Sturgeon Creek Watershed. The model simulated timing of the peak matches well, but the peak flow is overestimated by 21%. Furthermore, other small peaks were not well captured. The deviation of runoff volume (D_v) was −17.9%; the Nash–Sutcliffe coefficient of efficiency (N_s) was 0.87. Figure 7b presents the scatter plot of simulated flow vs. measured flow and shows a strong positive correlation (0.95). The line of correlation 1.0 (i.e., y = x line) is also shown in the plot.

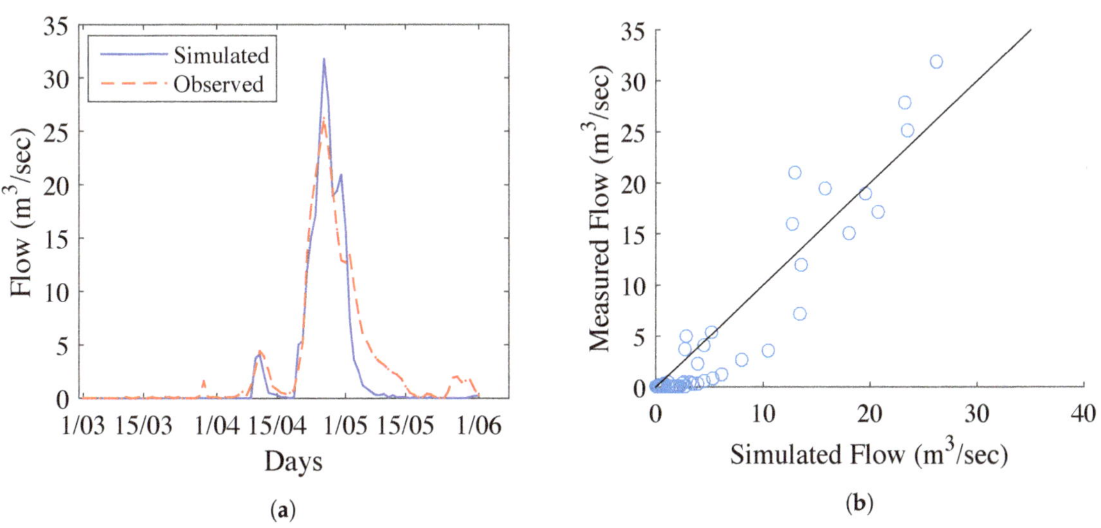

Figure 7. Continuous modelling of daily flow series for the Sturgeon Creek Watershed. The calibration is performed for 2014 from 1 March 2014–1 June 2014. (**a**) Comparison of observed and simulated hydrograph; (**b**) distributions of points are shown from the y = x line.

A second continuous model simulation was performed for 2011 using the 2014 parameters. Figure 8a depicts simulated flow data compared to observed flows. The 2011 simulation also shows that the timing of peak flow arrival is well captured, but the peak flow volume is over-estimated by 13%. The D_v and N_s were found to be −7.22% and 0.88, respectively. Figure 8b presents the scatter plot, which shows a strong positive correlation (0.94).

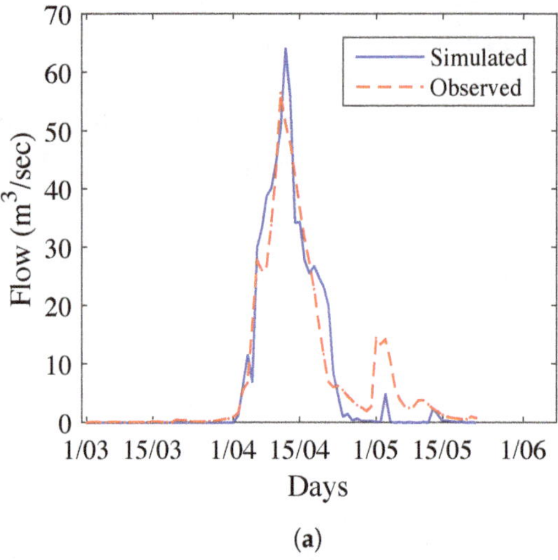

(a)

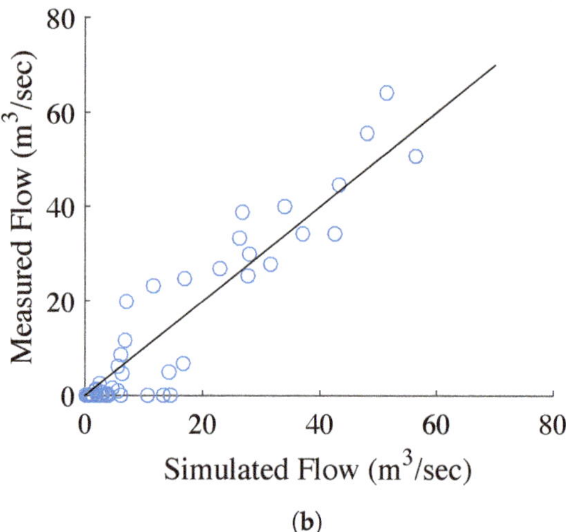

(b)

Figure 8. Continuous modelling of daily flow series for the Sturgeon Creek Watershed. The validation is performed for 2011 starting from 1 March 2011–1 June 2011. (**a**) Comparison of observed and simulated hydrograph; (**b**) distributions of points are shown from the y = x line.

The model performance results are summarized in Table 3. Event Simulation 2 showed improved performance over Simulation 1 due to well-defined initial soil moisture values from the RADARSAT-2 satellite data in all sub-basins. Performance indicators presented in Table 3 also reveal that the flow simulations generated by the HEC-HMS model are suitable for the Sturgeon Creek Watershed. Several authors [33,38,39] have successfully used these indicators for the performance evaluation of hydrological models. Henriksen et al. [37], Table 4, suggest that values of N_s obtained in this study are in very good agreement.

Table 3. Performance measures of the HEC-HMS model.

Model Simulation	Duration	Dev. of Runoff Volume, D_v (%)	Nash Co-efficient, N_s
Event (Simulation 2)	28 March 2015–6 April 2015	−4.83	0.74
Event (Simulation 1)	28 March 2015–6 April 2015	−34.25	0.31
Continuous	01 March 14–1 June 2014	−17.9	0.87
Continuous	01 March 11–1 June 2011	−7.22	0.88

The snow melt process is important for this Manitoba watershed due to spring flooding [3]. The snow depth data from the ECC station at Marquette were used for validation in this study. Following the method described by Strum et al. [40], the snow depth was converted to snow water equivalent (SWE) using a bulk density value of 0.312 gm/cm^3. The HEC-HMS simulates SWE as an internal variable. Figure 9a presents a comparison of simulated SWE from four different sub-watersheds with the measurements from the Marquette weather station. The simulated SWE from different sub-watersheds follows closely with the measurements at the Marquette station. The timing of the snow melt is considered a crucial process for hydrological models in order to capture the appearance of flood peak for forecasting. The timing of the snow melt in different sub-watersheds shows a good agreement with observed spring snow melt in Figure 9a. Figure 9b plots the relationship of observed snow depth and temperature, which shows that the snow accumulation and melt closely follow the observed temperature variation.

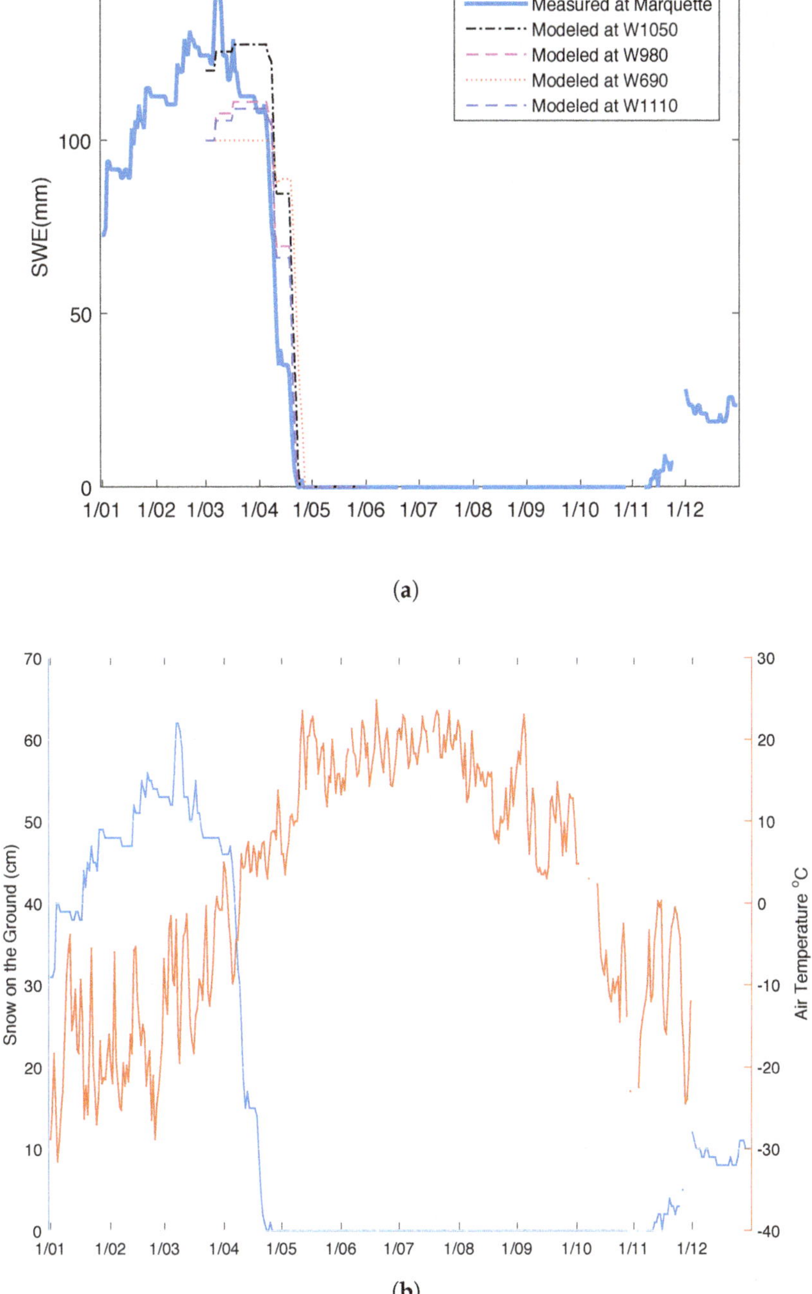

Figure 9. Validation of snow water equivalent (SWE) and timing of spring melt with temperature. (**a**) Comparison of measured and simulated snow water equivalent (SWE); (**b**) observed snow accumulation and melt pattern with temperature.

4.3. Snowmelt Model

4.3.1. Sensitivity of SMA Parameters

In the non-winter months, Prairie watersheds follow a sequence of wetting and drying periods. For a specific rainfall event, the initial moisture condition at the beginning of the rainfall event will have a major influence on a watershed's hydrological response. The initial soil moisture sensitivity on event modelling from this study revealed this fact as discussed in Section 4.3.2. Due to wetting and drying during a continuous simulation, the model state reaches a value that is no longer dependent on the models' initial soil moisture saturation. In this study, the influence of initial soil moisture in a continuous HEC-HMS simulation is examined and found not sensitive (not presented).

Other important SMA parameters such as soil storage, soil percolation, tension storage, maximum infiltration and maximum canopy storage were tested within ±40% variation. The sensitivity of each parameter was tested against the simulated runoff volume at the outlet of the watershed. The percent change of parameters was done individually by ±10% increments while keeping other parameters unchanged. Results from each parameter sensitivity test to the percent change in runoff volume are presented in Table 4. The slope of each parameter's sensitivity was estimated using Theil–Sen's [41,42] non-parametric slope estimator. This method chooses the median slope among all lines through pairs of two-dimensional sample points. Ranking of each SMA parameter is performed using Theil–Sen's slope. It is evident from the Table 4 that the soil storage is ranked the most sensitive parameter for simulated stream flow in this watershed.

Table 4. Sensitivity of SMA parameters tested in Sub-watersheds W1100 and W870.

Percent Deviation (%)	Soil Storage (mm)	Soil Percolation (mm)	Tension Storage (mm)	Maximum Infiltration (mm/h)	GW1 Storage (mm)	Max Canopy Storage (mm)
Sub-Watershed W1100						
40	−20.06	−16.07	17.02	−15.20	−0.28	−9.73
30	−15.83	−12.98	11.40	−12.11	−0.24	−9.02
20	−11.04	−9.14	6.93	−8.46	−0.16	−7.36
10	−5.74	−4.87	3.36	−4.71	−0.08	−4.63
0	0.00	0.00	0.00	0.00	0.00	0.00
−10	6.73	5.54	−2.89	5.32	0.12	5.70
−20	15.43	11.79	−6.02	11.74	0.28	12.74
−30	27.78	18.60	−8.94	18.84	0.44	20.74
−40	48.12	26.39	−11.83	28.08	0.67	28.41
Slope [a] (b)	0.68	0.52	−0.33	0.51	0.01	0.50
Ranking	1	2	5	3	6	4
Sub-Watershed W870						
40	−16.67	−11.34	5.92	−8.19	−0.07	−6.41
30	−12.87	−8.80	4.53	−6.84	−0.07	−2.82
20	−8.86	−5.89	3.38	−4.65	−0.04	−1.37
10	−4.62	−3.23	1.42	−2.56	−0.03	−0.49
0	0.00	0.00	0.00	0.00	0.00	0.00
−10	5.19	3.12	−0.72	3.13	0.00	0.52
−20	10.81	6.36	−1.13	8.39	0.04	4.56
−30	17.53	10.45	−1.77	14.03	0.07	9.65
−40	25.09	13.38	−2.12	21.97	0.11	12.12
Slope (b)	0.50	0.31	−0.12	0.32	0.00	0.21
Ranking	1	3	5	2	6	4

[a] Slope is estimated using Theil–Sen's method.

As shown in Table 4, other SMA parameters, such as maximum infiltration, soil percolation, maximum canopy storage and tension storage, can be ranked as sensitive. The SMA parameter sensitivity between Sub-watersheds W1100 and W870 also varies (i.e., soil percolation and maximum infiltration). A recent study presented by Roy et al. [43] demonstrated the variation of parameter sensitivity over different sub-watersheds. The sensitivity analysis of SMA parameters helps to understand the soil moisture accountability of the model.

4.3.2. Initial Soil Moisture

The event modelling of HEC-HMS was set up with the SMA loss method (Table 1) where the initial soil moisture condition was specified as the percentage of soil that is saturated with water at the beginning of simulation. The initial sub-watershed soil moisture (m^3m^{-3}) shown in Figure 4 is converted to soil saturation (%) using Equation (1) and used as the initial value for the HEC-HMS simulation.

In order to test the sensitivity of the initial soil moisture, the calibrated event model was used. Table 5 shows the sensitivity test of initial soil moisture over three sub-watersheds in the Sturgeon Creek. During calibration, the initial soil moisture of the W610 sub-watershed was set to 45.5% saturation, which is equivalent to 0.229 m^3m^{-3} volumetric moisture. For the purpose of the sensitivity test, the input soil moisture is set to 60% soil saturation (0.326 m^3m^{-3}) and 25% saturation (0.125 m^3m^{-3}).

Table 5. Sensitivity of peak flow (m^3/s) and cumulative flow (1000 m^3) at different initial soil moistures. Results from three sub-watersheds (W610, W690, and W1000) are shown.

	Sub-Watershed Sensitivity				**Watershed Sensitivity**			
Initial Soil Moisture	Peak Flow (m^3/s)	Diff. [a] Peak (%)	Cum. Outflow (1000 m^3)	Cum. Flow Diff. (%)	Peak Flow (m^3/s)	Diff. Peak (%)	Cum. Outflow (1000 m^3)	Cum. Flow Diff. (%)
Sub-watershed W610, bulk density 1.32 gm/cm^3								
Calibration; 0.229 m^3m^{-3} /45.5%.	0.4		42.9		3.9		1097	
0.326 m^3 m^{-3} /60%	0.8	+100	94.6	+121	4.3	+10	1148	+5
0.125 m^3 m^{-3} /25%	0.1	−75	12.3	−71	3.7	−5	1067	−3
Sub-watershed W960, bulk density 1.39 gm/cm^3								
Calibration; 0.291 m^3m^{-3} /61.1%.	0.8		101.1		3.9		1097	
0.215 m^3 m^{-3} /45%	0.4	−50	51.5	−49	3.8	−3	1048	−4
0.125 m^3 m^{-3} /25%	0.2	−75	22.9	−77	3.8	−3	1020	−7
Sub-watershed W1000, bulk density 1.04 gm/cm^3								
Calibration; 0.243 m^3m^{-3} /40%.	1.1		155.3		3.9		1097	
0.364 m^3 m^{-3} /60%	1.9	+73	313.8	+102	4.2	+8	1254	+14
0.151 m^3 m^{-3} /25%	0.5	−55	67.5	−57	3.8	−3	1010	−8

[a]. Differences are calculated based on calibration output. A (+) sign indicates percent increase and a (-) sign indicates percent decrease from the calibration results.

The result of the initial soil moisture change to 60% shows an increase in peak and cumulative flow by 100% and 121%, respectively, for the sub-watershed. Increasing the initial soil moisture to 60% also resulted in an overall increase in the peak discharge and cumulative flow by 10% and 5%, respectively. The result of initial soil moisture change to 25% shows a decrease in peak and cumulative flow by 75% and 71%, respectively, for the sub-watershed. Decreasing initial soil moisture to 25% also resulted in an overall decrease in the peak discharge and cumulative flow by 5% and 3%, respectively.

Similar sensitivity tests on the initial soil moisture setting on two other sub-watersheds (i.e., W960 and W1000) were also performed. These are presented in Table 5, where high sensitivity on individual sub-watersheds can be seen as peak flow changes are in a range of 50%–75%, and cumulative flow changes are in a range of 49%–102% due to the change of initial soil moisture saturation on a sub-watershed. The overall impact on peak and cumulative flow due to the change of initial soil moisture on sub-watersheds is within a range of peak difference of 3%–8% and cumulative flow difference of 4%–14%. It can be concluded from these tests that the modelled peak flow and cumulative output flow are very sensitive to the antecedent soil moisture condition. This confirms a similar study by [44,45], which concluded that HEC-HMS simulations are highly influenced by initial soil moisture on flood generation.

5. Conclusions

This study investigated the applicability of the Hydrologic Modelling System developed by the Hydrologic Engineering Center (HEC-HMS) in the Sturgeon Creek watershed; a snow melt-dominated watershed in Manitoba, Canada. Soil moisture was estimated from RADARSAT-2 satellite data and subsequently used to set the initial soil water content of the HEC-HMS model's event simulation. Event and continuous modelling of HEC-HMS has been performed in order to confirm the applicability of the model in Manitoba basins. Model performance measurements indicate that simulated flows are in good agreement with the observed results. The study also demonstrated that HEC-HMS and the temperature index method were able to accurately simulate the timing and magnitude of a spring snowmelt. Therefore, the model is well suited to determine runoff values for flood forecasting and other purposes.

Analysis of SMA parameters was performed to understand the sensitivity of each parameter on the movement and storage of water through different layers. Results from the analysis identified soil storage as the most sensitive parameter. A sensitivity analysis of initial soil moisture was performed to provide changes in peak flow and cumulative flow of the watershed and sub-watersheds under different saturations. The results confirm that peak flow in a snowmelt event is highly influenced by the initial soil moisture setting of the model.

Soil moisture information from RADARSAT-2 and the Fall Soil Moisture Survey of Manitoba were used to set the initial soil moisture states for two event simulations. Modelled flow data using the Fall Soil Moisture Survey did not agree well with observed flows. However, soil moisture data from RADARSAT-2 substantially improved the agreement between modelled and observed flows. These results demonstrate the ability of RADARSAT-2 to improve the performance of hydrology models over the Fall Soil Moisture Survey, which is based on field-collected data to define the soil moisture state. Satellites have a much greater ability to provide accurate soil moisture information with improved spatial/temporal resolution and coverage over the entire watershed.

This study supports the Hydrologic Forecast Center (HFC) and their efforts to find and to select an appropriate hydrology model for flood forecasting in Manitoba as recommended by the Flood Review Task Force [2]. The HFC has implemented MANAPI, which uses precipitation data from a sparse station network to define an antecedent precipitation index (API) map. MANAPI computes a single runoff value for a selected watershed from an event based on historical data. MANAPI is not capable of addressing complex watershed processes. Historical data for a graphical relationship of runoff-precipitation-API were not available for the Sturgeon Creek Watershed. Therefore, an MANAPI runoff value could not be compared with the event simulation.

This is the first attempt to use RADARSAT-2-derived soil moisture in hydrological modelling in an area where flooding events are caused by snowmelt. Despite the efforts to make the research comprehensive, further studies using RADARSAT-2-derived soil moisture for other Manitoba watersheds and for other years should be done to validate and enhance the findings of this study. Although this study only used one year (2014) of satellite soil moisture data for event modelling, the results indicate that the initial setting of RADARSAT-2-derived soil moisture can improve the performance of HEC-HMS and can be an appropriate tool for flood forecasting at the HFC in Manitoba.

Acknowledgments: This project was supported by the Science and Technology Branch (STB) of Agriculture and Agri-Food Canada and partially funded by the Government Related Initiatives Program (GRIP) of the Canadian Space Agency. We thank our colleagues from the Hydrologic Forecast Center (HFC) of Manitoba, who provided feedback that greatly assisted the research.

Author Contributions: H. McNairn, J. Powers, A. Merzouki and A. Bhuiyan conceived of and designed the experiments. A. Bhuiyan performed the experiments. A. Bhuiyan and J. Powers analysed the data. A. Merzouki contributed tools/processing RADARSAT-2 data. A. Bhuiyan wrote the paper.

Conflicts of Interest: The authors declare no conflict of interest.

Abbreviations

The following abbreviations are used in this manuscript:

AAFC	Agriculture and Agri-Food Canada
ECC	Environment and Climate Change Canada
MA	Manitoba Agriculture
SAR	Synthetic aperture radar
RADARSAT	Canadian remote sensing Earth observation satellite
HEC-HMS	Hydrologic Modelling System developed by the Hydrologic Engineering Center
SMA	Soil moisture accounting
MANAPI	Manitoba Antecedent Precipitation Index model
WSC	Water Survey Canada
HFC	Hydrologic Forecast Center
DEM	Digital elevation model
LiDAR	Light detection and ranging
IEM	Integral equation model
HH	Backscatter horizontal transmission and horizontal receive
VV	Backscatter vertical transmission and vertical receive
SCS	Soil Conservation Service
SWE	Snow water equivalent

References

1. Blais, E.L.; Clark, S.; Dow, K.; Ranniec, B.; Stadnyk, T.; Wazney, L. Background to flood control measures in the Red and Assiniboine River Basins. *Can. Water Resour. J.* **2015**, doi:10.1080/07011784.2015.1036123.
2. Flood Review Task Force. Manitoba 2011 Flood Review Task Force Report. Report to the Minister of Manitoba Infrastructure and Transportation, 2013. Availableonline:https://www.gov.mb.ca/asset_library/en/2011flood/flood_review_task_force_report.pdf (accessed on 7 April 2016).
3. Rannie, W. The 1997 flood event in the Red River basin: Causes, assessment and damages. *Can. Water Resour. J.* **2015**, doi:10.1080/07011784.2015.1004198.
4. Stadnyk, T.; Dow, K.; Wazney, L.; Blais, E.L. The 2011 flood event in the Red River Basin: Causes, assessment and damages. *Can. Water Resour. J.* **2015**, 65–73.
5. Bower, S.S. Natural and unnatural complexities: Flood control along Manitoba's Assiniboine River. *J. Hist. Geogr.* **2010**, *36*, 57–67.
6. Rasmussen, P.F. Evaluation of Flood Forecasting and Warning SSystem in Canada. 22nd Canadian Hyrotechnical Conference, 2015. Availableonline:http://www.nsercfloodnet.ca/files/Track_5_-_Presentation_-_Flood_forecasting_in_Canada_CSCE_2015.pdf (accessed on 7 April 2016).
7. Steenbergen, N.V.; Willems, P. Rainfall uncertainty in flood forecasting: Belgian case study of rivierbeek. *J. Hydrol. Eng.* **2014**, *19*, doi:10.1061/(ASCE)HE.1943-5584.0001004.
8. Dietrich, J.; Schumann, A.H.; Redetzky, M.; Walther, J.; Denhard, M.; Wang, Y.; Pfutzner, B.; Büttner, U. Assessing uncertainties in flood forecasts for decision making: prototype of an operational flood management system integrating ensemble predictions. *Nat. Hazards Earth Syst. Sci.* **2009**, *9*, 1529–1540.
9. Brocca, L.; Melone, F.; Moramarco, T.; Wagner, W.; Naeimi, V.; Bartalis, Z.; Hasenauer, S. Improving runoff prediction through the assimilation of the ASCAT soil moisture product. *Hydrol. Earth Syst. Sci.* **2010**, *14*, 1881–1893.
10. Sutanudjaja, E.H.; van Beek, L.P.H.; de Jong, S.M.; van Geer, F.C.; Bierkens, M.F.P. Calibrating a large-extent high-resolution coupled groundwater-land surface model using soil moisture and discharge data. *Water Resour. Res.* **2014**, *50*, 687–705.

11. Tramblay, Y.; Bouaicha, R.; Brocca, L.; Dorigo, W.; Bouvier, C.; Camici, S.; Servat, E. Estimation of antecedent wetness conditions for flood modelling in northern Morocco. *Hydrol. Earth Syst. Sci.* **2012**, *16*, 4375–4386.
12. Li, Y.; Grimaldi, S.; Walker, J.P.; Pauwels, V.R.N. Application of Remote Sensing Data to Constrain Operational Rainfall-Driven Flood Forecasting: A Review. *Remote Sens.* **2016**, *8*, doi:10.3390/rs8060456.
13. Massari, C.; Brocca, L.; Barbetta, S.; Papathanasiou, C.; Mimikou, M.; Moramarco, T. Using globally available soil moisture indicators for flood modelling in Mediterranean catchments. *Hydrol. Earth Syst. Sci.* **2014**, *18*, 839–853.
14. Xu, X.; Li, J.; Tolson, B.A. Progress in integrating remote sensing data and hydrologic modeling. *Appl. Meteorol. Climatol.* **2014**, *87*, 61–77.
15. McNairn, H.; Merzouki, A.; Pacheco, A. Estimating surface soil moisture ubing RADARSAT-2. In *International Archives of the Photogrammetry, Remote Sensing and Spatial Information Science*; Copernicus GmbH: Göttingen, Germany, 2010; pp. 576–579.
16. AAFC Information Bulletin 99-4. *Soils and Terrain. An Introduction to the Land Resource. Rural Municipality of Rosser. Information Bulletin 99-4*; Technical Report; Land Resources Unit, Brandon Research Centre, Research Branch, Agriculture and Agri-Food Canada: Winnipeg, MB, Canada, 1999.
17. AECOM Canada. *Sturgeon Creek Hydrodynamic Model and Economic Study, Project No.: F685 003 00 (4.6.1)*; Technical Report; Water Stewardship, Government of Canada: Winnipeg, MB, Canada, 2009.
18. Jacob, D.; Lorenz, P. Future trends and variability of the hydhydrologic cycle in different IPCC SRES emission scenarios—A case study for the Baltic Sea Region. *Boreal Environ. Res.* **2009**, *14*, 100–113.
19. Thornthwaite, C.W. An Approach toward a Rational Classification of Climate. *Geogr. Rev.* **1948**, *38*, 55–94.
20. Feldman, A.D. *Hydrologic Modeling System HEC-HMS, Technical Reference Manual*; U.S. Army Corps of Engineers, Hydrologic Engineering Center HEC: Davis, CA, USA, 2000.
21. Alvarez, J.; Verhoest, N.E.C.; Casali, J.; Gonzalez-Audicana, M.; Lopez, J.J. RADARSAT based surface soil moisture retrieval on agricultural catchments of Navarre (Spain). In Proceedings of the 2004 IEEE International Geoscience and Remote Sensing Symposium, Anchorage, AK, USA, 20-24 September 2004; Volume 5, pp. 3507–3510.
22. Merzouki, A.; McNairn, H. A Hybrid (Multi-Angle and Multipolarization) Approach to Soil Moisture Retrieval Using the Integral Equation Model: Preparing for the RADARSAT Constellation Mission. *Can. J. Remote Sens.* **2015**, *41*, 349–362.
23. Eilers, P. *Sturgeon Creek Soil Moisture Monitoring Stations (SMMS); Soil and Landscape Classification;* Contract Number 3000528851. Technical Report; Science and Technology Branch, Agriculture and Agri-Food Canada: Winnipeg, Manitoba, 2013.
24. Schaffenberg, W.A. *Hydrologic Modeling System HEC-HMS, User Manual: Version 4.0.* U.S. Army Corps of Engineers, Hydrologic Engineering Center HEC, 609 Second Street, Davis, CA, USA, 2013.
25. U.S. Army Corps of Engineers. *Hydrologic Modeling System (HEC-HMS) Application Guide: Version 4.0*; Institute for Water Resources, Hydrologic Engineering Center: Devis, CA, USA, 2015.
26. Knebl, M.; Yang, Z.L.; Hutchison, K.; Maidment, D. Regional scale flood modeling using NEXRAD rainfall, GIS, and HEC-HMS/RAS: A case study for the San Antonio River Basin Summer 2002 storm event. *J. Environ. Manag.* **2005**, *75*, 325–336.
27. Du, J.; Qian, L.; Rui, H.; Zuo, T.; Zheng, D.; Xu, Y.; Xu, C.Y. Assessing the effects of urbanization on annual runoff and flood events using an integrated hydrological modeling system for Qinhuai River basin, China. *J. Hydrol.* **2012**, *464–465*, 127–139.
28. Haberlandt, U.; Radtke, I. Hydrological model calibration for derived flood frequency analysis using stochastic rainfall and probability distributions of peak flows. *Hydrol. Earth Syst. Sci.* **2014**, *18*, 353–365.
29. Fleming, M.; Neary, V. Continuous hydrologic modeling study with the hydrologic modeling system. *ASCE J. Hydrol. Eng.* **2004**, *9*, 175–183.
30. Gebre, S.L. Application of the HEC-HMS model for runoff simulation of upper blue Nile River Basin. *Hydrol. Curr. Res.* **2015**, *6*, 1–8.
31. Singh, W.R.; Jain, M.K. Continuous Hydrological Modeling using Soil Moisture Accounting Algorithm in Vamsadhara River Basin, India. *J. Water Res. Hydraul. Eng.* **2015**, *4*, 398–408.
32. Gyawali, R.; Watkins, D.W. Continuous Hydrologic Modeling of Snow-Affected Watersheds in the Great Lakes Basin Using HEC-HMS. *ASCE J. Hydrol. Eng.* **2013**, *18*, 29–39.
33. Hall, M.J. How well does your model fit the data? *J. Hydroinf.* **2001**, *3*, 49–55.

34. Krause, P.; Boyle, D.P.; Base, F. Comparison of different efficiency criteria for hydrological model assessment. *Adv. Geosci.* **2005**, *5*, 89–97.
35. Bardsley, W.E. A goodness of fit measure related to r^2 for model performance assessment. *Hydrol. Process.* **2013**, *27*, 2851–2856.
36. Nash, J.E.; Sutcliffe, J.V. River flow forecasting through conceptual models part I–A discussion of principles. *J. Hydrol.* **1970**, *10*, 282–290.
37. Henriksen, H.J.; Troldborg, L.; Nyegaard, P.; Sonnenborg, T.O.; Refsgaard, J.C.; Madsen, B. Methodology for construction, calibration and validation of a national hydrological model for Denmark. *J. Hydrol.* **2003**, *280*, 52–71.
38. Legates, D.R.; McCabe, G.J. Evaluating the use of goodness-of-fit measures in hydrologic and hydroclimatic model validation. *Water Resour. Res.* **1999**, *35*, 233–241.
39. Moriasi, D.N.; Arnold, J.G.; Liew, M.W.V.; Bingner, R.L.; Harmel, R.D.; Veith, T.L. Model evaluation guidelines for systematic quantification of accuracy in watershed simulations. *Am. Soc. Agric. Biol. Eng.* **2007**, *50*, 885–900.
40. Sturm, M.; Taras, B.; Liston, G.E.; Derksen, C.; Jonas, T.; Lea, J. Estimating snow water equivalant using snow depth data and climate classes. *J. Hydrometeorol.* **2010**, *11*, 1380–1394.
41. Theil, H. A rank-invariant method of linear and polynomial regression analysis, I, II, III. In Proceedings of the Royal Netherlands Academy of Sciences, Amsterdam, the Netherlands, 30 September 1950; pp. 1397–1412.
42. Sen, P.K. Estimates of the regression coefficient based on Kendall's tau. *J. Am. Stat. Assoc.* **1968**, *63*, 1379–1389.
43. Roy, D.; Begam, S.; Ghosh, S.; Jana, S. Calibration and validation of HEC-HMS model for a river basin in Eastern India. *J. Eng. Appl. Sci.* **2013**, *8*, 40–56.
44. Czigany, S.; Pirkhoffer, E.; Geresdi, I. Impact of extreme rainfall and soil moisture on flash flood generation. *J. Hung. Meteorol. Serv.* **2010**, *114*, 79–100.
45. Hegedus, P.; Czigany, S.; Balatonyi, L.; Pirkhoffer, E. Analysis of soil boundary conditions of flash Floods in a small basin in SW Hungary. *Cent. Eur. J. Geosci.* **2013**, *5*, 97–111.

9

Evaluating Global Reanalysis Datasets as Input for Hydrological Modelling in the Sudano-Sahel Region

Elias Nkiaka *, N. R. Nawaz and Jon C. Lovett

School of Geography, University of Leeds, Leeds LS2 9JT, UK; N.R.Nawaz@leeds.ac.uk (N.R.N.); J.Lovett@leeds.ac.uk (J.L.)

* Correspondence: gyenan@leeds.ac.uk

Abstract: This paper investigates the potential of using global reanalysis datasets as input for hydrological modelling in the data-scarce Sudano-Sahel region. To achieve this, we used two global atmospheric reanalyses (Climate Forecasting System Reanalysis and European Center for Medium-Range Weather Forecasts (ECMWF) ERA-Interim) datasets and one global meteorological forcing dataset WATCH Forcing Data methodology applied to ERA-Interim (WFDEI). These datasets were used to drive the Soil and Water Assessment Tool (SWAT) in the Logone catchment in the Lake Chad basin. Model performance indicators after calibration showed that, at daily and monthly time steps, only WFDEI produced Nash Sutcliff Efficiency (NSE) and Coefficient of Determination (R^2) values above 0.50. Despite a general underperformance compared to WFDEI, CFSR performed better than the ERA-Interim. Model uncertainty analysis after calibration showed that more than 60% of all daily and monthly observed streamflow values at all hydrometric stations were bracketed within the 95 percent prediction uncertainty (95PPU) range for all datasets. Results from this study also show significant differences in simulated actual evapotranspiration estimates from the datasets. Overall results showed that biased corrected WFDEI outperformed the two reanalysis datasets; meanwhile CFSR performed better than the ERA-Interim. We conclude that, in the absence of gauged hydro-meteorological data, WFDEI and CFSR could be used for hydrological modelling in data-scarce areas such as the Sudano-Sahel region.

Keywords: reanalysis; SWAT; CFSR; ERA-Interim; WFDEI; Logone catchment; Sudano-Sahel

---

1. Introduction

Long-term and well distributed climate information is essential to enhance water resources management and to guide policies aimed addressing the consequences of climate variability and change from a local to global scale [1]. This data is needed because the quantitative estimation of water balance components is important to understand the variations taking place at catchment/global level [2]. However, in many developing and arid regions of the world, the assessment and management of water resources is still a major challenge due to data scarcity [3]. According to Gorgoglione et al. [4], the difficulty in collecting data in semi-arid and other remote regions can be attributed to several reasons: (i) lack of reliable equipment; (ii) absence of good archiving system and software to store and process the data, and lack of funds to organize data collection campaigns. Another challenge in these regions is that even when data is collected and archived, the effort and money required to access them can be quite substantial [5]. Hydrological models are designed to fill some of these gaps, and their application to enhance water resources management is widely acknowledged [6].

Rainfall is one of the most important inputs used to drive hydrological models; hence it is important to obtain rainfall data of sufficient temporal and spatial resolution. Nevertheless, due to the high spatiotemporal variability of rainfall, it can only be accurately captured by a dense network of rain gauge stations [7]. However, most often, rain gauges may be located outside the area of interest or could exhibit significant gaps in spatial coverage, especially in remote and ungauged areas [5].

Current advances in remote sensing offer many advantages, e.g., satellites observing the Earth have generated potentially useful data that can be used to improve water resources management. Even so, satellite data is usually developed for application in large areas, e.g., at continental or global scale. Therefore, its application at catchment scale for hydrological modelling requires further downscaling, transformation or interpolation which may increase uncertainties in the data [8].

To overcome this challenge, multiyear global gridded representations of weather known as reanalysis datasets are now available. Examples of widely used reanalysis datasets include: National Centers for Environmental Prediction (NCEP)/National Center for Atmospheric Research (NCAR), Climate Forecasting System Reanalysis (CFSR) [9], European Center for Medium-Range Weather Forecasts (ECMWF) ERA-Interim [10] and Modern-Era Retrospective Analysis for Research and Applications (MERRA) [11]. However, it has been shown that significant differences exist in precipitation estimates from these products [12]. Lorenz and Kunstmann [12] assert that the quality of precipitation estimates from reanalyses datasets depends on the geographic location, especially in tropical regions. Furthermore, a recent study by Essou et al. [13] demonstrated that the performance of reanalysis datasets may vary from one climatic zone to another.

To address the issue of bias inherent in reanalysis products; global forcing datasets have been developed using post processing techniques (e.g., bias correction) based on observations [14]. An example of such bias corrected dataset is the WATCH Forcing Data methodology applied to ERA Interim (WFDEI) [14].

Another issue often overlooked in most studies evaluating the performance of reanalysis datasets in hydrologica modelling is the impact of spatial resolution of each dataset on the quality of the simulated streamflow. In fact, the effect of rainfall spatial variability on streamflow and water balance components have been shown to be significant in catchments with high spatial variability [15]. Lobligeois et al. [16] in their study demonstrated the importance of spatial representation in areas subjected to high spatial variability in rainfall. Given that the distance between reanalysis grid points is quasi uniform, these datasets could be used to investigate the impact of rainfall spatial variability on hydrological processes such as streamflow and evapotranspiration in large catchments.

Recently, reanalysis datasets have been used as input for hydrological modelling in many studies with different degrees of successes recorded. For example, Essou et al. [13] used CFSR, ERA-Interim, MERRA and WFDEI as input for streamflow simulation using a conceptual model in several watersheds in the US and concluded that these datasets had good potential to be used for hydrological modelling. Monteiro et al. [17] used CFSR and WFDEI to drive the Soil and Water Assessment Tool (SWAT) for hydrological modelling in the Tocantins catchment in Brazil and asserted that WFDEI outperformed CFSR in their study area. Andersson et al. [18] used ERA-Interim and WFDEI as input to drive the hydrological catchment model (HYPE) in Europe and Africa. They concluded that WFDEI improved streamflow simulation compared to Watch Forcing data methodology applied to ERA-40. Krogh et al. [19] used CFSR and ERA-Interim to drive the Cold Regions Hydrological Model (CRHM) in the upper Baker river basin in Chile and concluded that CFSR simulated streamflow better than ERA-Interim. These numerous studies suggest that reanalysis datasets could be used for hydrological modelling in data scarce regions. Despite widespread hydro-meteorological data scarcity in Africa in general and the Sudano-Sahel region in particular, the use of reanalyses datasets for hydrological modelling in this area remains largely unstudied.

The Logone catchment presents special attributes for the evaluation of reanalysis datasets because it is located at the transition zone between the Sudano and Sahel areas where rainfall is highly variable both in space and time [20]. Furthermore, like most catchments in the region, the Logone suffers from acute observational data scarcity. Given that the performance of reanalysis products in hydrological modelling is largely determined by the quality of the precipitation estimates, Essou et al.; Monteiro et al. and Krogh et al. [13,17,19] recommend that the correlation between observed rainfall and reanalysis precipitation estimates should be assessed before the latter is used as input for hydrological modelling.

In a previous study in the catchment, the authors of [21] evaluated the quality of precipitation estimates from CFSR and ERA-Interim against observed monthly rainfall covering the period 1979–2002 and concluded that, precipitation estimates from both reanalyses products could reproduce the seasonal rainfall cycle in the catchment albeit significant variability in the data.

The objectives of this study were; (i) to evaluate the ability of two reanalysis datasets; CFSR and ERA-Interim and one bias corrected global meteorological forcing dataset WFDEI to be used as input to drive the SWAT model in the Logone catchment; and (ii) to evaluate the impact of reanalysis spatial resolution on the quality of simulated flows. This study will be useful in validating the use of reanalysis datasets in data scarce catchments subject to high spatial rainfall variability. In addition, Siam et al. [22] have argued that driving hydrological models with reanalyses datasets to reproduce observed streamflow represents one of the most accurate ways to evaluate how the hydrological cycle is simulated by reanalysis forecast models. Including WFDEI will permit us to assess the impact of bias correction on the performance of ERA-Interim. It is not our intention in this study to judge the quality of each reanalysis dataset or recommend the use of one product over another. This choice depends on personal preference because the performance of each reanalysis product varies from one region to another and one from climatic zone to another as mentioned earlier. A limitation of this study is the absence of daily rain gauge data that could also be used to drive SWAT to compare the performance of the reanalysis datasets against gauge data in simulating streamflow.

2. Materials and Methods

2.1. Study Area

The Logone catchment (Figure 1) is a transnational catchment shared by Cameroon, Chad and Central Africa Republic, with an estimated area of about 86,500 km^2 lying between latitude 6° and 12° N and longitude 13°–17° E. There are two National Parks in the catchment (Waza and Kalamaloue), with high concentration of wildlife [23]. The Logone River has its source in Cameroon through the Mbere and Vina rivers from the north eastern slopes of the Adamawa Plateau. In Lai, it is joined by the Pende River from Central Africa Republic and flows for about 1000 km in a South-North direction with an elevational range from 300 masl in the north to about 1200 masl in the south. The basin topography, apart from some local mountains in the south is very flat with an average slope of less than 1.3%. The catchment has a semi-arid climate in the north where annual rainfall varies between 600 and 900 mm/year and Sudano climate in the south where annual rainfall varies between 900 and 1400 mm/year. The climate is also characterized by high spatio-temporal variability in rainfall controlled by the oceanic regime from the south and the continental regime from the north [20]. Almost all rain falls during the rainy season from May/June to September/October with high spatial and temporal variability and mean annual temperature is about 28 °C [23].

2.2. Data Sources

2.2.1. Observed River Discharge Data

Daily river discharge measurements were obtained from the Lake Chad Basin Commission (LCBC) covering the period 1983–1997 at four discharge stations. Gaps in the river discharge data were filled using Artificial Neural Networks Self-Organizing Maps (ANN-SOM) [24].

2.2.2. Spatial Datasets

Digital Elevation Model (DEM) data obtained from Shuttle Radar Topographic Mission (SRTM) at a spatial resolution of 90 m was used for catchment delineation. Land cover/use maps were obtained from Climate Change Initiative Land Cover (CCI-LC) at a spatial resolution of 300 m. The land cover was reclassified in the ARCSWAT interface according to model input requirements. Soil data was obtained from the Food and Agricultural Organization (FAO), Harmonize World Soil Database (HWSD) at a spatial resolution of 1 km.

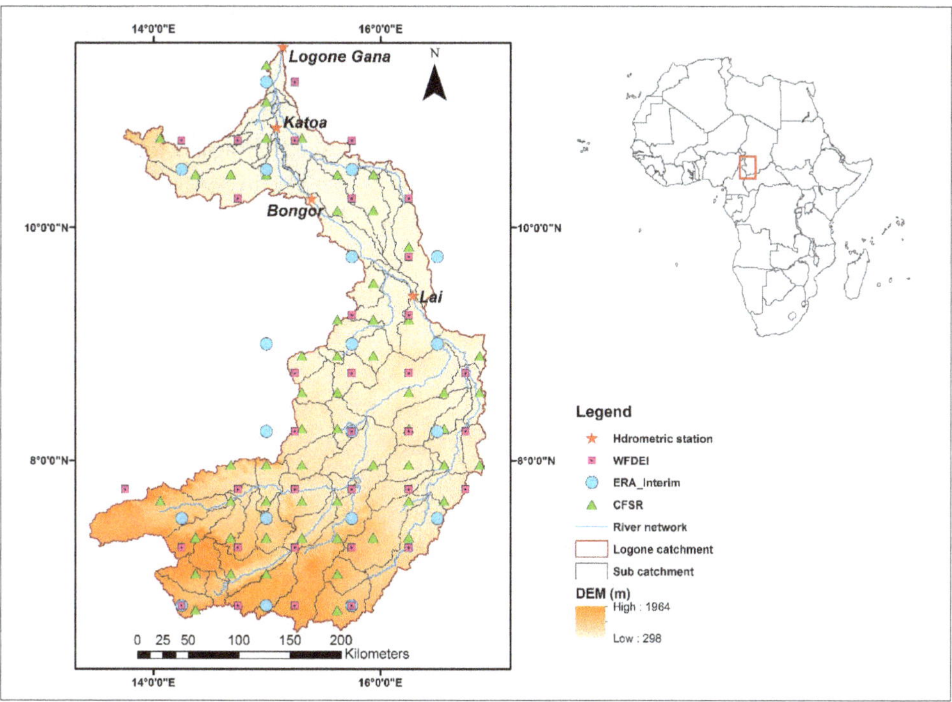

Figure 1. Map of the study area showing the Logone river network, sub catchments and reanalysis grid points used for streamflow simulation. DEM: Digital Elevation Model in metres.

2.2.3. Reanalysis Data

A reanalysis project involves the reprocessing of observational data spanning an extended historical period. "It makes use of a consistent modern analysis system, to produce a dataset, that to a certain extent can be regarded as a "proxy" for observation with the advantage of providing coverage and time resolution often unobtainable with normal observational network" [25]. It is generated with a data assimilation system combining observations with a numerical weather prediction model. For the entire reanalysis period, the model physics remain unchanged in the forecast model for consistency of the output data. The reanalysis consequently provides a physical picture of the global climate over a period during which observational data are available.

2.3. CFSR

The Climate Forecast System, NCEP version 2 is an upgraded version of CFS version one (CFSv1). It was first developed as part of the Climate Forecast System by NCEP in 2004 with quasi-global coverage, fully coupled atmosphere-ocean-land model used by NCEP for seasonal prediction [9]. CFSR has a 3D-variational analysis scheme of the upper-air atmospheric state with 64 vertical levels with a horizontal resolution of 38km spanning the period 1st January 1979 to present day [9].

2.4. ERA-Interim

ERA-Interim is the latest global atmospheric reanalysis produced by the European Centre for Medium-Wave Forecasts (ECMWF) and covers the period from 1 January 1979 to present day [10]. The core component of the ERA-Interim data assimilation system is the 12-h 4D-variational analysis scheme of the upper-air atmospheric state, which is on a spectral grid with triangular truncation of 255 waves (corresponding to approximately 80 km) spatial resolution and a hybrid vertical coordinate system with 60 vertical levels.

2.5. WFDEI

The WATCH Forcing Data methodology applied to ERA-Interim (WFDEI) dataset [14] is produced from Watch Forcing Data (WFD) and ERA-Interim reanalysis via sequential interpolation to a 0.5° resolution, elevation correction and monthly-scale adjustments based on CRU TS3.1/TS3.21 and GPCCv5/v6 monthly precipitation observations for 1979–2012.

Details of the three products can be found in [9,10,14] for CFSR, ERA-Interim and WFDEI respectively. For the Logone catchment, the reanalysis datasets were obtained for an area bounded by latitude 6°–12.0° N and longitude 13°–17.25°E from the Texas A&M University for CFSR, ECMWF for ERA Interim and Lund University for WFDEI. All variables were obtained at a daily time step with spatial resolution of 0.312° (~38 km), 0.50° (~55 km) and 0.75° (~80 km) for CFSR, WFDEI and ERA-Interim respectively hereafter referred to as high, medium and low resolution.

2.6. Model Setup

River discharge at various locations along the Logone River was simulated using the SWAT [26] in the ArcSWAT interface. SWAT is one of the most widely used river basin–scale models worldwide, applied extensively for solving a broad range of hydrologic and environmental problems [26].

In this study, we focus only on water quantity simulation accomplished through two steps: (i) the land phase of the hydrological cycle which controls the amount of water transferred to the main channel from each sub catchment; and (ii) the routing phase which involves the movement of water through the channel network to the outlet. The hydrologic cycle in the land phase of the model is simulated using the water balance equation:

$$SW_t = SW_0 + \sum_{i=1}^{n}(R_{day} - Q_{surf} - E_a - W_{seep} - Q_{gw}) \tag{1}$$

SW_t is the final soil water content (mm), SW_0 is the initial water content (mm), R_{day} is the amount of precipitation on day i (mm) Q_{surf} is the amount of surface water runoff on day i (mm), E_a is the amount of actual transpiration on day i (mm), W_{seep} is the amount of water entering the vadose zone from the soil profile on day I (mm) and Q_{gw} is the amount of return flow on day i (mm). Details of equations and methods used to estimate various hydrological components can be found in [27]. During model development, SWAT divides a catchment into sub catchments using digital elevation model (DEM) data. The spatial distribution of hydrological processes over each sub catchment is represented through hydrologic response units (HRUs), used to further divide the sub catchments into smaller units. The HRU can be defined as a land area within a sub catchment with the same land use class, soil type, slope class and management combinations.

While building the model, an attempt was made to maximize the number of grid points used for streamflow simulation using CFSR as the reference dataset because of its high spatial resolution (0.312°) compared to the other two. Different threshold areas were tested for catchment delineation. Reducing the threshold area to 500 km^2 did not increase the number of reanalysis grid points selected while increasing it 1000 km^2 reduced the number to only 45. An optimum threshold area of 750 km^2 was finally used to delineate the catchment into 66 sub catchments. Threshold values for creation of hydrological response units (HRUs) were set at 10%, 15%, and 15% for land use, soil and slope classes respectively creating 266 HRUs. A separate model was developed for each of the reanalysis datasets using the same threshold values.

The Hargreaves method for estimating potential evapotranspiration (PET) was applied owing to the less onerous data demands (only rainfall, minimum and maximum temperature) compared to the alternative Priestley-Taylor and Penman-Monteith methods. Surface runoff was calculated using the Soil Conservation Service's curve number (CN2) method while flow routing was accomplished through the variable storage method [27].

2.7. Model Calibration and Uncertainty Analysis

The model was calibrated in the SWAT Calibration and Uncertainty Program software (SWAT-CUP) using the Sequential Uncertainty Fitting algorithm (SUFI-2) [28]. During the calibration process in SUFI-2, parameters can be changed using either the relative or absolute parameter ranges. Each parameter value can be modified either by replacement of the initial value, addition of absolute change or multiplication by a relative change factor to obtain the optimum value. Given the multiple sources of uncertainties inherent in the use of hydrological models; the advantage of using SWAT-CUP is that these are taken into consideration during model calibration [28]. As model parameters often depend on the input data used to drive the model which is susceptible to seasonal variation [29]; the calibrated parameter values in SWAT-CUP are given within a range to represent this variability. Model calibration consisted of running 500 simulations in each iteration with the parameter set shown in Table 1. The best parameter range obtained in the first iteration was then substituted and used in the next iteration for each of the five iterations performed. This was done for the three different datasets at daily and monthly time steps. To obtain the values of the different water balance components such as evapotranspiration, the simulation number that produced the best model output was used to calculate the water balance for the whole catchment.

The model was evaluated using three different evaluation statistics: (i) the Nash Sutcliffe Efficiency (NSE); (ii) coefficient of determination (R^2); and (iii) Percent Bias (PBIAS). The NSE is used to assess the predictive capacity of the model and measures how well the observed and simulated flows match. Its value range from $-\infty$ to 1 with values close to 1 indicating high model performance. The R^2 measures how well the observed data is correlated to the simulated data and varies from 0 to 1 with values closer to 1 also indicating high model performance. PBIAS indicates the average tendency of the simulated flows to be over/underestimated than observed flows with absolute low values indicating accurate model simulation. Positive values indicate model underestimation while negative values indicate overestimation. According to [30], the results of the calibrated model may be considered to be satisfactory if NSE > 0.50, $R^2 > 0.60$ and PBIAS $\pm$ 25%.

The degree of uncertainty in the calibrated model(s) was quantified using the *p-factor* and *r-factor*. The *p-factor* represents the percentage of observed streamflow bracketed by the 95% prediction uncertainty (95PPU) while the *r-factor* is the average width of the 95PPU. The 95PPU is calculated at the 2.5% and 97.5% confidence interval of observed streamflow obtained through Latin hypercube sampling. In SUFI-2, the goal is to minimize the width of the uncertainty band and enclose as many observations as possible because these observations are a result of all processes taking place in the catchment [28]. The *p-factor* can vary between 0 and 1 while the ideal value for *r-factor* is 0, indicating that there is no uncertainty in the model outputs. However, an *r-factor* of 0 will indicate that fewer flow observations were included in the 95PPU band.

Given that the goal of this study was to evaluate how well each reanalysis dataset was able to simulate streamflow as closely as possible to the observed, all parameters that influence this process, were calibrated. Evapotranspiration (ESCO); surface runoff (CN2, Surlag, Ch_K2); groundwater exchange (Rchrg_DP, GWQMN, GW_REVAP, REVAPMN, GW_DELAY, ALPHA_BF) and infiltration (SOL_AWC). Furthermore, since this study objective did not include evaluation of alternative scenarios for which it would be necessary to establish the performance limits of different parameter sets e.g., by validating the parameter set(s) using independent observations, the entire period of the available streamflow record was used for calibration. The advantage of this approach is that, longer input time series are included in the simulation with the possibility of capturing long term trends and variability as simulated by reanalysis forecast models. Auerbach et al. [31] used a similar approach to evaluate the performance of CFSR dataset as input for hydrological modelling in the tropics. Furthermore, the parameters range obtained during model over this long time scale could be used for climate change impact assessment in the catchment. The model was calibrated from 1980 to 1997 at daily and monthly time steps using the first three years as warm-up period. This calibration was done at Logone Gana, Katao, Bongor and Lai hydrometric stations (Figure 1).

Table 1. Description of model parameters, parameter ranges used for calibration. The ranges are given for the three datasets used in the study.

Parameter	Description	Model Process	Parameter Range Used
CN2 [a]	Curve number for moisture condition II	Surface runoff generation. High values lead to high surface flow	−0.5–0.15
GW_Delay	Groundwater delay	Groundwater (affects groundwater movement). It is the lag between the time water exits the soil profile and enters the shallow aquifer	30–250
GW_REVAP	Groundwater "revap" coefficient	Affects the movement of water from the shallow aquifer to the unsaturated soil layers. Low values lead to high baseflow	0.10–0.40
GWQMN	Threshold depth of water in the shallow aquifer required for return flow to occur	Groundwater (when reduced streamflow increases)	20–95
Revapmn	Threshold depth of water for "revap to occur" (mm)	Groundwater (when increased, base flow will increase)	0–20
Rchrg_DP	Deep aquifer percolation	Groundwater (the fraction of percolation from the root zone which recharges the deep aquifer. Higher values lead to high percolation).	0.05–0.50
Ch_K2	Hydraulic conductivity of main channel	Channel infiltration	1.69–6.0
ESCO	Soil evaporation compensation factor	Controls the soil evaporative demand from different soil depth. High values lead to low evapotranspiration	0.25–0.95
SOL-AWC [a]	Available Water Capacity or available is calculated as the difference between field capacity the wilting point	Groundwater, evaporation. When increased less water is sent to the reach as more water is retained in the soil thus increasing evapotranspiration	−0.04–0.04
ALPHA_BF	Base flow alpha factor	Shows the direct index of groundwater flow response to changes in recharge	0.3–0.9
Surlag	Surface runoff lag coefficient	Surface runoff	1.5–5.0

[a] Parameter value is multiplied by (1 + a given value). For example if CN2 = 85 then the calibrated CN2 value will be $(1 + (-0.5)) \times 85 = 0.5 \times 85 = 42.5$.

3. Results

The optimum threshold area used for delineating the catchment into different sub catchments was 750 km^2. Using this area 57, 34 and 19 reanalysis grid points were selected for CFSR, WFDEI and ERA Interim respectively (Figure 1).

Figure 2 shows the variability in annual rainfall from the three datasets used in this study. It can be observed from the figure that the variability in the datasets is not the same because maximum and minimum rainfall occur in almost different years except in a few cases when all the datasets produced maximum/minimum rainfall in the same year e.g., 1985, 1988 and 1992. Annual rainfall from WFDEI varies between 1000 and 1300 mm/year, CFSR varies between 900–1550 mm/year and ERA-Interim varies between 750 and1650 mm/year. Overall the analysis showed that the variability is highest for ERA-Interim followed by CFSR while WFDEI has lowest variability in annual rainfall. The annual average rainfall in the catchment as simulated by SWAT model for the three datasets was 1237 mm, 1240 mm and 1047 mm for WFDEI, CFSR and ERA-Interim respectively indicating that ERA-Interim recorded the lowest amount of rainfall in the catchment for the period under study.

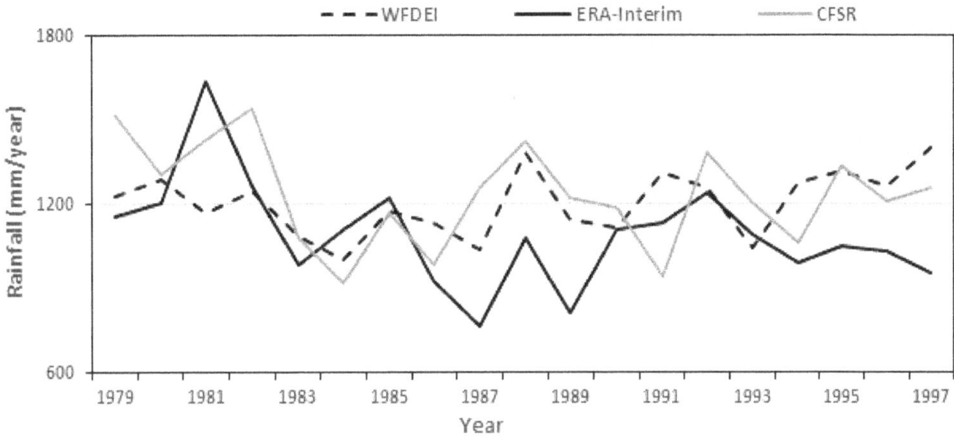

Figure 2. Reanalysis annual rainfall variability in the Logone catchment.

Results of model calibration are shown in Table 2 for daily and monthly time steps respectively. It can be observed from the table that only WFDEI dataset produced NSE and R^2 values considered to be satisfactory according to Moriasi et al. [30] model evaluation criteria at both time steps for most hydrometric stations. CFSR and ERA-Interim both produced unsatisfactory results because most NSE values fall below the minimum threshold although the performance of the former was better compared to the latter. Generally, it was observed that there was a considerable improvement in NSE values at the monthly time step compared to daily for all datasets. For example, NSE values for WFDEI data improved from a range of 0.05–0.66 to 0.43–0.77 while that of CFSR improved from a range of −0.67–0.43 to −0.43–0.59. Despite a general under performance compared to WFDEI, CFSR registered negative NSE values at both time steps only at one hydrometric station (Logone Gana).

The PBIAS values obtained also showed that only WFDEI was able to produce values that fall within the acceptable limits while results from the other two datasets show a consistent over estimation of annual discharge throughout the simulation period at all hydrometric stations.

Results further show that all the datasets were able to replicate the streamflow seasonal cycle at all hydrometric stations. This follows the finding of Nkiaka et al. [21] who showed that CFSR and ERA-Interim precipitation estimates could replicate the seasonal cycle of rainfall in the catchment. However, from the streamflow hydrographs shown in Figures 3–6 it can be observed that WFDEI and CFSR were able to simulate low flows (baseflow) throughout the period under study while ERA-Interim overestimated low flows in most years during the same period. Apart from a few cases of overestimation, the WFDEI dataset was able to simulate peak discharges at Logone Gana hydrometric station but consistently underestimated at other stations. Although there were a few cases of overestimation and a general underperformance compared to WFDEI; CFSR was able to simulate peak flows at most hydrometric stations compared to WFDEI and ERA-Interim (Figure 5). Only daily streamflow hydrographs for WFDEI are shown herein. Comparing the results of the other two reanalysis products showed that CFSR outperformed ERA-Interim. ERA-Interim consistently underestimated streamflow in 1987, 1989 and from 1994–1997 (Figure 6). This underestimation of discharge by ERA-Interim follows the general underestimation of average rainfall in the catchment during these years.

From Table 2 and Figures 3–6, the *p-factor* values obtained indicate that more than 60% of observed streamflow values at all the hydrometric stations were bracketed within the 95PPU band at both time steps for all the datasets although CFSR outperformed WFDEI and ERA-Interim at daily time step. At the monthly time step, WFDEI outperformed the other two datasets with more than 80% of observed streamflow bracketed within the 95PPU band. Nevertheless, *r-factor* values obtained for CFSR and ERA-Interim as shown by Figures 5 and 6 and Table 2 indicate that the uncertainty band for these datasets was much wider compared to that of WFDEI. This suggest that streamflow simulated using WFDEI dataset had the lowest level of uncertainty followed by CFSR while ERA-Interim produced the highest uncertainty.

Regarding the impact of spatial resolution of reanalysis datasets on streamflow simulation, results showed that WFDEI which has a lower spatial resolution (0.5°) compared to CFSR (0.312°) performed better than the latter in streamflow simulation given that the calibration results produced by the WFDEI are better than those of CFSR. Furthermore, the PBIAS values showed that CFSR with fine resolution compared to WFDEI consistently overestimated simulated streamflow during the period under study.

Analysis of water balance components showed that 74%, 65% and 58% of total rainfall received in the catchment was lost through evapotranspiration for WFDEI, CFSR and ERA-Interim respectively. Compared to the amount of rainfall received in the catchment, the evapotranspiration estimates from WFDEI compare well with the results of [32,33] obtained in the Ouemé river basin which is located in the same latitudinal zone with the Logone catchment.

4. Discussion

4.1. Selection of Grid Points

During the selection of grid points used as meteorological stations input in SWAT, the model selects each grid point depending on its proximity to the centroid of the sub catchment [27]. When low resolution reanalysis data is used to drive SWAT, the possibility of the model locating a grid point in each sub catchment may be reduced. This explains why many grid points (57) were selected for CFSR because of its high spatial resolution which is almost two times that of WFDEI with (34) grid points and three times that of ERA Interim (19) grid points. Even so, not every sub catchment had a different grid point because only 57 grid points were selected for CFSR instead of 66 to correspond to the number of sub-catchments in the catchment.

4.2. Model Evaluation

Results of model evaluation indices showed that WFDEI had the best performance among the three datasets. This is not surprising given that WFDEI had the best rainfall input among three datasets evaluated because of reduced variability in rainfall estimates. This demonstrates the importance of post-processing or bias correcting global reanalysis datasets before using them for hydrological modelling. The post-processing reduces the uncertainty in the rainfall data thus leading to better streamflow simulation. We therefore conclude that WFDEI outperformed the other two datasets (CFSR and ERA-Interim) in simulating streamflow in the Logone catchment due to reduced uncertainty in the rainfall estimates from this dataset as shown in Figure 2. These results are similar to those obtained by [18,19] who reported that WFDEI improved streamflow simulation compared to other global reanalysis datasets in their respective study areas. Meanwhile, [19] asserted that CFSR outperformed ERA-Interim in streamflow simulation in the Patagonia basin in South America.

Table 2. Results of model calibration at daily and monthly time steps.

Time Step	Evaluation Criteria	WFDEI				CFSR				ERA Interim			
		Gana	Katoa	Bongor	Lai	Gana	Katoa	Bongor	Lai	Gana	Katoa	Bongor	Lai
Daily	NSE	0.05	0.58	0.66	0.57	−0.67	0.17	0.43	0.31	−3.97	−1.54	−0.59	−0.56
	R2	0.64	0.68	0.68	0.6	0.65	0.62	0.57	0.51	0.47	0.44	0.38	0.31
	PBIAS (%)	−15.2	2.7	16.6	22.7	−74.5	−51.7	−32.3	−42.0	−146.1	−109.6	−81	−78.7
	p-factor	0.61	0.64	0.6	0.68	0.78	0.80	0.81	0.78	0.63	0.65	0.66	0.62
	r-factor	1.69	1.3	1.02	0.89	2.47	1.87	1.48	1.46	3.78	2.58	2.01	1.73
Monthly	NSE	0.43	0.75	0.77	0.67	−0.28	0.39	0.59	0.49	−3.12	−1.17	−0.38	−0.31
	R2	0.73	0.77	0.8	0.73	0.74	0.71	0.68	0.61	0.52	0.48	0.44	0.37
	PBIAS (%)	−16.2	3.5	17.7	23.6	−66.9	−45.4	−26.8	−36.6	−163.3	−125.5	−94.8	−91.4
	p-factor	0.86	0.88	0.81	0.83	0.68	0.73	0.78	0.74	0.64	0.66	0.67	0.63
	r-factor	1.65	1.26	1	0.87	2.04	1.55	1.25	1.23	3.41	2.6	2.09	1.86

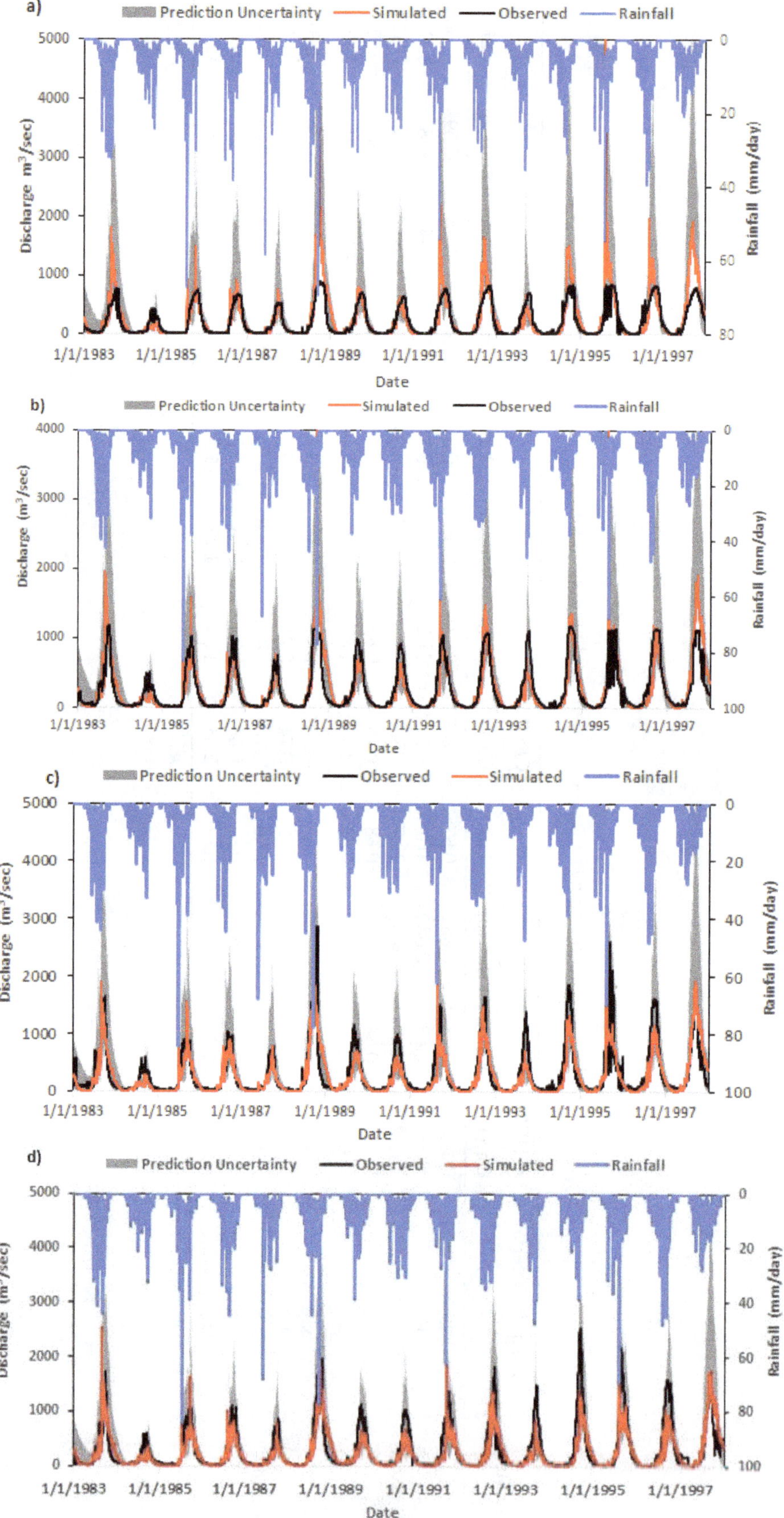

Figure 3. WATCH Forcing Data methodology applied to ERA-Interim (WFDEI) daily hydrographs for observed and simulated flows at (**a**) Logone Gana; (**b**) Katoa and (**c**) Bongor (**d**) Lai.

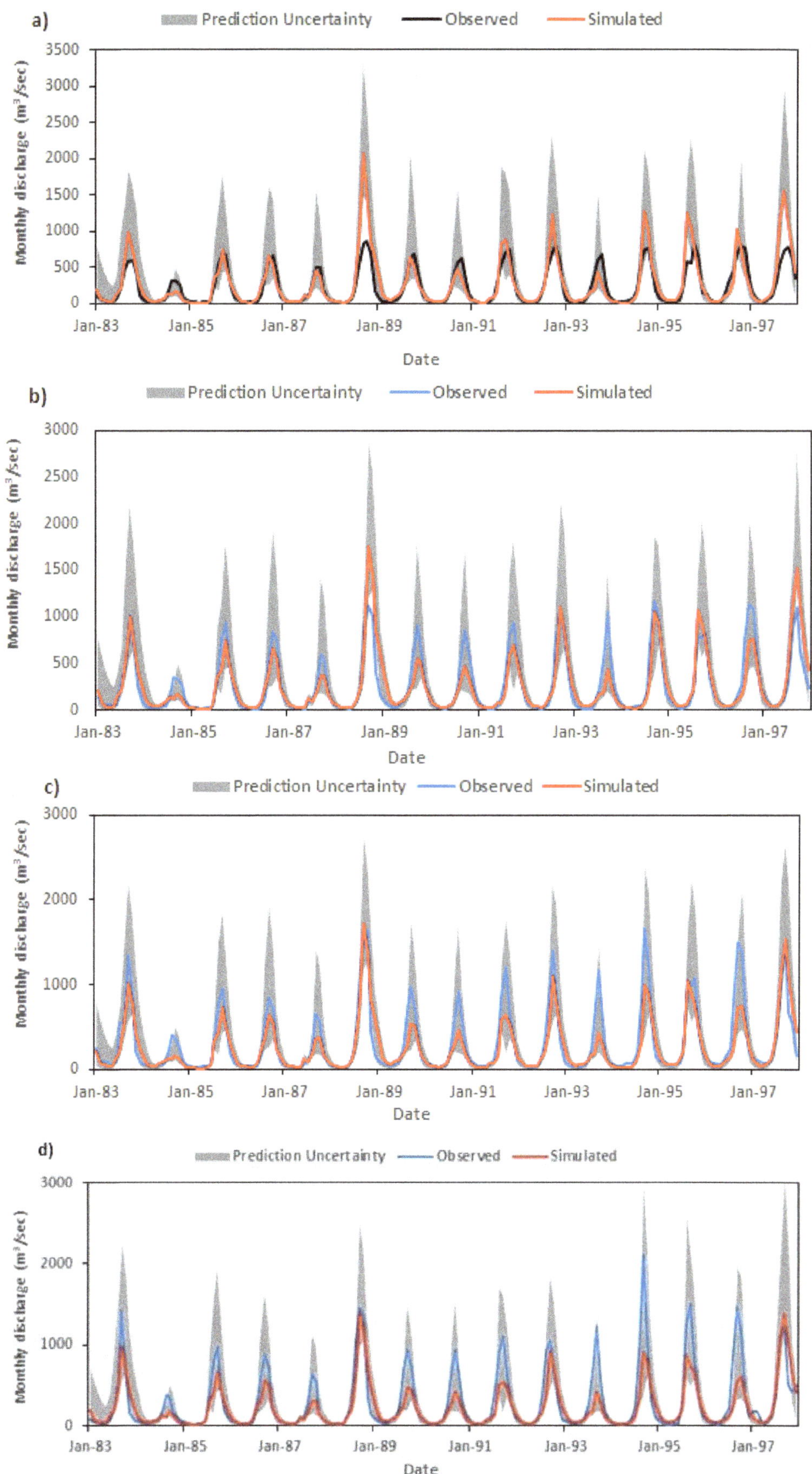

Figure 4. WFDEI monthly hydrographs for observed and simulated flows at (**a**) Logone Gana; (**b**) Katoa and (**c**) Bongor (**d**) Lai.

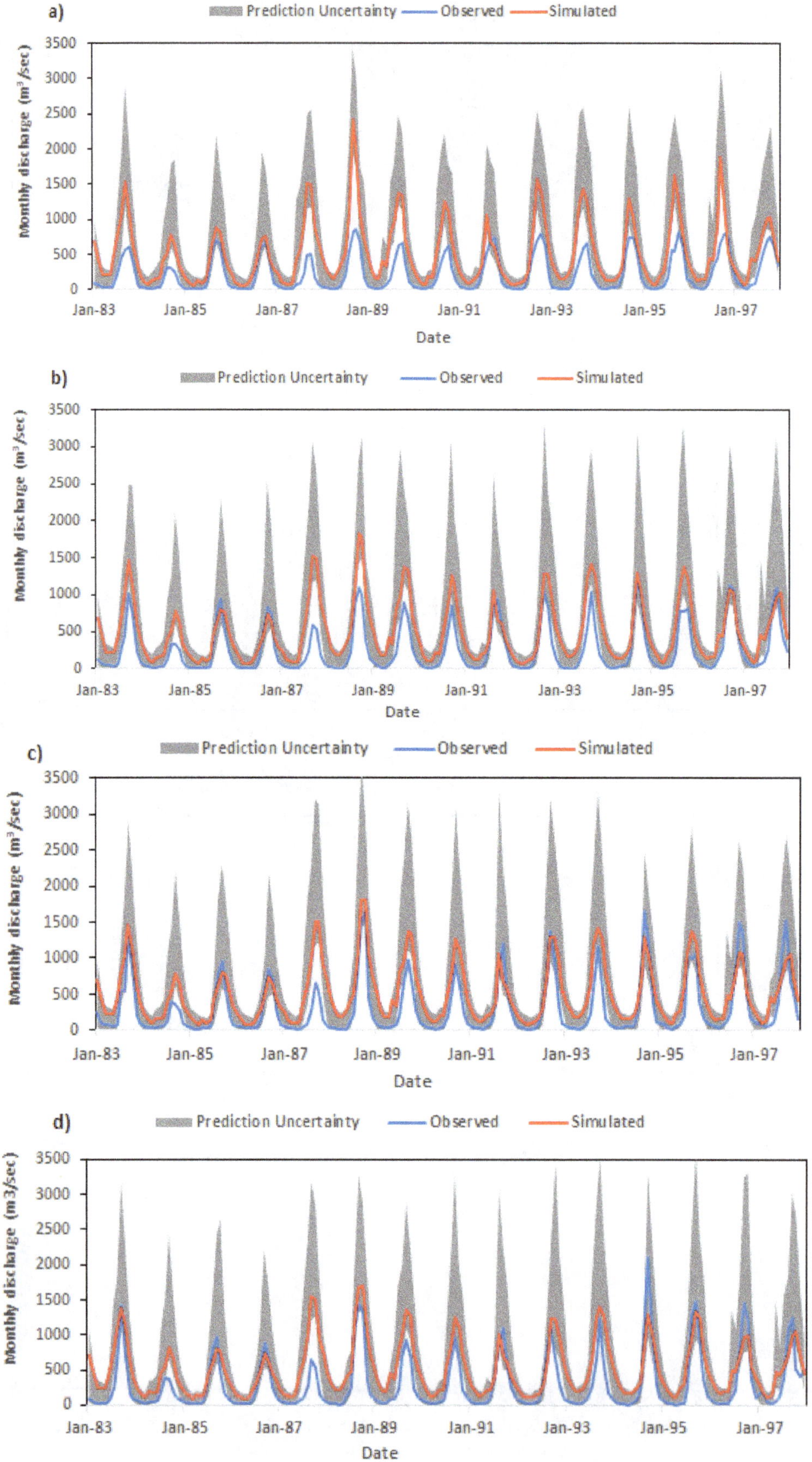

Figure 5. Climate Forecasting System Reanalysis (CFSR) monthly hydrographs for observed and simulated flows at (**a**) Logone Gana; (**b**) Katoa and (**c**) Bongor (**d**) Lai.

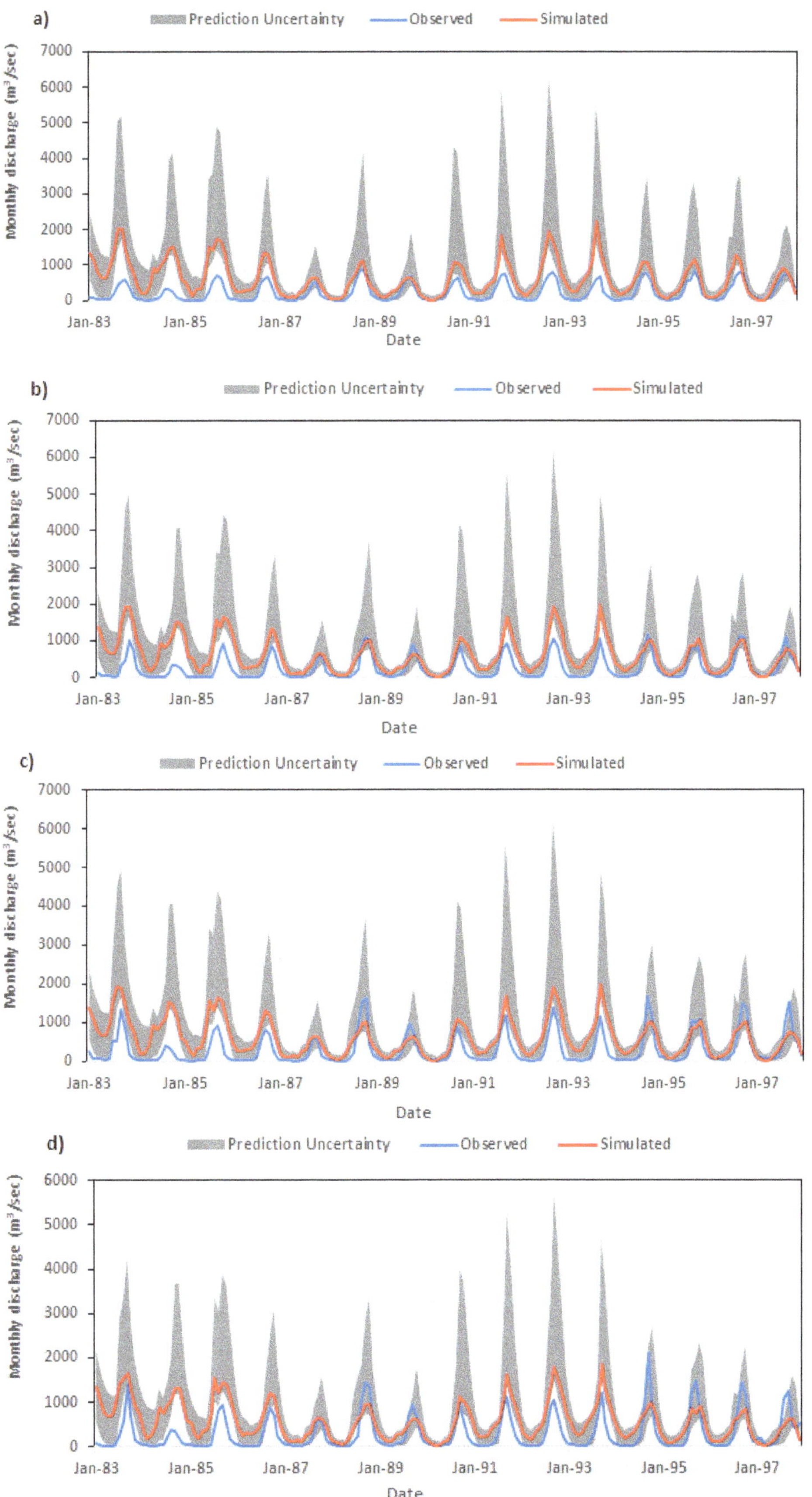

Figure 6. European Center for Medium-Range Weather Forecasts (ECMWF) ERA-Interim monthly hydrographs for observed and simulated flows at (**a**) Logone Gana; (**b**) Katoa and (**c**) Bongor (**d**) Lai.

Given that low flows and peak discharges were adequately simulated by WFDEI and CFSR datasets indicate that the parameter range(s) used to calibrate the model(s) can be considered to be satisfactory and cases of streamflow under/overestimation may be attributed to the uncertainty in the rainfall input used in calibrating the models or to parameter conditionality. This is because the same parameter set was used to calibrate the model in both semi-arid and Sudano areas; although the amount of rainfall received by each zone is different which has implications for parameter values used in calibrating the model.

The poor performance of ERA-Interim can be attributed to the high variability in annual rainfall produced by this dataset compared to the other two datasets. For example, Figure 2 shows that annual rainfall produced by ERA-Interim was consistently lower compared to the other two datasets in 1987, 1989 and 1994–1997 leading to a systematic underestimation of simulated streamflow by ERA-Interim during these years. This high variability in ERA-Interim rainfall estimates may have offset the interaction among the different model parameters making it difficult to find a parameter range that could simulate streamflow above the minimum threshold limit. This suggest that rainfall input plays a significant role in model calibration because it has the potential to influence calibrated parameters as reported by [29]. Nevertheless, the significant variability in CFSR and ERA-Interim datasets in this study follow the findings of [21] in the Logone catchment.

4.3. Prediction Uncertainty

The daily streamflow hydrographs with corresponding prediction uncertainty band and rainfall input shown in Figure 3 indicate that, as the variability in rainfall input increases, the values of *r-factor,* which measures the prediction uncertainty band increases as well. This indicates that rainfall input contributes significantly to increase the level of uncertainty in the simulated streamflow because as the variability in rainfall increases, the uncertainty band also increases. This suggests that, reducing the variability in the rainfall input or accurately estimating the rainfall data used for driving the model could lead to a significant improvement in simulated streamflow thereby reducing the level of uncertainty in the latter. This follows the findings of [13] who asserted that performance of reanalysis datasets in hydrological modelling depends largely on the quality of the rainfall data.

The improved performance of model evaluation indices and improvement in model uncertainty at monthly time steps compared to daily can be attributed to the fact that, monthly rainfall is a cumulative measurement in which all the daily variability within the month is summed, thus reducing the variability in the input data which leads to an overall improvement in model performance.

Nevertheless, it is worth noting that the contribution of model parameters to the overall uncertainty in the simulated streamflow cannot be overlooked since it is difficult to decouple the uncertainty inherent in model parameters from that of input data [9]. It should be noted that assessing the uncertainty of model parameters was not part of this study.

The high variability in rainfall estimates from the reanalysis datasets in the study area can be attributed to the low rain gauge density and few radiosonde coverages in Central Africa [11,34] used for optimization and data-assimilation in the reanalysis forecast models thus increasing the forcing uncertainty. According to Sperna Weiland [35], when using global reanalysis datasets for hydrological modelling, the forcing uncertainty decreases with increasing number of sampling points available for optimization and data-assimilation, suggesting that a limited number of sampling points will lead to an increased level of forcing uncertainty. Although WFDEI has fewer grid points compared to CFSR, the improvement of WFDEI rainfall estimates compared to CFSR can be attributed to the fact that this dataset is bias corrected.

4.4. Effects of Spatial Resolution

The results obtained from this study also suggest that streamflow simulation may not represent an important factor to be used for evaluating the impact of spatial resolution of reanalysis datasets as long as the average rainfall over the modelled catchment is accurately estimated. This is because in moderate size catchments such as the Logone, the model integrates rainfall data from a very large area which dampens and smooths the impact of rainfall spatial variability on the catchment outflow, hence limiting the effect of spatial resolution on streamflow. Under such circumstances, the ability of the reanalysis product to produce accurate precipitation estimates in the catchment is more important than its spatial resolution. Gascon et al. [36] also demonstrated that the spatial resolution of rainfall datasets had no significant impact on streamflow simulation in the Ouémé basin. These authors used rainfall datasets at spatial resolutions of 0.05°, 0.1°, 0.25° and 0.5°. Results from this study also corroborate the findings of Fu et al. [37], where the authors demonstrated that, the impact of rainfall spatial resolution was insignificant for catchment sizes above 250 km^2 and negligible for catchments larger than 1000 km^2. A recent study, [15] demonstrated that, the impact of spatial resolution on flow simulation was scale, catchment and event characteristic-dependent. We conclude that, the ability of the reanalysis dataset(s) to accurately produce good quality rainfall estimates in the study area can significantly improve streamflow simulation compared to the spatial resolution of the rainfall. Nevertheless, Tramblay et al. [38] have shown that spatial rainfall representation is important in the simulation of flood events.

4.5. Simulation of Evapotranspiration

The values of actual evapotranspiration estimates obtained showed that there were significant differences in the values produced by the threedatasets. These differences in actual evapotranspiration values from the different datasets can be attributed to the different amounts of rainfall input used in simulating each model. WFDEI and CFSR produced almost the same amount of average rainfall during the simulation period but there is a significant discrepancy between the actual evapotranspiration values from the two datasets. This discrepancy can be attributed to the uncertainty inherent in each of the r datasets. This is because precipitation estimates can strongly influence the parameter values that control the rates and threshold of hydrological processes taking place in the catchment [39]. Furthermore, as pointed out by Remesan and Holman [39] our results show that the uncertainty in the rainfall estimates is conserved and propagated into streamflow and other water balance components including evapotranspiration. Although ground data was not available to compare the evapotranspiration estimates in this study, the estimates from WFDEI are acceptable given that similar values have been obtained in other catchments the region [32,33]. Furthermore, the importance of temperature in influencing actual evapotranspiration cannot be overstated implying that the minimum and maximum temperature estimates used in the simulations could also interact to strongly influence the results obtained.

We conclude that the estimation of actual evapotranspiration and other water balance components by the model is influenced by the precipitation estimates and other input data used in driving the model.

5. Conclusions

The objectives of this study were to evaluate the ability of two global reanalysis datasets; CFSR and ERA-Interim and one bias corrected global meteorological forcing dataset WFDEI to be used as input to drive the SWAT model in the Logone catchment, and to evaluate the impact of reanalysis spatial resolution on the quality of simulated flows.

The results of our study showed that the WFDEI out-performed the other two datasets in simulating streamflow in the study area. This highlights the importance of bias correcting global reanalysis datasets before using them for hydrological modelling. As seen in the hydrographs of WFDEI, the bias correction reduces the uncertainty in precipitation estimates used to drive the hydrological model which has a direct positive impact in reducing the overall uncertainty in the simulated streamflow.

The results obtained also showed that the ability of reanalysis forecast models to produce accurate precipitation estimates is more important than its spatial resolution. This is because accurate streamflow simulation and hydrological modelling in general depend on accurate rainfall input given that the impacts of spatial resolution on streamflow simulation are not significant in medium to large catchments.

From the result of evapotranspiration estimates obtained, we conclude that the estimation of actual evapotranspiration depends on the input data used in driving the model. This is because rainfall, temperature and other variables play a significant role in influencing model parameters that interact to control hydrological processes, e.g., evapotranspiration at catchment scales.

Finally, we conclude that in the absence of gauged hydro-meteorological data, WFDEI and CFSR could be used for hydrological modelling in data-scarce areas such as the Sudano-Sahel and other remote locations with poor data availability.

This study is part of an on-going research aimed at understanding the hydrological dynamics of the Logone catchment with the aim of improving water resources management. Future research in the catchment will use the WFDEI dataset, which has been shown to out-perform the other two datasets in this study, for detailed hydrological analysis of the catchment to determine the main processes and feedback mechanisms driving the response of the catchment to natural and anthropogenic changes.

Acknowledgments: This study was supported through a Commonwealth Scholarship award to the first author. The authors are indebted to Lund University, IIASA, LCBC, Jet Propulsion Laboratory, NCEP/NCAR, ESA and FAO for providing the data used in the study.

Author Contributions: Elias Nkiaka conducted this research as a part of his Ph.D. thesis. Elias Nkiaka conceived and designed the research; prepared the data, performed the simulations; interpreted the results and wrote the paper. N.R. Nawaz and Jon C. Lovett supervised the work and helped in improving the manuscript.

Conflicts of Interest: The authors declare no conflict of interest.

References

1. Van de Giesen, N.; Hut, R.; Selker, J. The trans-African hydro-meteorological observatory (TAHMO). *WIREs Water.* **2014**, *1*, 341–348. [CrossRef]
2. Buma, W.G.; Lee, S.; Seo, J.Y. Hydrological evaluation of Lake Chad basin using space borne and hydrological model observations. *Water* **2016**, *8*, 205. [CrossRef]
3. Buytaert, W.; Friesen, J.; Liebe, J.; Ludwig, R. Assessment and management of water resources in developing, semi-arid and arid regions. *Water Resour. Manag.* **2012**, *26*, 841–844. [CrossRef]
4. Gorgoglione, A.; Gioia, A.; Iacobellis, V.; Piccinni, A.F.; Ranieri, E. A rationale for pollutograph evaluation in ungauged areas, using daily rainfall patterns: Case studies of the Apulian region in Southern Italy. *Appl. Environ. Soil Sci.* **2016**, *2016*, 9327614. [CrossRef]
5. Liu, Y.; Gupta, H.; Springer, E.; Wagener, T. Linking science with environmental decision making: Experiences from an integrated modeling approach to supporting sustainable water resources management. *Environ. Model. Softw.* **2008**, *23*, 846–858. [CrossRef]
6. Worqlul, A.W.; Maathuis, B.; Adem, A.A.; Demissie, S.S.; Langan, S.; Steenhuis, T.S. Comparison of rainfall estimations by TRMM 3B42, MPEG and CFSR with ground-observed data for the Lake Tana basin in Ethiopia. *Hydrol. Earth Syst. Sci.* **2014**, *18*, 4871–4881. [CrossRef]
7. Fuka, D.R.; MacAllister, C.A.; Degaetano, A.T.; Easton, Z.M. Using the Climate Forecast System Reanalysis dataset to improve weather input data for watershed models. *Hydrol. Process.* **2013**, *28*, 5613–5623. [CrossRef]
8. Skinner, C.J.; Bellerby, T.J.; Greatrex, H.; Grimes, D.I.F. Hydrological modelling using ensemble satellite rainfall estimates in a sparsely gauged river basin: The need for whole ensemble calibration. *J. Hydrol.* **2015**, *522*, 110–122. [CrossRef]
9. Saha, S.; Moorthi, S.; Wu, X.; Wang, J.; Nadiga, S.; Tripp, P.; Behringer, D.; Hou, Y.-T.; Chuang, H.-Y.; Iredell, M. The NCEP climate forecast system version 2. *J. Clim.* **2014**, *27*, 2185–2208. [CrossRef]
10. Dee, D.P.; Uppala, S.M.; Simmons, A.J.; Berrisford, P.; Poli, P.; Kobayashi, S.; Andrae, U.; Balmaseda, M.A.; Balsamo, G.; Bauer, P.; et al. The ERA-Interim reanalysis: Configuration and performance of the data assimilation system. *Q. J. R. Meteorol. Soc.* **2011**, *137*, 553–597. [CrossRef]

11. Rienecker, M.M.; Suarez, M.J.; Gelaro, R.; Todling, R.; Bacmeister, J.; Liu, E.; Bosilovich, M.G.; Schubert, S.D.; Takacs, L.; Kim, G.-K.; et al. MERRA: NASA's modern-era retrospective analysis for research and applications. *J. Clim.* **2011**, *24*, 3624–3648. [CrossRef]
12. Lorenz, C.; Kunstmann, H. The hydrological cycle in three state-of-the-art reanalyses: Intercomparison and performance analysis. *J. Hydrometeorol.* **2012**, *13*, 1397–1420. [CrossRef]
13. Essou, G.R.C.; Sabarly, F.; Lucas-Picher, P.; Brissette, F.; Poulin, A. Can precipitation and temperature from meteorological reanalyses be used for hydrological modelling? *J. Hydrometeorol.* **2016**, *17*, 1929–1950. [CrossRef]
14. Weedon, G.P.; Balsamo, G.; Bellouin, N.; Gomes, S.; Best, M.J.; Viterbo, P. The WFDEI meteorological forcing data set: WATCH forcing data methodology applied to ERA-Interim reanalysis data. *Water Resour. Res.* **2014**, *50*, 7505–7514. [CrossRef]
15. Zhao, F.; Zhang, L.; Chiew, F.H.S.; Vaze, J.; Cheng, L. The effect of spatial rainfall variability on water balance modelling for south-eastern Australian catchments. *J. Hydrol.* **2013**, *493*, 16–29. [CrossRef]
16. Lobligeois, F.; Andréassian, V.; Perrin, C.; Tabary, P.; Loumagne, C. When does higher spatial resolution rainfall information improve streamflow simulation? An evaluation using 3620 flood events. *Hydrol. Earth Syst. Sci.* **2014**, *18*, 575–594. [CrossRef]
17. Monteiro, J.A.F.; Strauch, M.; Srinivasan, R.; Abbaspour, K.; Gücker, B. Accuracy of grid precipitation data for Brazil: Application in river discharge modelling of the Tocantins catchment. *Hydrol. Process.* **2016**, *30*. [CrossRef]
18. Andersson, J.C.M.; Pechlivanidis, I.G.; Gustafsson, D.; Donnelly, C.; Arheimer, B. Key factors for improving large-scale hydrological model performance. *Eur. Water.* **2015**, *49*, 77–88.
19. Krogh, S.A.; Pomeroy, J.W.; McPhee, J. Physically based mountain hydrological modelling using reanalysis data in Patagonia. *J. Hydrometeorol.* **2015**, *16*, 172–193. [CrossRef]
20. Nkiaka, E.; Rizwan, N.R.; Lovett, J.C. Analysis of rainfall variability in the Logone catchment, Lake Chad basin. *Int. J. Climatol.* **2016**. [CrossRef]
21. Nkiaka, E.; Rizwan, N.R.; Lovett, J.C. Evaluating global reanalysis precipitation datasets with rain gauge measurements in the Sudano-Sahel region: Case study of the Logone catchment, Lake Chad Basin. *Meteorol. Appl.* **2016**, *24*, 9–18. [CrossRef]
22. Siam, M.S.; Demory, M.; Eltahir, E.A.B. Hydrological cycles over the Congo and upper Blue Nile basins: Evaluation of general circulation model simulations and reanalysis products. *J. Clim.* **2013**, *26*, 8881–8894. [CrossRef]
23. Loth, P. *The Return of the Water: Restoring the Waza Logone Floodplain in Cameroon*; IUCN: Cambridge, UK, 2004.
24. Nkiaka, E.; Rizwan, N.R.; Lovett, J.C. Using self-organizing maps to infill missing data in hydro-meteorological time series from the logone catchment, Lake Chad basin. *Environ. Monit. Assess.* **2016**, *188*, 400. [CrossRef] [PubMed]
25. Morse, A.; Caminade, C.; Tompkins, A.; McIntyre, K.M. *The QWeCI Project (Quantifying Weather and Climate Impacts on Health in Developing Countries)*; Final Report; Liverpool: University of Liverpool, UK, 2013.
26. Gassman, P.W.; Sadegh, I.A.M.; Srinivasan, R. Applications of the SWAT model special section: Overview and insights. *J. Environ. Qual.* **2014**, *43*, 1–8. [CrossRef] [PubMed]
27. Neitsch, S.L.; Arnold, J.G.; Kiniry, J.R.; Williams, J.R. *Soil and Water Assessment Tool Theoretical Documentation Version 2009*; Technical Report 406; Texas Water Resources Institute: Temple, TX, USA, 2011.
28. Abbaspour, K.C. *SWAT-CUP: SWAT Calibration and Uncertainty Programs: A User Manual*; Swiss Federal Institute of Aquatic Science and Technology: Eawag, Switzerland, 2015.
29. Pathiraja, S.; Marshall, L.; Sharma, A.; Moradkhani, H. Hydrologic modeling in dynamic catchments: A data assimilation approach. *Water Resour. Res.* **2016**, *52*, 3350–3372. [CrossRef]
30. Moriasi, D.N.; Arnold, J.G.; van Liew, M.W.; Bingner, R.L.; Harmel, R.D.; Veith, T.L. Model evaluation guidelines for systematic quantification of accuracy in watershed simulations. *Trans. Asabe* **2007**, *50*, 885–900. [CrossRef]
31. Auerbach, D.A.; Easton, Z.M.; Walter, M.T.; Flecker, A.S.; Fuka, D.R. Evaluating weather observations and the climate forecast system reanalysis as inputs for hydrologic modelling in the tropics. *Hydrol. Process.* **2016**, *30*, 3466–3477. [CrossRef]

32. Sintondji, L.O.; Zokpodo, B.; Ahouansou, D.M.; Vissin, W.E.; Agbossou, K.E. Modelling the water balance of Ouémé catchment at the Savè outlet in Benin: Contribution to the sustainable water resource management. *Int. J. Agric. Sci.* **2014**, *4*, 74–88.
33. Giertz, S.; Diekkrugerm, B.; Jaeger, A.; Schopp, M. An interdisciplinary scenario analysis to assess the water availability and water consumption in the upper Oueme catchment in Benin. *Adv. Geosci.* **2006**, *9*, 3–13. [CrossRef]
34. Maidment, R.I.; Grimes, D.I.F.; Allan, R.P.; Greatrex, H.; Rojas, O.; Leo, O. Evaluation of satellite-based and model re-analysis rainfall estimates for Uganda. *Meteorol. Appl.* **2013**, *20*, 308–317. [CrossRef]
35. Sperna Weiland, F.C.; Vrugt, J.A.; van Beek, R.P.H.; Weerts, A.H.; Bierkens, M.F.P. Significant uncertainty in global scale hydrological modeling from precipitation data errors. *J. Hydrol.* **2015**, *529*, 1095–1115. [CrossRef]
36. Gascon, T.; Vischel, T.; Lebel, T.; Quantin, G.; Pellarin, T.; Quatela, V.; Leroux, D.; Galle, S. Influence of rainfall space-time variability over the Ouémé basin in Benin. *Proc. Int. Assoc. Hydrol. Sci.* **2015**, *368*, 102–107. [CrossRef]
37. Fu, S.; Sonnenborg, T.O.; Jensen, K.H.; He, X. Impact of precipitation spatial resolution on the hydrological response of an integrated distributed water resources model. *Vadose Zone J.* **2011**, *10*, 25–36. [CrossRef]
38. Tramblay, Y.; Bouvier, C.; Ayral, P.-A.; Marchandise, A. Impact of rainfall spatial distribution on rainfall-runoff modelling efficiency and initial soil moisture conditions estimation. *Nat. Hazards Earth Syst. Sci.* **2011**, *11*, 157–170. [CrossRef]
39. Remesan, R.; Holman, I.P. Effect of baseline meteorological data selection on hydrological modelling of climate change scenarios. *J. Hydrol.* **2015**, *528*, 631–642. [CrossRef]

10

Catchment Hydrology during Winter and Spring and the Link to Soil Erosion

Torsten Starkloff [1,*], Rudi Hessel [2], Jannes Stolte [1] and Coen Ritsema [2]

1 Norwegian Institute of Bioeconomy Research (NIBIO), Fredrik A Dahlsvei 20, N-1431 Ås, Norway; Jannes.Stolte@nibio.no
2 Wageningen Environmental Research (Alterra), P.O. Box 47, 6700 AA Wageningen, The Netherlands; rudi.hessel@wur.nl (R.H.); Coen.Ritsema@wur.nl (C.R.)
* Correspondence: torsten.starkloff@nibio.no

Abstract: In the Nordic countries, soil erosion rates in winter and early spring can exceed those at other times of the year. In particular, snowmelt, combined with rain and soil frost, leads to severe soil erosion, even, e.g., in low risk areas in Norway. In southern Norway, previous attempts to predict soil erosion during winter and spring have not been very accurate owing to a lack of catchment-based data, resulting in a poor understanding of hydrological processes during winter. Therefore, a field study was carried out over three consecutive winters (2013, 2014 and 2015) to gather relevant data. In parallel, the development of the snow cover, soil temperature and ice content during these three winters was simulated with the Simultaneous Heat and Water (SHAW) model for two different soils (sand, clay). The field observations carried out in winter revealed high complexity and diversity in the hydrological processes occurring in the catchment. Major soil erosion was caused by a small rain event on frozen ground before snow cover was established, while snowmelt played no significant role in terms of soil erosion in the study period. Four factors that determine the extent of runoff and erosion were of particular importance: (1) soil water content at freezing; (2) whether soil is frozen or unfrozen at a particular moment; (3) the state of the snow pack; and (4) tillage practices prior to winter. SHAW performed well in this application and proved that it is a valuable tool for investigating and simulating snow cover development, soil temperature and extent of freezing in soil profiles.

Keywords: SHAW; soil freezing; snow; infiltration; modelling; soil erosion

---

1. Introduction

In the Nordic countries, soil erosion rates in winter and early spring can exceed those occurring during other seasons of the year. A factor of particular importance is the incidence of frozen soil, which modifies surface runoff generation and also the erosivity of the soil material [1]. In addition, water infiltration into frozen soils is more complicated than water infiltration into unfrozen soils, because it involves coupling water and heat transport (temperature of the infiltrating water) with phase change (from liquid to ice and vice versa) [2].

A large number of laboratory studies has investigated different processes occurring in soils during freezing and thawing. Using a rain simulator, Edwards and Burney [3] showed that soil freezing and thawing can significantly increase soil erosivity. They concluded that only plant cover is effective in reducing soil losses due to rain and overland flow on frozen ground. Other more recent studies, e.g., Ban et al. [4], have shown that water flows much faster over a frozen slope than over a thawed slope. Watanabe et al. [5] found that the speed of snowmelt and/or rain infiltration into frozen soils is largely dependent on initial water content, frost depth and temperature of the soil. In addition,

Yami et al. [6] showed that increasing soil moisture and finer soil structure advance the speed and depth of the freezing front.

Al-Houri et al. [7] added to knowledge about water transport in frozen soils by showing that the amount of time available for soil water redistribution before freezing affects the infiltration capacity of frozen soil, with more time resulting in better infiltration capacity under frozen conditions. In addition to laboratory studies, a great number of studies has investigated infiltration processes under field conditions at the plot or point scale. In a study examining nine different plots in North Dakota, Willis et al. [8] showed that soils that were dry in autumn freeze faster and deeper than wet soils and that a dry profile thaws upward, while a wet soil thaws both upward and downward. They also recorded less runoff from dry soils.

Stähli et al. [9] and Nyberg et al. [10] concluded that frost has little effect on runoff from forest soils, in contrast to reported effects on agricultural soils [11], and that forests probably do not increase runoff episodes in winter and spring. Furthermore, they predicted that critical initial conditions, such as high water content and early frost penetration combined with heavy rain on still frozen soil, could have a decisive effect on the amount of runoff. Iwata et al. [12,13] showed that a frozen soil layer can significantly impede snowmelt infiltration and thus increase runoff of spring snowmelt water. Zhao et al. [13] demonstrated that soil freezing can reduce hydraulic conductivity by blocking pores and retaining water in the profile, thereby reducing the infiltration capacity during snowmelt.

In a study at five locations in Finland, Sutinen et al. [14] found that a snow pack with a thickness exceeding 30 cm can reduce or even prevent soil freezing. In addition, Zhao et al. [15] showed that snow pack less than 20 cm deep can cause deeper soil freezing than no snow cover, due to an increase in ground albedo caused by the snow. The effects of different tillage practices on soil freezing was investigated by Parkin et al. [16] in a soil profile over 10 years. They found that conventional autumn tillage resulted in lower soil temperatures than no-till.

In a recent five-year field study in Canada, He et al. [17] presented results that confirmed many of the above-mentioned effects and demonstrated the complexity of the interaction between the topsoil and snow cover, especially during snowmelt. They also pointed out that only a limited amount of field studies to date has taken all of these processes into account in a series of measurements that covers several winter periods. The complexity of the different processes occurring in soil during winter is amplified when all of these interacting processes have to be monitored and interpreted at the catchment scale, where different soil types, terrain and water flow at the surface and in the soils interact [18,19]. In addition, detailed observations have to be made over several years to identify processes that can only occur during certain conditions or are masked by confounding factors in a catchment [20].

In Norway, the incidence of soil erosion from agricultural land is greatest during spring [21,22], and the severity of the erosion is often amplified by preceding winter conditions. Snowmelt, combined with rain and soil frost, can lead to severe gully and rill erosion, even in low risk areas in Norway [23]. In southern Norway, previous attempts to predict soil erosion during winter and spring have not been very accurate [24], probably owing to a lack of catchment-based studies covering several winters, resulting in a lack of knowledge about the interacting processes described above. In the present study, field measurements covering three winter periods (2013, 2014 and 2015) were carried out, with the aim of improving overall understanding of how soil hydraulic properties behave during winter and affect surface runoff caused by snowmelt and rain and how these processes are linked to soil erosion. Furthermore, the data collected were used to calibrate and validate a hydrodynamic model (Simultaneous Heat and Water (SHAW) model; [25], in order to acquire better insights into the complex interactions between freezing, thawing, snowpack and runoff and erosion dynamics.

The focus of this study lies in Norway; however, severe soil erosion on agricultural areas during winter and spring is a problem in many other countries around the world (e.g., USA [26], Belgium [27], the U.K. [28], Germany [29], Russia [30]). In these areas, like in Norway, soil erosion during winter and spring depletes the irreplaceable nutrient-rich top layer of agricultural soils and results in a major part of the annual input of phosphorous and nitrogen from agricultural catchments to fresh water bodies [31]. It is therefore hoped that this study will also contribute to the understanding of winter processes outside of Norway.

2. Methodology

2.1. Study Area

The study area is located in the Skuterud catchment (4.5 km^2) in Ås and Ski municipalities, approximately 30 km south of Oslo, Norway. For the field investigations and sampling, a sub-catchment, Gryteland (0.29 km^2), in the southeastern part of the main catchment (Figure 1), was chosen. This area has been used for different hydrological studies in the past, and it can easily be reached under all weather conditions. A monitoring station was installed at the outlet of the sub-catchment in 2008. This station measures precipitation, air temperature, surface runoff and drainage discharge. In addition, five stations (one at the outlet) were installed along a transect in the catchment [32] (Figure 1), in order to measure soil moisture and soil temperature at three depths (5, 10 and 20 cm).

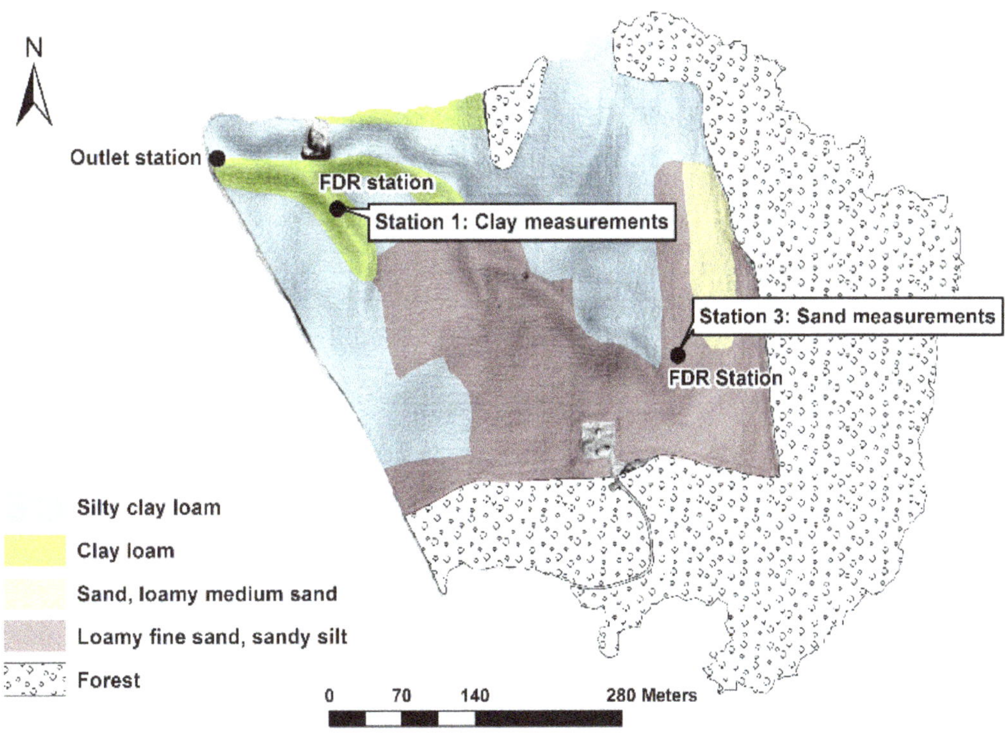

Figure 1. Soil and hill shade map of the Gryteland catchment in southern Norway. FDR, Frequency Domain Reflectometry.

The sub-catchment is characterized by an undulating landscape (elevation 106–141 m, slope 2%–10%) covered by approximately 60% arable land and 40% coniferous forest. Soil types for the arable land are a levelled clay loam (Stagnosol) and a silty clay loam (Albeluvisol) (Group 1), as well as a sandy silt on clay (Umbrisol) and a sand to loamy medium sand (Histic Gleysol) (Group 2). The two soil groups are often not clearly distinguishable in the field. Within the groups, the soils have similar physical properties. Hereafter, Groups 1 and 2 are referred to as clay and sand, respectively (Figure 1).

Mean annual temperature in the study area is 5.3 °C, with an average minimum of −4.8 °C in January/February and an average maximum of 16.1 °C in July. Mean annual precipitation is 785 mm, with a minimum monthly amount of 35 mm in February and a maximum of 100 mm in October [33]. Winter is usually relatively unstable, with alternating periods of freezing and thawing and several snowmelt events [34].

There was no tillage (no-till system) after harvest in 2013, leaving the fields covered in stubble. In 2014 and 2015, secondary tillage was performed after harvest with a cultivator on the slopes, leaving the depressions still covered with stubble.

2.2. Weather Data

A weather station was installed in the catchment outlet at the end of 2013, providing hourly data on net solar radiation, air temperature, wind speed and wind direction for the winters (December–March) of 2014 and 2015. For winter 2013 (January–April), data from a station 6 km away from the catchment were used ([33].

2.3. Soil Temperature and Soil Moisture Measurements

To obtain more detailed measurements of soil water content and temperature during winter, the measuring Stations 1 (clay measurements) and 3 (sand measurements) were upgraded to measure soil water content and soil temperature at four depths, 5, 20, 30 and 40 cm, using Decagon 5 TM temperature and Frequency Domain Reflectometry (FDR) sensors. Measurements from these two stations and the outlet station were used in the present study. However, it should be noted that the soil water content, calculated from the dielectricity of the soil, measured with the FDR probes, represents only the liquid soil water content, not water in the form of ice, and therefore, only the liquid soil water content

2.4. Discharge Measurements

To estimate how the catchment reacted to precipitation and to analyze the infiltration capacity of the soils in the catchment, data on discharge measured at the outlet were analyzed. Besides measuring how much discharge was produced during the winter periods, the runoff coefficient was calculated as:

$$D_{ro} = 100 \times D_M / P_A$$

where D_{ro} = runoff coefficient (%); D_M = discharge (m^3); P_A = precipitation on area (m^3).

2.5. Snow Cover Properties

Snow has a significant influence on changes in soil temperature and soil water content [14]. Therefore, snow properties (depth and density) were monitored in the catchment during the three winters. The measurements taken at the outlet, Stations 1 and 3, are presented in this study. Snow Water Equivalent (SWE) was sampled after weather changes expected to result in changes in SWE [35]. The measured snow depth data were used to validate snow depth values simulated with the SHAW model.

2.6. Erosion Mapping

In addition to the other measurements carried out in the field, soil erosion features were documented. Minor erosion damage was recorded by taking pictures. The extent of any large features observed was mapped using a differential GPS, and the depth and width were measured at several points using a ruler.

2.7. SHAW Model Setup and Calibration

The SHAW model, which was originally developed to simulate soil freezing and thawing [36], simulates heat, water and solute transfer within a one-dimensional profile extending downwards from the vegetation canopy to a specified depth within the soil. A layered system is established through the plant canopy, snow, residue and soil, and each layer is represented by an individual node [25]. Infiltration is calculated using a Green–Ampt approach for a multi-layered soil. Water flow in frozen soil is assumed to be similar to flow in unsaturated soil. Therefore, the relationships for matric potential and hydraulic conductivity of unsaturated soils are assumed to be valid for frozen soils. However, hydraulic conductivity is reduced linearly with ice content, assuming zero conductivity at an available porosity of 0.13 [37]. A detailed description of the model can be found in Flerchinger [38].

Input to the SHAW model includes: initial conditions for snow, soil temperature and water content profiles; daily or hourly weather conditions (temperature, wind speed, humidity, precipitation and solar radiation); general site information; and parameters describing the vegetative cover, snow, plant residues and soil. General site information includes slope, aspect, latitude and surface roughness parameters. Input soil parameters are bulk density, saturated conductivity, albedo and coefficient for the soil water potential-water content relationship [38].

To obtain the necessary soil input data (Table 1), undisturbed samples were taken in April 2014 at three different depths (0, 25 and 35 cm) at Stations 1 and 3. These depths corresponded to the depth between FDR probes, avoiding the disturbed area around the probes. For determination of saturated soil hydraulic conductivity (K_{sat}) and saturated water content (θ_S), two samples (volume of sample ring 250 cm^3) were taken. Two additional samples (volume of sample ring 98 cm^3) were used for determination of bulk density and soil organic matter. In total, 12 samples were taken at each of the two stations.

Saturated soil hydraulic conductivity was determined using the constant head method [39]. The two soil profiles defined for SHAW are presented in Table 1, with the corresponding depths of the simulation nodes (same as the installation depth of the FDR/temperature probes). For the simulations of winter 2013, only the clay was included, as a three-layered soil profile with the location of the nodes at 5, 10 and 20 cm, due to missing data for the sand and only three FDR/temperature probes in the clay.

Table 1. Input parameters for the SHAW model. Only the three layers marked with an asterisk (*) were used for the 2013 simulation, and the depth was reduced from 15 down to 10 cm and from 25 cm down to 20 cm.

Soil Type	Clay				Sand			
Location	59°40′ N, North Facing (22.5°), Slope 12°, Elevation ASL 100 m				59°40′ N, Northwest Facing (330.5°), Slope 0°, Elevation ASL 140 m			
Surface	Albedo of dry soil: 0.15 Wind profile surface roughness: 0.1 cm							
Depth	5 cm *	15 cm *	25 cm *	35 cm	5 cm	15 cm	25 cm	35 cm
Campbell's b	20	20	20	20	3	1	1	1
Air entry potential (hPa)	−31	−31	−34	−35	−31	−31	−34	−35
K_{sat} (cm·h^{-1})	2.60	1.86	1.00	0.60	16.80	18.00	22.00	24.00
Bulk density (kg·m^{-3})	1331	1400	1535	1537	1190	1346	1346	1347
θ_S	0.40	0.40	0.40	0.35	0.43	0.40	0.40	0.31
Sand (%)	13				70			
Silt (%)	58				13			
Clay (%)	29				7			
Organic matter content (%)	4.5	4.0	3.7	3.5	3.4	3.4	2.7	2.0

The model was calibrated to fit the measured snow depth and soil temperature by adjusting the site-specific parameters maximum temperature at which precipitation is snow (T_{max}) (only for winter 2015) and the wind profile roughness for momentum transfer of the snow cover (z_m) (for all three winters).

3. Results

3.1. Weather Measurements

Measured air temperature and precipitation are presented in Figure 2. Periods with continuous snow cover, indicated with blue bands in the diagram, were of differing duration in the different years.

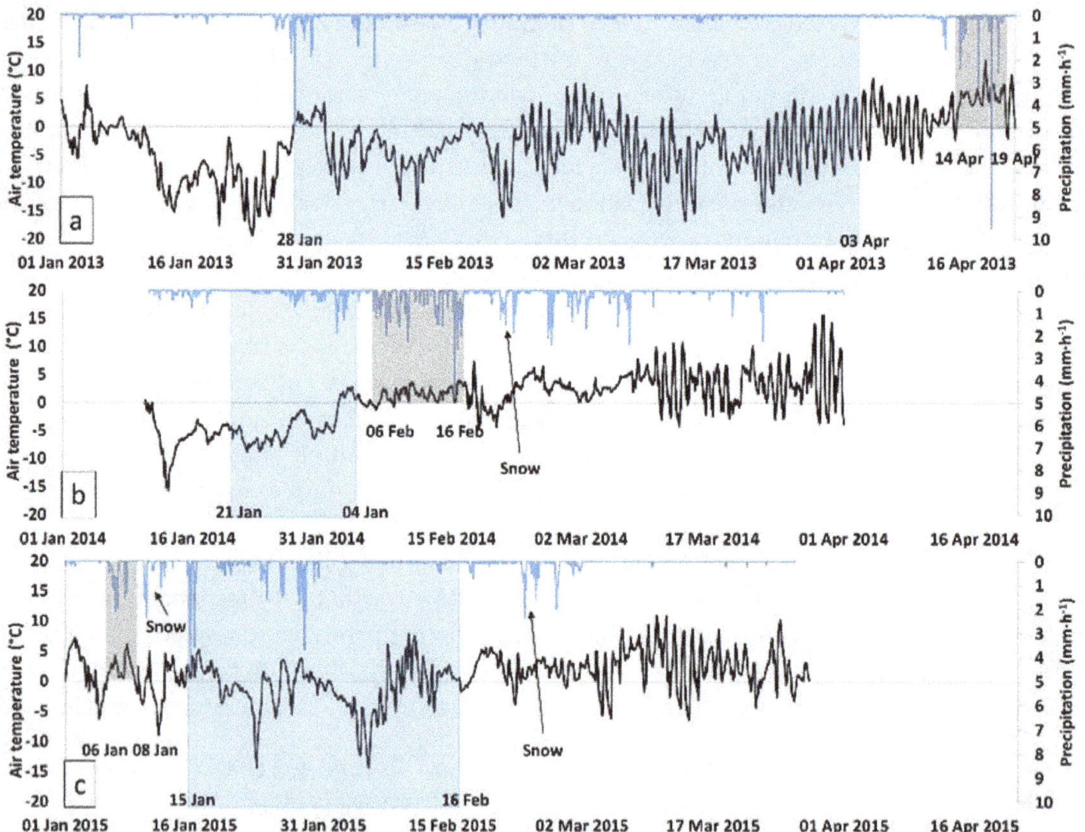

Figure 2. Measured air temperature and precipitation for the three winter periods: 2013 (**a**); 2014 (**b**); and 2015 (**c**). Rain events of interest for this study with start and end dates are marked with grey bands. The duration of unbroken snow cover with start and end dates is indicated with blue bands. Snow events outside the unbroken snow cover period are marked as 'snow'.

The number, duration and intensity of rain events during the three winter periods also differed considerably. For each winter period, rain events of interest were selected (grey bands in Figure 2) for detailed analysis. A rain event was classified as 'interesting' for this study when the precipitation fell as rain on completely or partially frozen ground. The total amount of rain and the duration of the 'main event' when the highest measured intensities occurred are shown in Table 2.

Table 2. Start and end dates of the rain events of interest, with measured total amount of precipitation, highest measured intensity and the duration and amount of precipitation of the main event within the whole event.

Date	Total (mm)	Main Event	Highest Intensity (mm·h^{-1})
14–19 April 2013	42.1	15.8 mm in 2 h	9.3
6–16 February 2014	105	10 mm in 4 h	4.6
6–8 January 2015	28	7 mm in 4 h	2.0

In terms of intensity, the 2013 event listed in Table 2 would be classified as heavy rain and the other events as moderate rain. The rain event in 2014 occurred on top of ongoing snowmelt after 4 February.

3.2. *Surface Discharge Measurements*

Measured surface discharge for the selected rain events is presented in Table 3, together with the estimated discharge coefficients. The main discharge events in the three winter periods occurred during these rain events, rather than as a result of snowmelt. While the rain event in 2013 was the largest of the three events, the rain event in 2015 produced the highest amount of discharge and had the highest discharge coefficient. During winter 2014, little discharge was produced by the rain event of interest (1008 m^3) compared with the other two years (2594–3096 m^3) (Table 3). However, it should be noted that some tunneling below the flume was observed, resulting in the by-pass of water and too low discharge values. This was repaired and did not happen in 2015.

Table 3. Start and end date of the rain events of interest with measured surface discharge, precipitation per area and percentage of precipitation water, which reached the outlet (discharge coefficient) for these rain events. Values in parenthesis show the discharge coefficient for the agricultural area only (no forest). Values marked with an asterisk (*) are incorrect measurements due to the by-pass of water below the flume.

Date	Discharge (m^3)	Precipitation (m^3)	Discharge Coefficient (%)
14–19 April 2013	3096	12,180	25 (50)
6–16 February 2014	1008 *	30,450	4 *
6–8 January 2015	2594	8120	32 (63)

By comparing the discharge measured at the Gryteland sub-catchment outlet with discharge measured at the outlet of the main Skuterud catchment, it was determined that a surface discharge coefficient of about 12%, rather than 4%, was more realistic for the event in 2014.

3.3. Liquid Soil Water Content and Soil Temperature

A malfunction in Stations 1 and 3 created a gap in the data for winter 2013. The two stations were repaired just before the extreme event at the end of the winter period, so that soil moisture and soil temperature measurements for the clay and sand were available for this event (insert diagram in Figure 3). For the clay, measured data were taken close by the outlet station, where undisturbed soil moisture and soil temperature data were obtained throughout winter 2013 (Figure 3).

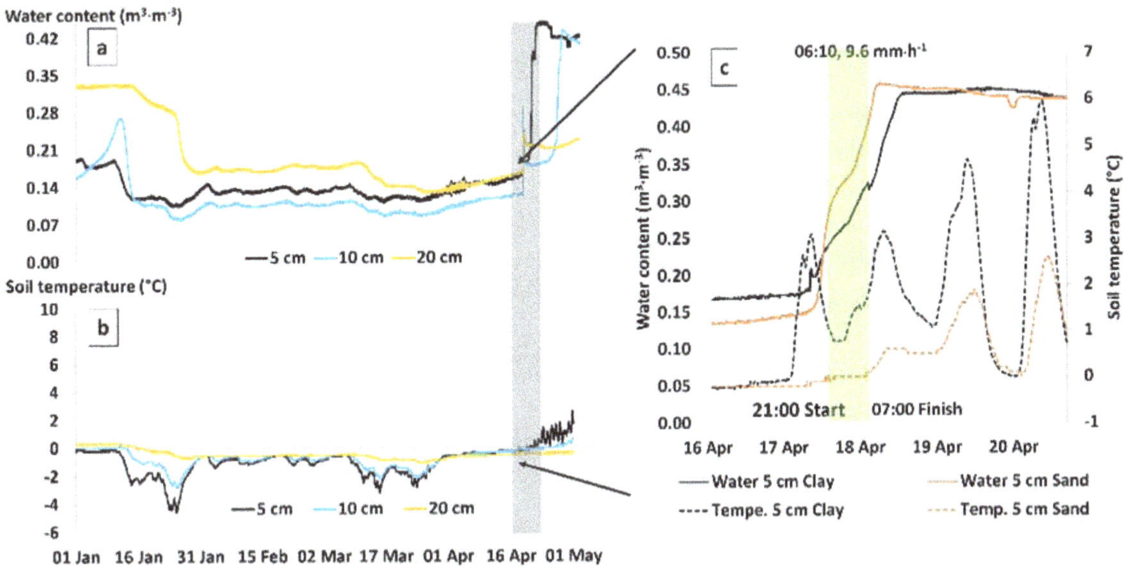

Figure 3. Measured soil water content (**a**) and soil temperature (**b**) in the clay during winter 2013 with the insert diagram (**c**) showing measured soil water content and soil temperature at 5 cm depth in the clay and sand during the rain event of interest in 2013 (duration marked with grey band in (**a**,**b**)). The period with the highest precipitation intensities is marked with a green band in (**c**).

During winter 2013, the soil temperature and liquid soil water content measurements were characterized by a long period of soil frost at all soil depths (Figure 3). Shortly after 2 December, the soil started to freeze at 5 and 10 cm and stayed frozen until the highest precipitation intensities (9.6 mm·h^{-1}) occurred on 18 March (Figure 3). Frozen soil was indicated by a low liquid soil water content (~0.12 m^3·m^{-3}) compared with pre-freezing (~0.38 m^3·m^{-3} in November 2012) and soil temperature below 0 °C. As can be seen in Figure 3, the soil was still frozen after the first rain event on 14 April. The liquid soil water content was still low in both soils (~0.17 m^3·m^{-3} in the clay and 0.13 m^3·m^{-3} in the sand), and the measured soil temperature was below 0 °C in both soils. On 17 April, the soil temperature in the top 5 cm in the clay soil started to rise above 0 °C, followed by a continuous increase in liquid soil water content at 5 and 10 cm depth. The soil temperature in the topsoil dropped to about 0.8 °C when the major rain event started in the evening of 17 April. However, the liquid soil water content continued to increase steadily and more rapidly in the sand than in the clay. When the highest precipitation rates occurred, the clay soil had reached a liquid soil water content of about 0.25 m^3·m^{-3}. The temperature in the sand soil rose much more slowly than in the clay soil, but the liquid soil water content increased rapidly in the sand topsoil layer at the same rate as in the clay soil, reaching a liquid soil water content of 0.34 m^3·m^{-3} when the highest precipitation intensities occurred. During the final period of the high intensity rain, the soil water content continued to rise to about 0.45 m^3·m^{-3} in both the clay and the sand. No further increase in soil water content occurred during the rest of the rain event, probably because both soils were fully saturated.

Measured soil water content during winter 2014 for the clay and sand is shown in Figure 4, and measured soil temperature is presented in Figure 5.

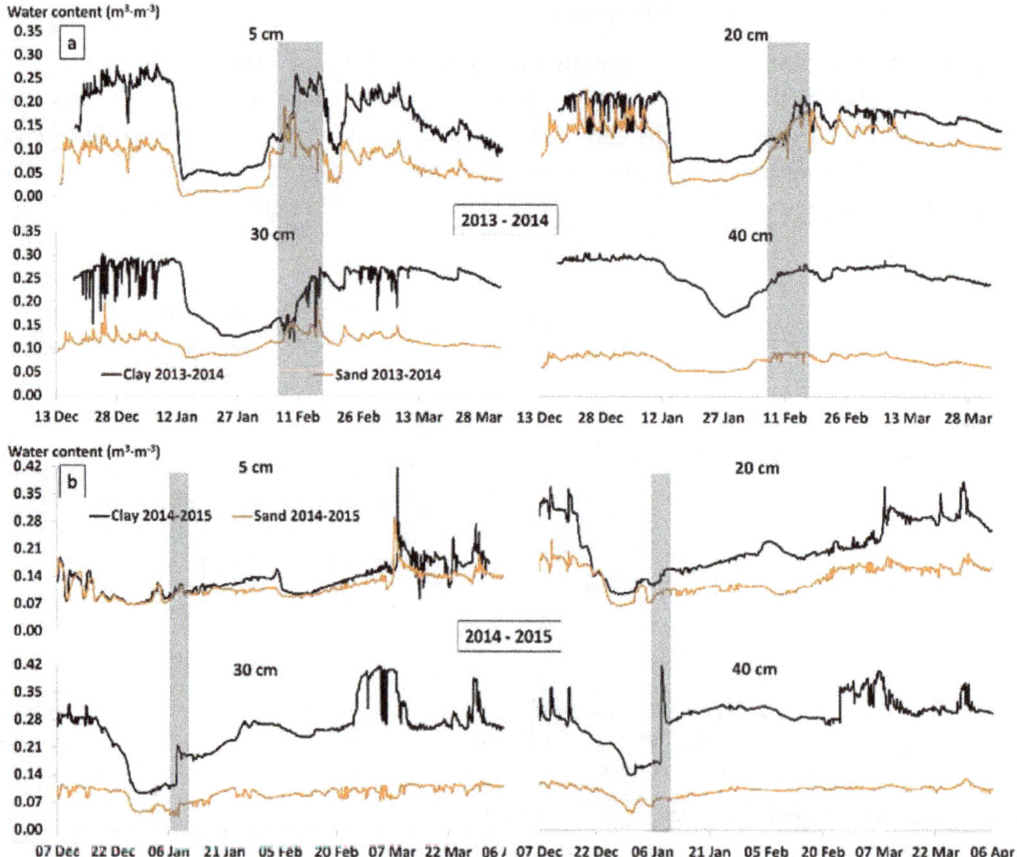

Figure 4. Measured soil water content in the clay and sand for the winter periods 2014 (**a**) and 2015 (**b**) at four depths: 5, 20, 30 and 40 cm. Durations of the rain events of interest (according to Table 2) are marked with grey bands.

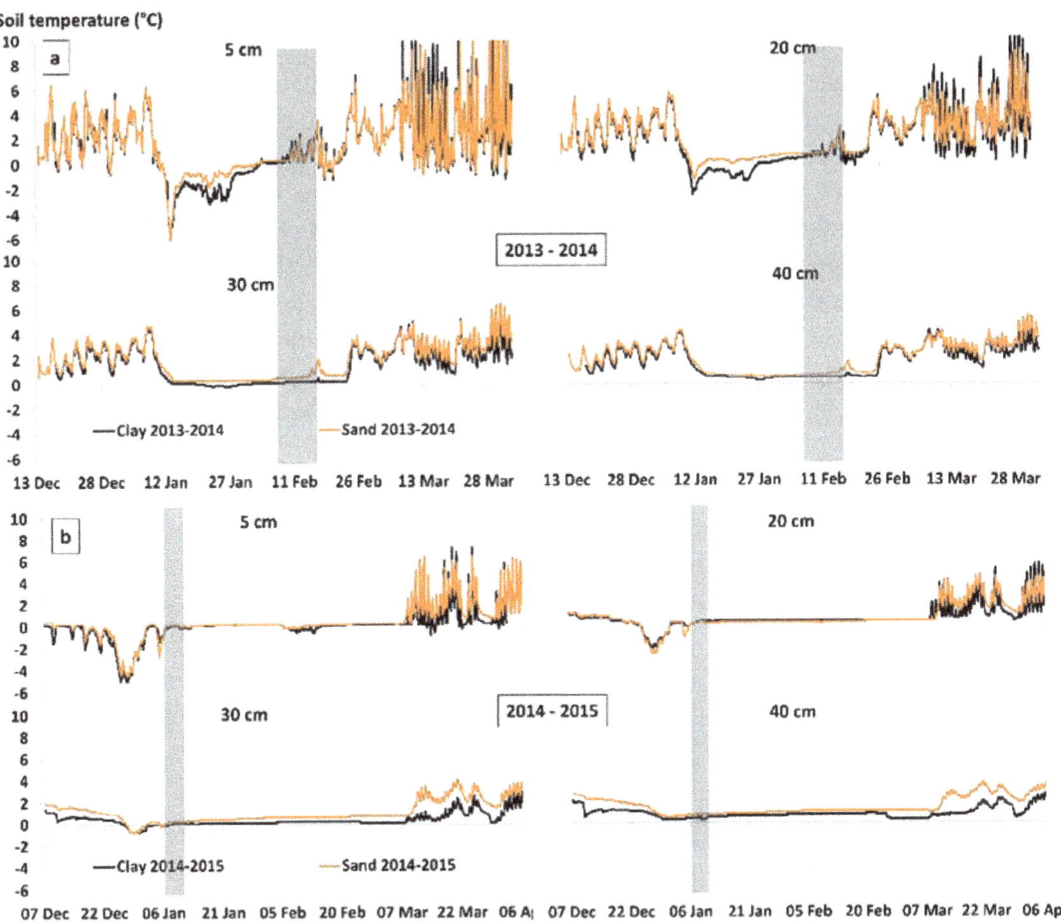

Figure 5. Measured soil temperature in the clay and sand for the winter periods 2014 (**a**) and 2015 (**b**) at four depths: 5, 20, 30 and 40 cm. Durations of the rain events of interest (according to Table 2) are marked with grey bands.

Measured soil water content and temperature declined significantly in both the clay and sand immediately after the air temperature started to drop below zero (Figures 4 and 5) around 6 January. Measured soil water content did not change much after 13 January, whereas soil temperature followed the changes in air temperature, although this effect decreased with increasing depth for both soil types. After 27 January, the liquid soil water content rose slowly, due to an increase in air temperature (Figure 2), resulting in thawing of ice in the soil. After 3 February, the liquid soil water content increased rapidly due to incoming snowmelt water. While the soil water content did not change much in the sand at 30 and 40 cm depth, a decrease in liquid soil water content was observed at these depths in the clay. Both soils showed similar temperature profiles for all four depths (Figure 5). The soil water content at all depths was much lower in the sand than in the clay, reducing the penetration depth of freezing [6]. Furthermore, at 30 and 40 cm depth in the sand, there was little water left to freeze (30 cm: 0.08 $m^3 \cdot m^{-3}$; 40 cm: 0.06 $m^3 \cdot m^{-3}$).

When the combined rain and snowmelt event occurred in 2014 (Table 2), both soils had just started to thaw. During the event, similar soil water contents as measured before freezing (~0.25 $m^3 \cdot m^{-3}$) were reached in the clay. In the sand, soil water content at 5 cm depth initially rose rapidly above (~0.18 $m^3 \cdot m^{-3}$) pre-freezing values (~0.10 $m^3 \cdot m^{-3}$), but decreased to pre-freezing values during the event, while at the other three depths studied, the soil water content continued to increase during the whole event. At all four soil depths in both soils, the soil temperature rose significantly to above 0 °C, reaching the highest temperature (3.6 °C) at 5 cm.

The change in liquid soil water content in winter 2015 (Figure 4) was characterized by low values at 5 cm in both soils (minimum of 0.07 $m^3 \cdot m^{-3}$) due to early freezing in the beginning of December. This was followed by a decrease in liquid soil water content at 20, 30 and 40 cm in the clay on 17 December, reaching a minimum of 0.10 $m^3 \cdot m^{-3}$ at 20 and 30 cm and 0.14 $m^3 \cdot m^{-3}$ at 40 cm. A sudden increase in liquid soil water content occurred on 7 March at all depths in the clay and at 5 cm in the sand, when the air (Figure 2) and soil temperature (Figure 5) rose significantly to above 0 °C. In winter 2015, the rain event (Table 2) was preceded by freezing temperatures at 5–30 cm soil depth and low liquid soil water content at all soil depths in both soils (Figures 5 and 6). During the event, the liquid soil water content rose rapidly at 30 and 40 cm depth in the clay, reaching its highest value (0.42 $m^3 \cdot m^{-3}$) at 40 cm depth. At 5- and 20 cm depth, the change in liquid soil water content was less pronounced. Soil temperature rose in the upper three depths in both soils to about 0 °C, while at 40 cm depth, the soil temperature stayed above 0 °C at all times in both soils.

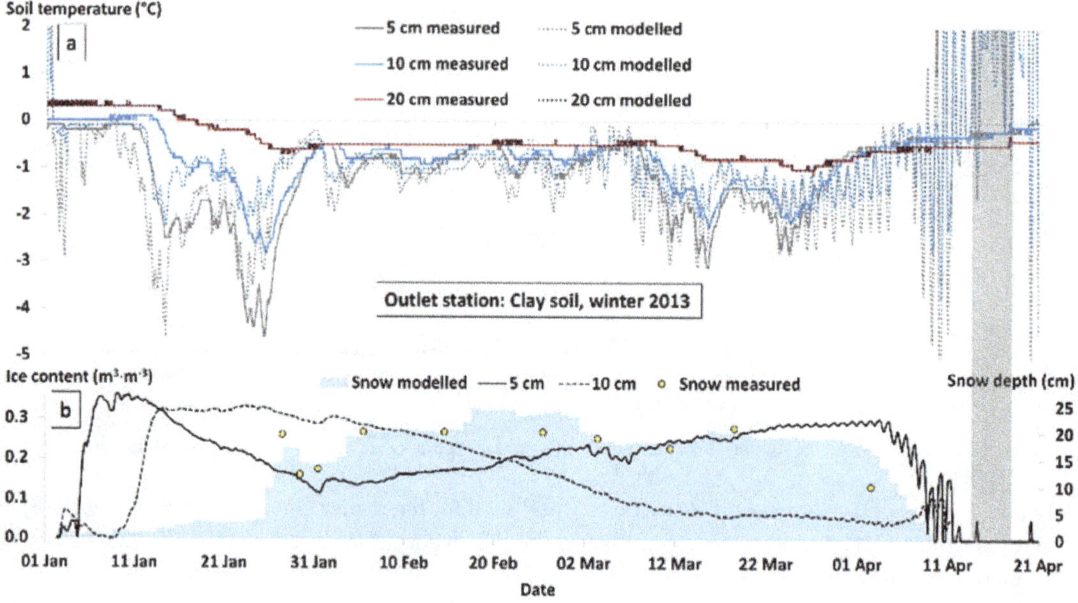

Figure 6. Diagram of (**a**) modelled and measured soil temperature at three depths (5, 10 and 20 cm) in the clay soil (at outlet station) during winter 2013; diagram of (**b**) measured and modelled snow depths at the outlet station and modelled ice content at the three depths in the clay soil (was zero at 20 cm). The duration of the rain event of interest is marked with a grey band.

3.4. SHAW Modelling

The simulated snow depth, soil temperature and ice content values for the clay during winter 2013 are presented in Figure 6, for the clay and sand during winter 2014 in Figures 7 and 8 and for winter 2015 in Figures 9 and 10. As mentioned, due to missing measurements for the sand during winter 2013, the sand was not modelled for that year.

As can be seen from Figure 6, the SHAW model simulated the changes in the snow pack very accurately for 2013. To fit the simulated snow depth to the measured values, the z_m parameter was set to 0.15 cm. The simulated soil temperature at 5 and 10 cm showed some fluctuations around the measured temperature, which increased after the snowpack disappeared in the end of March, when the air temperature fluctuated between −7 and 8 °C on a daily basis (Figure 2). However, it should be noted that the soil temperature was measured below a thick grass layer at the outlet station, and this grass layer probably acted as insulation [40], buffering the soil from the fluctuating air temperature. The model results show soil temperatures for bare ground, for comparison with the other winters. Due to the satisfactory simulation of snow depth and soil temperature, it was assumed that SHAW also performed well in simulating ice content in the soil profile. Simulated ice content was high (0.35 $m^3 \cdot m^{-3}$) at 5 and 10 cm depth and zero at 20 cm depth, indicating that almost all of the water at these depths was frozen. The ice at these depths was apparently formed during the low soil temperatures in January, before the snow pack was established. SHAW was able to simulate the rise in soil temperature after the snow pack reached a depth of about 20 cm in February and predicted a decrease in ice content to zero at all depths by 12 April. The SHAW model was also able to simulate the observed decrease in soil temperature in the middle of April.

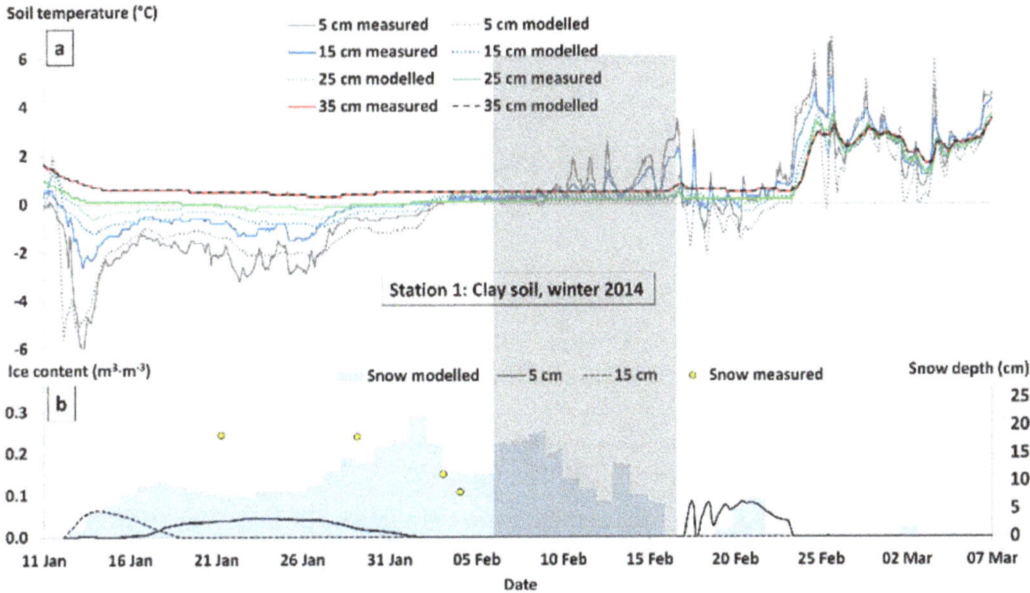

Figure 7. Diagram of (**a**) modelled and measured soil temperature at four depths (5, 10 and 20 cm) in the clay soil (at Station 1) during winter 2014; diagram of (**b**) measured and modelled snow depths at Station 1 and modelled ice content at four depths in the clay soil (was zero at 25 and 35 cm). The duration of the rain event of interest is marked with a grey band.

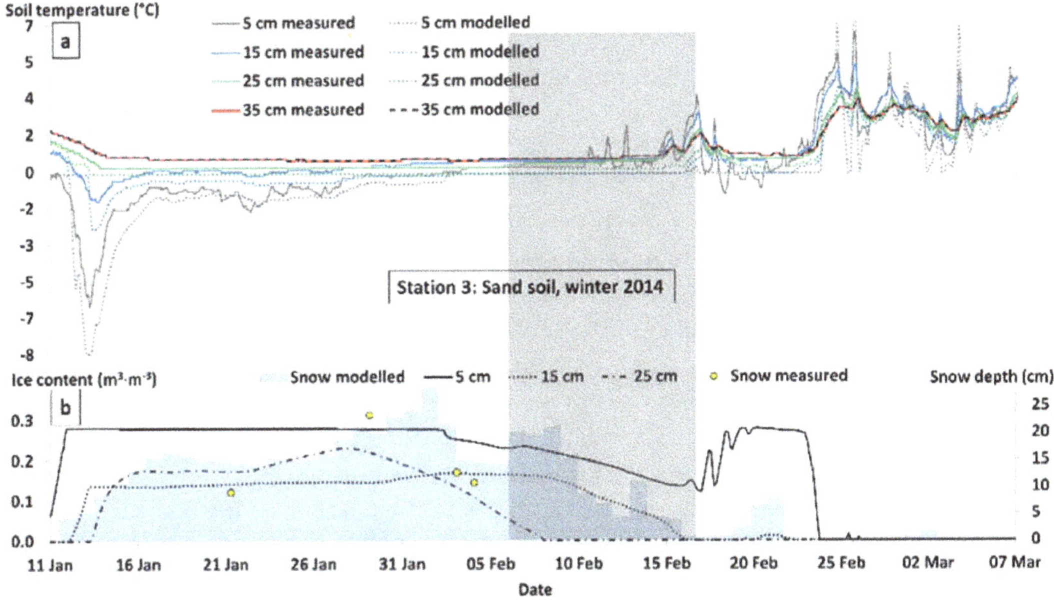

Figure 8. Diagram of (**a**) modelled and measured soil temperature at four depths (5, 15, 25 and 35 cm) in the sand soil (at Station 3) during winter 2014; diagram of (**b**) measured and modelled snow depths at Station 3 and modelled ice content at four depths in the sand soil (was zero at 35 cm). The duration of the rain event of interest is marked with a grey band.

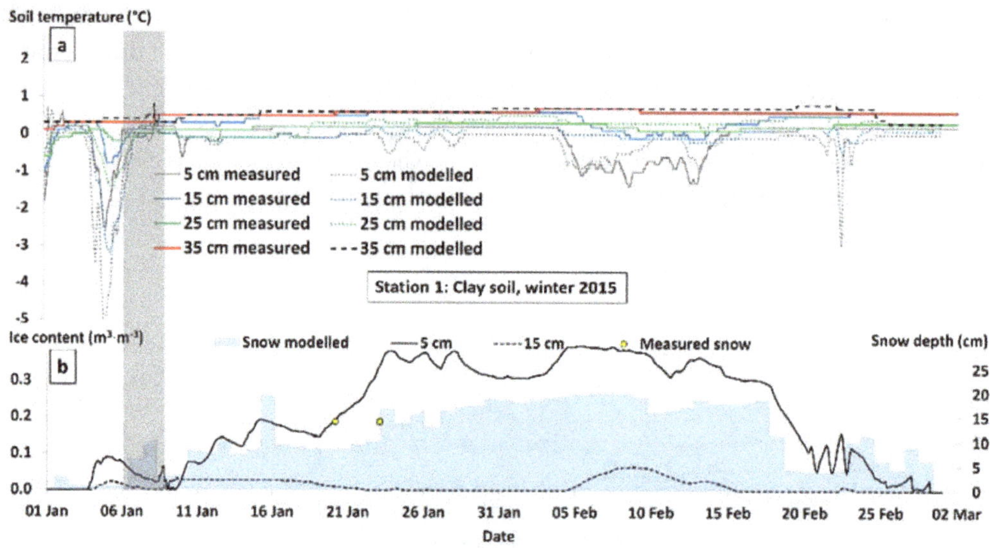

Figure 9. Diagram of (**a**) modelled and measured soil temperature at four depths (5, 15, 25 and 35 cm) in the clay soil (at Station 1) during winter 2015; diagram of (**b**) measured and modelled snow depths at Station 1 and modelled ice content at four depths in the clay soil (was zero at 25 and 35 cm). The duration of the rain event of interest is marked with a grey band.

Because only four snow depth measurements were possible in 2014, it was difficult to compare the simulated and measured snow depth values. However, it can be seen that SHAW underpredicted the snow depth above the clay soil during January and overpredicted it above the sand soil (Figures 7 and 8). The z_m parameter was adjusted to 0.5 cm for the clay and 1.5 cm for the sand.

The simulated soil temperature in the clay soil (Figure 7) followed the measured temperature rather closely during the freezing period in January, but during the start of the snowmelt in the middle of February, the measured temperature showed a higher fluctuation than the simulated temperature. For the sand, SHAW simulated slightly lower temperature (−8 °C at 5 cm) than the measured values (−6 °C at 5 cm), but in general, the simulated temperature followed the trend in the measured data quite well (Figure 8). SHAW predicted low ice content for the clay at 5 and 15 cm (max. 0.06 $m^3 \cdot m^{-3}$ at 15 cm) and no ice at lower depths, while for the sand, it predicted an ice content of 0.28 $m^3 \cdot m^{-3}$ at 5 cm and 0.22 $m^3 \cdot m^{-3}$ at 25 cm depth. At 5 and 15 cm, the model predicted that all the available water was frozen, which can be seen as the constant ice contents between 12 January and 2 February. Moreover, the period with frozen water in the sand profile was considerably longer (11 January–23 February at 5 cm) than that in the clay (12 January–1 February at 5 cm).

Except for a slight underprediction of the snow depth for the period when the first two snow measurements were taken in 2015, SHAW simulated the snow depth well for winter 2015 (Figure 10). To obtain this fit, z_m was set to 0.05 cm for the clay and 0.08 cm for the sand. In addition, T_{max} had to be set to 3.8 °C to fit the modelled snow thicknesses to the measured, which resulted in the simulation of a thin snow cover, which did not occur in the field, during the rain event of interest in 2015. This adjustment of T_{max} was in accordance with [32], who used a comparable snow model for the study area and had to make a similar adjustment to this parameter.

The simulated soil temperature showed similar low fluctuations as the measured (Figures 9 and 10), and the first negative peak was also simulated, with a good fit in the sand and slightly lower temperature in the clay (−5.8 °C simulated compared with −3 °C measured at 5 cm).

For both the clay and the sand, SHAW predicted high ice content at 5 cm depth, mainly during the period when the snow pack was between 10 and 20 cm thick (Figures 9 and 10). The predicted values were 0.39 $m^3 \cdot m^{-3}$ in the clay and 0.45 $m^3 \cdot m^{-3}$ in the sand. For both soils, SHAW simulated an increase in ice content at 5 cm with growing snow thickness and a decrease in ice content with decreasing snow thickness.

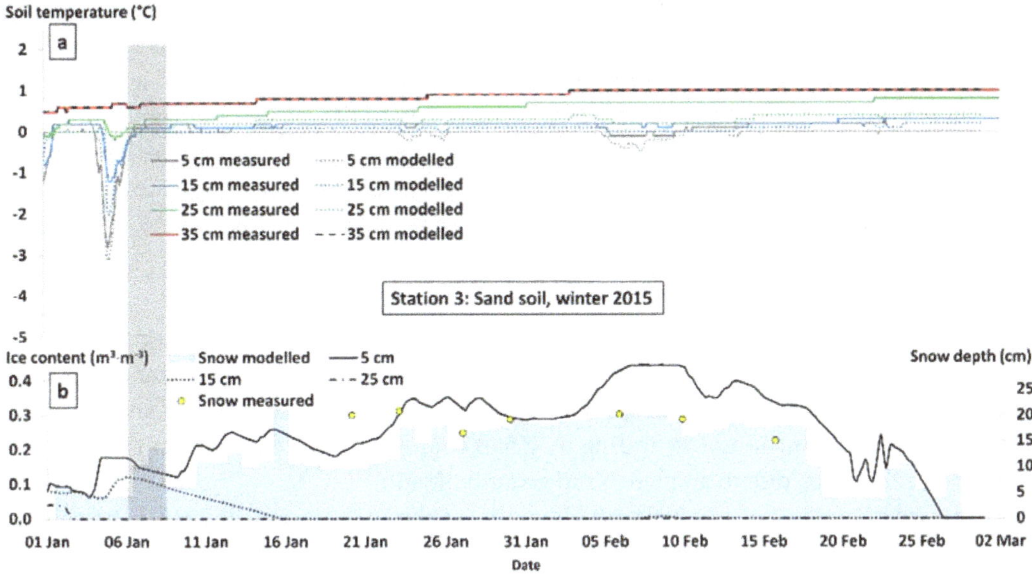

Figure 10. Diagram of (**a**) modelled and measured soil temperature at four depths (5, 15, 25 and 35 cm) in the sand soil (at Station 3) during winter 2015; diagram of (**b**) measured and modelled snow depths at Station 3 and modelled ice content at four depths in the sand soil (was zero at 35 cm). The duration of the rain event of interest is marked with a grey band.

3.5. Erosion Mapping

During winter 2013, no erosion was observed anywhere in the catchment. The combined snowmelt and rain event in 2014 formed several small rills with a maximum width of 20 cm and 5–10 cm deep on two slopes in the catchment (Figure 11). The observed rills were formed mainly in the tracks created by the cultivator.

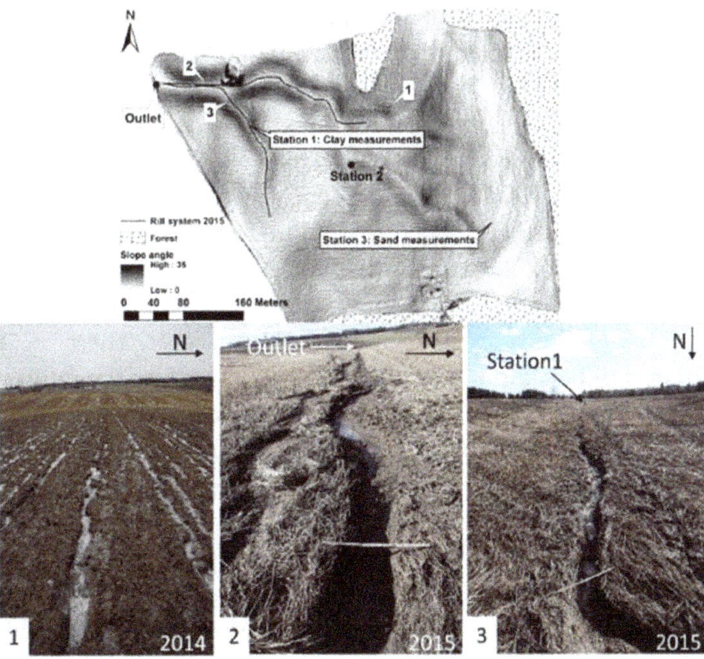

Figure 11. Slope angle map showing; the extent of the rill system, which formed during winter 2015 and the location where images 1–3 were taken. Image 1 shows rills which formed during the combined snowmelt and rain event in 2014 and images 2,3 show rills that formed during winter 2015.

In winter 2015, a continuous rill system (total length 493 m) was formed in depressions, where surface runoff was concentrated, during the main rain event. The rills varied in width and depth (Figure 11, Image 2,3), with measured maximum width of 40 cm and a depth of 20 cm. Several sedimentation areas were observed where the surface runoff was slowed down, due to reduced slope angle and increased flow width, indicating that not all of the eroded soil reached the outlet. At the other slope where rills were observed in 2014 (Figure 11, Image 1), the rills were connected to the extensive rill system in 2015.

4. Discussion

The winters of 2013, 2014 and 2015 differed significantly from each other in terms of the number of freezing periods, the length of period with a continuous snow cover and the number of rain events (Figure 2). The soils stayed frozen throughout the whole time with a continuous snow cover in the three winter periods, confirming the finding by Zhao et al. [15] that a thin (<25 cm) snow pack can increase soil freezing, e.g., due to an increase in ground albedo.

The amount of water that can infiltrate depends on the water permeability of the soil and the speed of surface runoff. In frozen soil, the water permeability depends on pre-freezing conditions [41], i.e., what the soil water content was before freezing started. In saturated soils the macropores are filled with water, which when frozen, clogs the pores. In the sand studied here, the initial soil water content was similar in 2013, 2014 and 2015, at between 0.10 and 0.25 $m^3{\cdot}m^{-3}$, which is far below the measured saturation of 0.40–0.43 $m^3{\cdot}m^{-3}$. Therefore, it can be assumed that the sand areas in the catchment contributed little to surface runoff during these three winters. This could explain why no erosion occurred in the depression on the sand soil in the catchment (Figure 1).

In the clay, however, the conditions differed between the years. In clays, water transport and infiltration capacity are highly dependent on macropores, particularly in the levelled clays found in the study area [42]. When freezing started in 2014, the clay had a soil water content of 0.25–0.30 $m^3{\cdot}m^{-3}$, which according to the soil hydraulic characteristic curve for the clay represents a matric potential of 32 kPa (5 cm depth) to 315 kPa (20 cm depth), at which macropores are filled with air [43]. Furthermore, the rain event of interest in 2014 was less intensive than that in 2013 (Table 2), and the thawed soil allowed more water to infiltrate, resulting in a smaller amount of water reaching the outlet (Table 3). The SHAW results suggested that the clay was completely thawed by the time the rain event occurred (Figure 7).

However, despite a smaller discharge coefficient compared with 2013, the 2014 event caused erosion (Figure 11), in the tracks created by the cultivator.

Similarly to 2013, in winter 2015 the soil water content was 0.40–0.45 $m^3{\cdot}m^{-3}$ at all soil depths in the clay when freezing started in November 2014, resulting in macropores filled with ice. Both events had a high discharge coefficient, of ~25% (50% when forest area was excluded) in 2013 and 32% (63%) during the first event in 2015 (Table 3). This, together with the low liquid soil moisture values, suggests that infiltration was restricted due to frozen soil during both events. Therefore, the fact that erosion did not occur in 2013, but did occur in 2015, cannot be explained by the infiltration capacity of the soil during the events. Moreover, the stability of the soils cannot explain the differences between 2013 and 2015, as confirmed by shear strength measurements carried out on the clay immediately before the event in 2013, which revealed low shear strength of 5–10 kPa at the soil surface (vane shear test, Eijkelkamp, The Netherlands).

Another process that could explain the differences between 2013 and 2015 is the speed of surface runoff, which determines the erosivity of surface runoff [44]. In 2013, the whole catchment was covered with stubble, but in 2015, only the depressions had intact stubble, while secondary tillage with a cultivator reduced the amount of plant residues on slopes. This tillage created a loose and smoother surface, probably causing higher speed of surface runoff, in freezing conditions in particular, as previously shown by Ban et al. [4]. This assumption was supported by the occurrence of soil erosion in the form of rills on the tilled slopes in 2014 (Figure 11). These findings were also in agreement with

Edwards and Burney [3], who concluded that only a plant cover can significantly reduce soil losses by rain and overland flow on frozen ground, e.g., through reduced runoff speeds and increased soil stability by roots. Our finding that the selected rain events on saturated and frozen soil produced a large amount of surface discharge confirmed the prediction by Stähli et al. [9] and Nyberg et al. [10] that high water saturation and early frost penetration, combined with heavy rain on still frozen soil, cause a marked increase in the amount of runoff.

Contrary to observation made by other studies (e.g., [21–23]), in all three winters studied, snowmelt played no significant role in terms of soil erosion. During snowmelt in 2014, the snow layer acted as a buffer for incoming rain. The rain infiltrated into the snow pack and surface runoff was delayed by the snow, reducing the erosive forces of the rain event. In 2013, no erosion occurred, and in 2015, the major soil erosion features had occurred before the first snow fell (Figure 2).

The performance of the SHAW model was satisfactory for all three winter periods. The change in soil temperature and snow pack was well reproduced, and the simulated ice content was in agreement with the measured liquid soil water content. In general, SHAW predicted ice in the soil for the periods when the FDR probes measured low liquid soil water contents. With the adjustment of two snow-related parameters in the model, it was possible to obtain reasonable results for the three different winter periods. Adjustment of T_{max} for winters when snow falls at temperatures above 0 °C, as was the case in 2015, allows the model to partition incoming precipitation into snow and rain based on field observations, rather than using linear interpolation. The performance of SHAW in this study proved that it can be a valuable tool for investigating and predicting: (1) water content at freezing; (2) whether soil is frozen or unfrozen at a particular moment; and (3) the state of the snow cover. These are three important factors that control the amount of runoff during winter and are indispensable for predicting when soil erosion can be expected.

5. Conclusions and Implications

Field observations carried out during three winters in a catchment in southern Norway showed how soil hydraulic properties changed due to freezing-thawing, affecting surface runoff caused by snowmelt and rain, and how these processes are linked to soil erosion. The largest amount of soil erosion was caused by a small rain event on frozen ground, before the snow cover was established, while snowmelt played no significant role in terms of soil erosion. Four factors that determine the extent of runoff and erosion were of particular importance: (1) soil water content at freezing; (2) whether soil was frozen or unfrozen at a particular moment; (3) the state of the snow cover; and (4) tillage operations prior to winter. The simulation results showed that the SHAW model, with its accurate snow pack routine, is a useful tool that can help to investigate and identify non-tillage factors (e.g., 1, 2 and 3) influencing erosion.

Acknowledgments: This research was funded by the Catchy project of NIBIO, Norwegian Institute of Bioeconomy Research.

Author Contributions: T.S. conducted this research as part of his Ph.D. and conceived and designed this study, acquired the data, performed the modelling, analyzed and interpreted the data and wrote the paper. R.H., J.S. and C.R. supervised the work and helped with improving the manuscript.

Conflicts of Interest: The authors declare no conflict of interest.

References

1. Ollesch, G.; Sukhanovski, Y.; Kistner, I.; Rode, M.; Meissner, R. Characterization and modelling of the spatial heterogeneity of snowmelt erosion. *Earth Surf. Process. Landf.* **2005**, *30*, 197–211. [CrossRef]
2. Iwata, Y.; Nemoto, M.; Hasegawa, S.; Yanai, Y.; Kuwao, K.; Hirota, T. Influence of rain, air temperature, and snow cover on subsequent spring-snowmelt infiltration into thin frozen soil layer in northern Japan. *J. Hydrol.* **2011**, *401*, 165–176. [CrossRef]
3. Edwards, L.M.; Burney, J.R. Soil erosion losses under freeze/thaw and winter ground cover using a laboratory rainfall simulator. *Can. Agric. Eng.* **1987**, *29*, 109–115.

4. Ban, Y.; Lei, T.; Liu, Z.; Chen, C. Comparison of rill flow velocity over frozen and thawed slopes with electrolyte tracer method. *J. Hydrol.* **2016**, *534*, 630–637. [CrossRef]
5. Watanabe, K.; Kito, T.; Dun, S.; Wu, J.Q.; Greer, R.C.; Flury, M. Water Infiltration into a Frozen Soil with Simultaneous Melting of the Frozen Layer. *Vadose Zone J.* **2013**, *12*. [CrossRef]
6. Yami, E.R.; Khalili, A.; Rahimi, H.; Etemad, A. Investigation of moisture on soil temperature regimes and frost depths in a laboratory model. *Int. J. Agric. Sci.* **2012**, *2*, 717–732.
7. Al-Houri, Z.M.; Barber, M.E.; Yonge, D.R.; Ullman, J.L.; Beutel, M.W. Impact of frozen soils on the performance of infiltration treatment facilities. *Cold Reg. Sci. Technol.* **2009**, *59*, 51–57. [CrossRef]
8. Willis, W.O.; Carlson, C.W.; Alessi, J.; Haas, H.J. Depth of freezing and spring run-off as related to fall soil-moisture level. *Can. J. Soil Sci.* **1961**, *41*, 115–123. [CrossRef]
9. Stähli, M.; Nyberg, L.; Mellander, P.; Jansson, P.; Bishop, K.H. Soil forst effects on soil water and runoff dynamics along a boreal transect: 2. Simulations. *Hydrol. Process.* **2001**, *15*, 927–941. [CrossRef]
10. Nyberg, L.; Stähli, M.; Mellander, P.; Bishop, K.H. Soil frost effect on soil water and runoff dynamics along a boreal forest transect: 1. Field investigations. *Hydrol. Process.* **2001**, *15*, 909–926. [CrossRef]
11. Lindström, G.; Löfvenius, M.O. *Tjäle Och Avrinning i Svartberget—Studier Med HBV-Modellen*; SMHI hydrologi Nr 84; SMHI: Norrköping, Sweden, 2000.
12. Iwata, Y.; Hayashi, M.; Suzuki, S.; Hirota, T.; Hasegawa, S. Effects of snow cover on soil freezing, water movement, and snowmelt infiltration: A paired plot experiment. *Water Resour. Res.* **2010**, *46*, 1–11. [CrossRef]
13. Zhao, Y.; Nishimura, T.; Hill, R.; Miyazaki, T. Determining Hydraulic Conductivity for Air-Filled Porosity in an Unsaturated Frozen Soil by the Multistep Outflow Method. *Vadose Zone J.* **2013**, *12*. [CrossRef]
14. Sutinen, R.; Hänninen, P.; Venäläinen, A. Effect of mild winter events on soil water content beneth snowpack. *Cold Reg. Sci. Technol.* **2008**, *51*, 56–67. [CrossRef]
15. Zhao, Y.; Huang, M.B.; Horton, R.; Liu, F.; Peth, S.; Horn, R. Influence of winter grazing on water and heat flow in seasonal frozen soil of Inner Mongolia. *Vadose Zone J.* **2013**, *12*. [CrossRef]
16. Parkin, G.; von Bertoldi, A.P.; McCoy, A.J. Effect of tillage on Soil Water Content and Temperature under Freeze-Thaw Conditions. *Vadose Zone J.* **2013**, *12*. [CrossRef]
17. He, H.; Dyck, M.F.; Si, B.C.; Zhang, T.; Lv, J.; Wang, J. Soil freezing—Thawing characteristics and snowmelt infiltration in Cryalfs of Alberta, Canada. *Geodermal. Reg.* **2015**, *5*, 198–208. [CrossRef]
18. Kormos, P.R.; Marks, D.; Williams, C.J.; Marshall, H.P.; Aishlin, P.; Chandler, D.G.; McNamara, J.P. Soil, snow, weather, and sub-surface storage data from a mountain catchment in the rain-snow transition zone. *Earth Syst. Sci. Data* **2014**, *6*, 165–173. [CrossRef]
19. Williams, C.J.; Mcnamara, J.P.; Chandler, D.G. Controls on the temporal and spatial variability of soil moisture in a mountainous landscape: The signature of snow and complex terrain. *Hydrol. Earth Syst. Sci.* **2009**, *13*, 1325–1336. [CrossRef]
20. Shanley, J.B.; Chalmers, A. The effect of frozen soil on snowmelt runoff at Sleepers River, Vermont. *Hydrol. Process.* **1999**, *13*, 1843–1857. [CrossRef]
21. Lundekvam, H.; Skøien, S. Soil erosion in Norway. An overview of measurements from soil loss plots. *Soil Use Manag.* **1998**, *14*, 84–89. [CrossRef]
22. Deelstra, J.; Kværnø, S.H.; Granlund, K.; Sileika, A.S.; Gaigalis, K.; Kyllmar, K.; Vagstad, N. Runoff and nutrient losses during winter periods in cold climates—Requirements to nutrient simulation models. *J. Environ. Monit.* **2009**, *11*, 602–609. [CrossRef] [PubMed]
23. Øygarden, L. Rill and gully development during an extreme winter runoff event in Norway. *Catena* **2003**, *50*, 217–242. [CrossRef]
24. Grønsten, H.A.; Lundekvam, H. Prediction of surface runoff and soil loss in southeastern Norway using the WEPP Hillslope model. *Soil Tillage Res.* **2006**, *85*, 186–199. [CrossRef]
25. Flerchinger, G.N.; Xiao, W.; Sauer, T.J.; Yu, Q. Simulation of within-canopy radiation exchange. *NJAS Wagening. J. Life Sci.* **2009**, *57*, 5–15. [CrossRef]
26. Hansen, N.C.; Gupta, S.C.; Moncrief, J.F. Snowmelt runoff, sediment, and phosphorous losses under three different tillage systems. *Soil Tillage Res.* **2000**, *57*, 93–100. [CrossRef]
27. Govers, G. Rill erosion on arable land in central Belgium: Rates, controls and predictability. *Catena* **1991**, *18*, 133–155. [CrossRef]

28. Boardman, J.; Shepheard, M.L.; Walker, E.; Foster, I.D.L. Soil erosion risk-assessment for on- and off-farm impacts: A test case using the Midhurst area, West Sussex, UK. *J. Environ. Manag.* **2009**, *90*, 2578–2588. [CrossRef] [PubMed]
29. Weigert, A.; Schmidt, J. Water transport under winter conditions. *Catena* **2005**, *64*, 193–208. [CrossRef]
30. Yakutina, O.P.; Nechaeva, T.V.; Smirnova, N.V. Consequences of snowmelt erosion: Soil fertility, productivity and quality of wheat on Greyzemic Phaeozem in the south of West Siberia. *Agric. Ecosyst. Environ.* **2015**, *200*, 88–93. [CrossRef]
31. Su, J.J.; van Bochove, E.; Thériault, G.; Novotma, B.; Khaldoune, J.; Denault, J.T.; Zhou, J.; Nolin, M.C.; Hu, C.X.; Bernier, M.; et al. Effects of snowmelt on phosphorus and sediment losses from agricultural watersheds in Eastern Canada. *Agric. Water Manag.* **2011**, *98*, 867–876. [CrossRef]
32. Kramer, G.J.; Stolte, J. *Cold-Season Hydrologic Modeling in the Skuterud Catchment. An Energy Balance Snowmelt Model Coupled with a GIS-Based Hydrology Model*; Bioforsk Report 4(126); Bioforsk: Ås, Norway, 2009; Volume 46.
33. Thue-Hansen, V.; Grimenes, A.A. *Meteorologiske Data for Ås*; Norges Miljø og Biovitenskaplige Universitet: Ås, Norway, 2013–2014.
34. Kværnø, S.H.; Øygarden, L. The influence of freeze-thaw cycles and soil moisture on aggreagte stability of three soils in Norway. *Catena* **2006**, *67*, 175–182. [CrossRef]
35. Starkloff, T.; Stolte, J.; Hessel, R.; Ritsema, C. Understanding snowpack development at catchment scale, using comprehensive field observations and spatially distributed snow modelling. *Hydrol. Res.* **2017**, in press.
36. Flerchinger, G.N.; Saxton, K.E. Simultaneous heat and water model of a freezing snow-residue-soil system. I. Theory and development. *Trans. Am. Soc. Agric. Eng.* **1989**, *32*, 565–571. [CrossRef]
37. Li, R.; Shi, H.; Flerchinger, G.N.; Akae, T.; Wang, C. Simulation of freezing and thawing soils in Inner Mongolia Hetao Irrigation District China. *Geoderma* **2012**, *173*, 28–33. [CrossRef]
38. Flerchinger, G.N. *The Simultaneous Heat and Water (SHAW) Model: Technical Documentation*; Technical Report NWRC 2000-09; Northwest Watershed Research Center, USDA Agricultural Research Service: Boise, ID, USA, 2000.
39. Black, C.A.; Evans, D.D.; White, J.L.; Ensminger, L.E.; Clark, F.E. *Methods of Soil Analysis Physical and Mineralogical Properties, Including Statistics of Measurement and Sampling*; American Society of Agronomy: Madison, WI, USA, 1965; Part 1, Volume 9, pp. 1–770.
40. Flerchinger, G.N.; Sauer, T.J.; Aiken, R.A. Effects of crop residue cover and architecture on heat and water transfer at the soil surface. *Geoderma* **2003**, *116*, 217–233. [CrossRef]
41. McCauley, C.A.; White, D.M.; Lilly, M.R.; Nyman, D.M. A comparison of hydraulic conductivities, permeabilities and infiltration rates in frozen and unfrozen soils. *Cold Reg. Sci. Technol.* **2002**, *34*, 117–125. [CrossRef]
42. Øygarden, L.; Kværner, J.; Jenssen, P.D. Soil erosion via preferential flow to drainage systems in clay soils. *Geoderma* **1997**, *76*, 65–86. [CrossRef]
43. Lundberg, A.; Ala-Aho, P.; Eklo, O.; Klöve, B.; Kværner, J.; Stumpp, C. Snow and frost: Implications for spatiotemporal infiltration patterns—A review. *Hydrol. Process.* **2015**, *30*, 1230–1250. [CrossRef]
44. Boardman, J.; Poesen, J. *Soil Erosion in Europe*; John Wiley & Sons Inc.: Hoboken, NJ, USA, 2006; p. 855.

11

Analysis and Predictability of the Hydrological Response of Mountain Catchments to Heavy Rain on Snow Events: A Case Study in the Spanish Pyrenees

Javier G. Corripio [1,*] and Juan Ignacio López-Moreno [2]

1 Meteoexploration, Höttinger Gasse 21/17, A-6020 Innsbruck, Austria
2 Department Procesos Geoambientales y Cambio Global, Instituto Pirenaico de Ecología, C.S.I.C, E-50059 Zaragoza, Spain; nlopez@ipe.csic.es
* Correspondence: jgc@meteoexploration.com

Abstract: From 18 to 19 June 2013, the Ésera river in the Pyrenees, Northern Spain, caused widespread damage due to flooding as a result of torrential rains and sustained snowmelt. We estimate the contribution of snow melt to total discharge applying a snow energy balance to the catchment. Precipitation is derived from sparse local measurements and the WRF-ARW model over three nested domains, down to a grid cell size of 2 km. Temperature profiles, precipitation and precipitation gradient are well simulated, although with a possible displacement regarding the observations. Snowpack melting was correctly reproduced and verified in three instrumented sites, and according to satellite images. We found that the hydrological simulations agree well with measured discharge. Snowmelt represented 33% of total runoff during the main flood event and 23% at peak flow. The snow energy balance model indicates that most of the energy for snow melt during the day of maximum precipitation came from turbulent fluxes. This approach forecast correctly peak flow and discharge during normal conditions at least 24 h in advance and could give an early warning of the extreme event 2.5 days before.

Keywords: ROS; snow; rain; flood; WRF; numerical weather forecast; energy balance; discharge estimation; early alert system

---

1. Introduction

From June 18 to June 19, 2013, the Ésera river in the Pyrenees, Northern Spain, caused widespread damage due to flooding. Damage to public properties, such as roads or bridges, and to private property was estimated in excess of seven million euros. Over 300 people had to be evacuated from their homes, but fortunately there were no fatalities. The increased flow came as a result of torrential rains at a time when the catchment presented an anomalously extensive snow cover above 2000 m [1], and the 0 °C isotherm was above 4000 m a.s.l.

The role of rain on snow events (ROS) is interesting from a scientific and applied point of view, since it is widely recognised that they dominate much of the flood generation in mountainous and boreal regions [2,3]. Most of the largest floods in British Columbia, Washington, Oregon and California have been associated with ROS [4–6]. In Germany, ROS have much larger hydrological influence than rain alone in catchments above 400 m a.s.l. ROS have been identified as a primary cause of changes in channel morphology due to erosion [7–11]. For example, at a site in the Oregon Cascades, 85% of all landslides which could be accurately dated were associated with snowmelt during rainfall [7]. Sandersen et al. [12] reported that triggering of debris flows in Norway is caused by the combination of rainfall and snow melt. ROS events are also considered a major cause in the release of some type of avalanches [2].

The importance of ROS events is highlighted above, however there is uncertainty about the physical processes that control the runoff generation during their occurrence and their sensitivity to temperature and other meteorological variables. McCabe et al. [6] made a comprehensive study of several ROS events in the Pacific Northwest of the United States (PNW) concluding that the severity of a ROS events does not only depend on the magnitude of the precipitation but also on the elevation of the freezing level and the extent and characteristics of the existing snowpack. Marks et al. [13] studied a heavy ROS event (winter 1996) in the Pacific North West of the United States, where they found that 60–90% of the energy for snowmelt came from turbulent heat exchange. Condensation during the event was a significant contributor to the flood. Van Heeswijk et al. [14] concluded that wind speed together with vapour pressure and temperature gradients are the most important climate variables that control snow melt. However, in a later study Mazurkiewicz et al. [15] found that radiation was the largest contributor to melt during ROS in the Pacific North West. Other studies show the importance of the low permeability of the snowpack when it is saturated, leading to a quick generation of excess runoff. This indicates that heavy rain water moves several times faster than the natural snowmelt [2,11,16]. During ROS events the potential for flooding is increased if the soil is frozen [17,18].

Quantifying the actual contribution of snow melt and rain to runoff in a particular ROS event is challenging, as well as it is the extrapolation of results to other events or geographical areas. Such complexity makes it difficult to anticipate the hydrological response of a catchment, even when detailed meteorological forecasts are available. In the last decades the spatial resolution and the accuracy of atmospheric mesoscale models have improved rapidly. These models are now able to produce reliable meteorological fields over complex terrain when forced with realistic initial and boundary conditions [19–21]. Such fields can be used as inputs for hydrological models enabling the forecasting and analysis of extreme weather events, such as ROS, and their hydrological consequences. However, a small deviation in determining initial conditions, or biases in the input data for snow models may lead to considerable errors in the timing, magnitude and spatial distribution of the simulated flood [22,23].

In this study, we investigate the skill of a high resolution weather model—the weather research and forecasting model, advanced research WRF core, WRF-ARW [24]—to simulate and forecast the occurrence of this extreme precipitation event. The output of the WRF combined with land observations and remote sensing data are used to feed a physically-based snow energy balance model. This approach is useful to improve our understanding of ROS in several aspects:

1. To assess the suitability of snow models coupled with atmospheric models to simulate ROS events
2. To estimate the spatial distribution of discharge in several sub-catchments
3. To estimate the relative forcing of meteorological variables to snow melt
4. To estimate the relative contribution of snowmelt and rainfall to the final river discharge
5. To estimate how much time in advance a weather model can give an early warning of dangerous floods.

All these questions are key to understanding the nature and predictability of ROS events in the Pyrenees and other similar mountain regions.

2. Materials and Methods

2.1. Site of Study

The Ésera river is one of the most important tributaries of the Ebro river in North Eastern Spain. Most of its runoff is generated in the headwater area located in the central Spanish Pyrenees. The flood event described in this study affected mostly the upper portion of the catchment, therefore, we restrict ourselves to the area above the gauging station of Eriste, at 1050 m a.s.l. (Figure 1). It was near this location that the most destructive effects of the flood were experienced. The catchment area is 323 km^2

and contains five small sub-catchments: Eriste, Estós, Maladeta, Ramascaró and Vallibierna. They drain one of the highest ranges in the Pyrenees, with many peaks exceeding 3000 m a.s.l., including the highest summit in the whole mountain range: Aneto (3404 m a.s.l.). In this catchment 65.6% of the area is located above 2000 m a.s.l., and 26.4% over 2500 m a.s.l. [25]. The geology exhibits a rather complex structure with granites, slates, schist and limestones heavily fractured and folded [26]. There is a mismatch in the extent of the topographic and the hydrological catchments, since several sinkholes divert part of the flow to the Garona basin (French Pyrenees). These diverted flow could represent 12–25% of the annual runoff of the catchment [25,27]. The upper portion of the catchment corresponds to a high mountain landscape, with the highest concentration of small glaciers and ice fields of the mountain range [28].

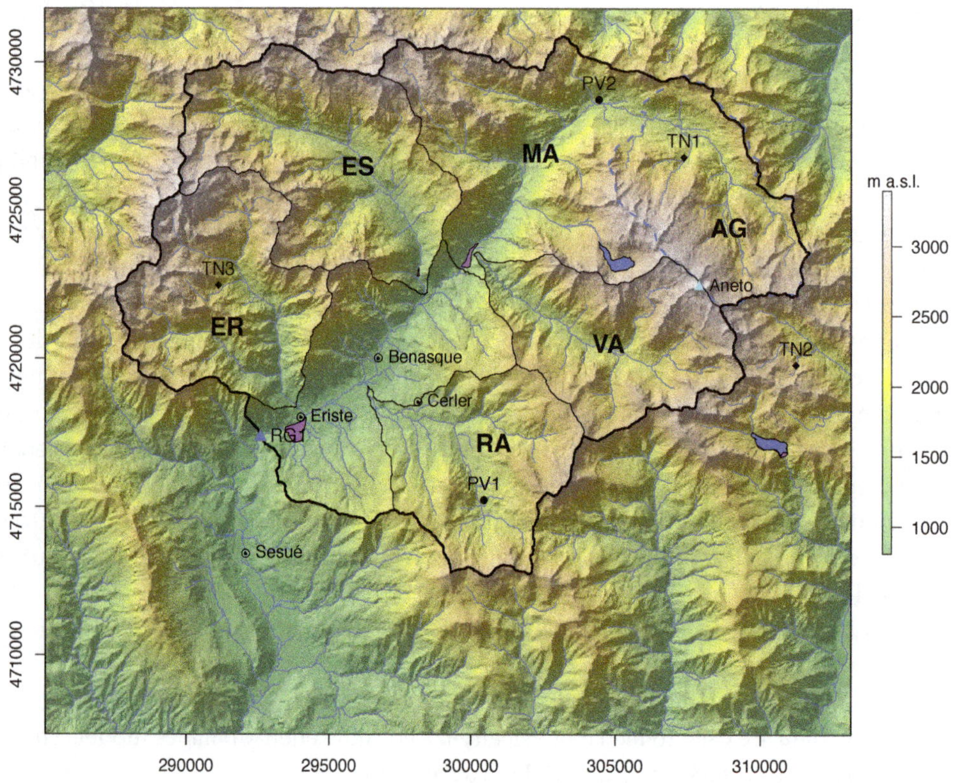

Figure 1. Map of the area of study. Coordinates are Universal Transverse Mercator UTM zone 31. The Ésera catchment and sub-catchments are shown. ER: Eriste, ES: Estós, MA: Maladeta, RA: Ramascaró and VA:Vallibierna, the Aiguallut endorheic sub-catchment is indicated by AG. Aneto, to the northeast is the highest peak in the range. Main towns are indicated by their full name and sensors listed in Table 1 by their acronyms. River network and natural lakes are drawn in blue and artificial dams in purple.

The estimated mean annual precipitation is 1840 mm [25], exceeding 2500 mm per year above 3000 m a.s.l. [29] Spring and autumn are the wettest seasons, while summer exhibits a marked dry period. Mean lapse rates have been estimated between −0.0049 and −0.0056 °C m^{-1} [30,31], and the 0 °C isotherm at around 1700 m a.s.l. during the cold season (November-April). This leads to a deep and extensive snow cover from late autumn to late spring. Consequently, the Ésera river exhibits a marked nival river regime with high flows in late spring and early summer. The lowest runoff occur in winter as consequence of snow accumulation, followed by late summer (August and September) in response to the lowest precipitation [32]. The annual runoff of the catchment is 309.2 hm^3 (9.8 m^3s^{-1} on average).

2.2. Data

There are two pluviometers from the Automatic Hydrologic Information System of the Ebro river basin (SAIH) in the whole catchment. Temperature measurements are recorded at three snow gauges ("telenivometers"), which also record snow height and snow water equivalent (SWE). Two of the snow gauges are inside the catchment and a third one very close to the catchment boundary. The location of measuring stations is indicated in Table 1. Rain gauges are tipping bucket pluviometers, which are not suitable for solid precipitation and need careful calibration (WMO 2008). Snow height is recorded by an ultrasonic sensor and snow water equivalent using a cosmic-ray snow gauge. All data are public but additional information on sensor type and specifications is limited. These measured data were used to evaluate the model output, but the snow model was fed with WRF data directly. Precipitation distribution is difficult to evaluate as there were only two measurement sites, insufficient for a correct spatial extrapolation.

Table 1. Coordinates of the measuring stations indicated in Figure 1. SWE is snow water equivalent, pcp is precipitation, temp is temperature and P is atmospheric pressure.

Station	Code	Longitud	Latitud	Elevation	variables
Pluviómetro en Cerler (Ampriu)	PV1	0.5690	42.5637	1900	pcp, temp
Llanos del Hospital	PV2	0.6127	42.6863	1752	pcp, temp
Río Esera en Eriste	RG1	0.4728	42.5807	1050	discharge
Telenivómetro en Renclusa	TN1	0.6490	42.6696	2180	SWE, depth, temp, P
Telenivómetro en Salenques	TN2	0.6987	42.6075	2600	SWE, depth, temp, P
Telenivómetro en Eriste	TN3	0.4529	42.6264	2350	SWE, depth, temp

Initial snow extent and albedo were derived from a Landsat image acquired from the U.S. Geological Survey and readily available through their online site http://earthexplorer.usgs.gov/. For the derivation of catchments and to calculate terrain parameters in the snow model, a Digital Elevation Model (DEM) was used. The original DEM at 5 m resolution was Lidar-generated and provided by the Spanish Geographical Office (IGN, Instituto Geográfico Nacional), this DEM was resampled at 30 m resolution in order to minimise artefacts, to make it compatible with the resolution of the Landsat images and to increase the computation speed of the snow model.

We approached the publicly funded Spanish Met Office (AEMET) for additional data, but were denied. We also tried to get data from the hydropower company operating the river dams, although this was an unsuccessful effort.

2.3. Remote Sensing

Initial albedo and snow cover extent were derived from a Landsat 8 image of 10 June 2013. The snow extent was obtained by estimating the Normalised Difference Snow Index (NDSI), according to Equation (1):

$$NDSI = \frac{\rho_2 - \rho_5}{\rho_2 + \rho_5} \tag{1}$$

where ρ_n is TOA (Top Of the Atmosphere) reflectance of Landsat band number n (see for example: [33]).

For the albedo we used the Landsat surface reflectance and the conversion from narrow band to broadband was done according to Wang et al. [34]. Additional corrections were made for the angle of incidence of the sun on the slope surface, which can vary greatly on mountain areas. The solar-terrain geometry was derived from the DEM following Corripio [35] and using the R package *insol* [36]. The resulting initial albedo is shown in Figure 2 and compared the original false colour Landsat image. The northern fringes of the image are partially covered in clouds, in these areas the snow cover limit was set according to the elevation. The limit was derived from the surrounding cloud-free areas as the mean lower elevation with snow. This limit was dependent on the orientation of the slope, and was calculated in steps of 15° from 0° to 345°.

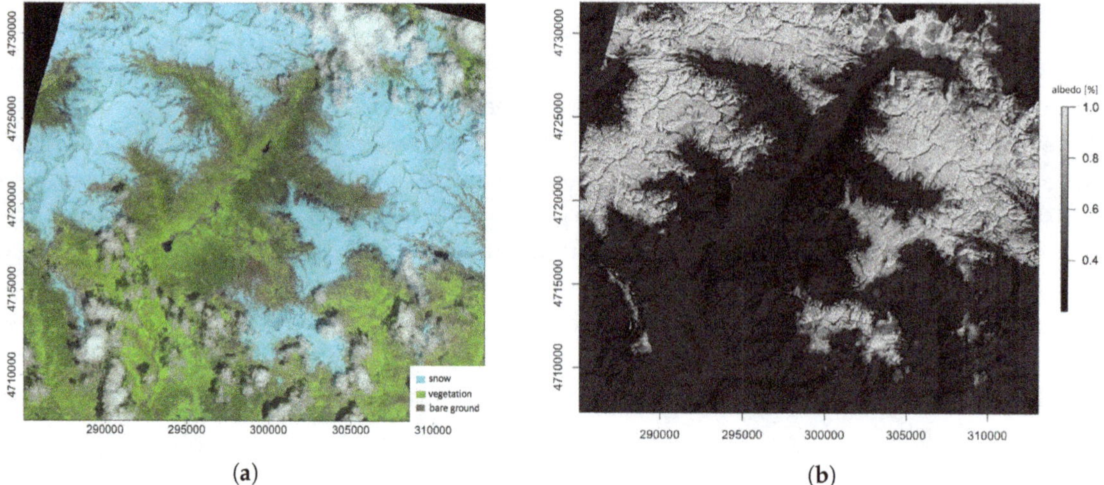

Figure 2. Initial conditions for snow cover and albedo derived from Landsat. (**a**) Landsat "natural look" image of 10 June 2013; (**b**) Derived albedo superimposed on a shaded relief of the DEM.

2.4. Snow Model

We use a multi-layered, distributed snow energy balance developed by Corripio [37], and applied successfully to previous studies in the Andes [38] and the Alps [39]. The energy balance of the snow pack can be described as in Equation (2):

$$EB = I_{\downarrow}(1-\alpha) + L_{W\downarrow} + L_{W\uparrow} + Q_S + Q_L + Q_r + Q_g \quad (2)$$

EB is the net energy balance at the snow pack, $I_{\downarrow}$ is incoming short-wave radiation, α is snow albedo, $L_{W\downarrow}$ is received long-wave radiation, both from the atmosphere and surrounding slopes, $L_{W\uparrow}$ is emitted long-wave radiation, Q_S is turbulent sensible heat transfer with the atmosphere, Q_L is latent heat transfer, Q_r is heat transfer due to precipitation, and Q_g is subsurface heat transfer within the snow pack.

If the snow pack reaches the melting point temperature EB will be the energy available for melting the snow, otherwise the net energy is used to change the snow pack temperature. Sublimation depends on the specific humidity gradient, whether the snow pack reaches melting point or not. Short wave incoming radiation was modified locally to account for shading, sky view factor and reflected radiation. A detailed description of the radiation components and the effect of surrounding topography is explained in Corripio [37]. Long Wave was derived from the WRF model, an additional cloud factor following Iziomon et al. [40] was applied to discriminate between the direct and diffuse radiation fraction depending on cloud cover. The WRF model calculates incoming long wave radiation independent of the local topography, so a further modification was applied to account for terrain influences such as sky view factor. Turbulent fluxes with the atmosphere were calculated according to the bulk aerodynamic flux method [41–44]. This method seems to be one of the most reliable according to a comparative study by Sexstone et al. [45]. Heat transfer due to precipitation was estimated according to Brun et al. [46]. The subsurface component of the model is a simplified three layer scheme where heat transfer is computed following Oke [47] with a simplified solar radiation snow extinction following Fukami et al. [48] and Warren [49]. Initial Albedo is derived from a Landsat image and local measurements, the temporal evolution of albedo is estimated according to an ageing parameterization considering local air temperature [50,51]. The temperature field was interpolated to the high resolution model according to local elevation and lapse rates from the WRF model. Snow-rain limit was prescribed by a wet bulb temperature of 1.3 °C [52,53]. This gives a satisfactory result but an estimation that includes the precipitation rate could be an improvement. Finding the correct solution is not easy as the identification of the correct snow line is difficult during heavy precipitation and on mountain ranges with sparse instrumentation. The detailed equations of the model and derived variables are given in the Appendix A.

2.5. Atmospheric Model

We run a nested WRF model ARW core [24], with three domains at 18, 6 and 2 km over the Pyrenees. The model initial and boundary conditions were derived from the GFS global model at 00Z every day and run for 30 h to overlap the next run. The available resolution of the GFS at the time was 0.5 degrees, while the current resolution is 0.25 degrees. We did not consider using data from the European Center for Medium-Range Weather Forecast model (ECMWF) for the initial conditions as they are prohibitively expensive and beyond the budget available for this study. The nested model was run daily from the 11th to the 26th of June.

The WRF model allows for different physical and dynamical settings and different radiative schemes, and these have an impact in the model ability to predict intense precipitation [54,55]. It would be desirable to make extensive tests to find which combination of settings perform better during heavy precipitation events in the Pyrenees. To our knowledge this is still to be tested. In the absence of this information, we decided to use the settings shown on Table 2, based on information from previous work in the region and other mountain areas using the WRF model [56–58]. For a detailed description of the settings see the WRF User's Guide [59].

Table 2. WRF physics settings.

Parameter	Domain 1	Domain 2	Domain 3	Scheme
Microphysics	6	6	6	WRF Single-Moment 6-class scheme
Cumulus Parameterization	1	0	0	Kain-Fritsch scheme, none
Long wave Radiation	5	5	5	New Goddard scheme
Short wave Radiation	5	5	5	New Goddard scheme

The output of the WRF model is in NetCDF format and the projection is in geographical longitude and latitude. To make it compatible with the DEM in Universal Transverse Mercator UTM projection, the files were reprojected using a Linux implementation of the GDAL - Geospatial Data Abstraction Library [60].

2.6. Discharge

From snowmelt and rainfall inputs river discharge was estimated using a parallel Single Linear Reservoir (SLR) model for every catchment. The SLR model assumes that storage S is linearly related to outflow Q by

$$S = KQ, \tag{3}$$

where K is the storage coefficient and has units of time [61].

Combining this equation with the hydrologic continuity equation e.g., [61,62] and integrating gives:

$$Q(t) = Q(t-1)e^{(-\Delta t/K)} + u(t)(1 - e^{(-\Delta t/K)}), \tag{4}$$

where u is inflow at time t.

Hannah and Gurnell [63] suggest deriving the parameter K from the gradient of a semilogarithmic plot of discharge over time when there is no recharge, following Equation (5):

$$K = \frac{-(t - t_0)}{log\, e\frac{Q_t}{Q_0}}, \tag{5}$$

However, deriving the parameter K is problematic without accurate discharge measurements. The problem in this case is that the only river gauge is located after a dam used for hydroelectric generation. Thus, the river flow may change considerably depending on whether turbines are operating or not. If electricity is generated, the stored water in the dam is evacuated directly through a pipeline and bypass the river gauge. To complicate things further, there are two additional dams upriver that channel water to a power station located at the lower dam. The outflow of the lower damn for electricity generation is 30 m^3s^{-1}; the capacity of the lower damn (Linsoles) is 3.5 hm^3; the two upper dams (Estós and Paso Nuevo) are linked to the lower dam by a pipe with a maximum flow of of 36 m^3s^{-1}. This situation makes difficult a proper calibration of the storage coefficient and therefore we have made an approximate calculation by trial and error. We used six different linear reservoirs corresponding to every sub-catchment, with different storage coefficients for snowmelt and rainfall. Travel times for every sub-catchment are taken into account and calculated with the GRASS module *r.traveltime* [64]. Singh et al. [2] recalls previous work showing that heavy rain may increase the speed of water drainage through snow, as more efficient channels develop in the snow pack. To account for this change we have modified the storage coefficient linearly with time.

The Aiguallut sub-basin on the north east of the main catchment (AG in Figure 1) is an endorheic basin, which drains to the Garona river, north of the area of study and already in French territory. It is a well developed karstic network with several intakes and an outflow at Uelhs deth Joeu to the Garona river [65,66]. Maximum measured outflow is 11–12 m^3s^{-1} [65], at the time of the flood event the main intake at Aiguallut was saturated and discharge overflew to the Ésera river, thus this maximum value was subtracted form the main discharge of the sub-basin.

3. Results

Figure 3 shows a good agreement between measured temperature and temperature simulated by WRF, with an overall mean error of 0.21 °C and a mean absolute error of 1.51 °C at La Renclusa (point TN1 in reference table and map). However, maximum measured temperature is higher (up to 3.5 °C) than simulated temperature during days with high incoming solar radiation. The temperature probes are shielded but not ventilated, which may result in an overestimation of the maximum measured value [67]. The WRF model simulates the amount of precipitation approximately, as seen in Figure 4. It shows the maximum simulated precipitation to the west of maximum measured precipitation, but the sensor network is too sparse to decide whether there is a displacement bias or it is a realistic simulation. The simulated north–south gradient in precipitation is also observed in the measured data, between Llanos del Hospital (PV2) and Ampriu (PV1) The geomorphological evidence also suggest a strong gradient as the Vallibierna sub-catchment (VA) was showing less sediment transport and deposition than in previous floods events in the area (Santiago Somolinos, personal communication).

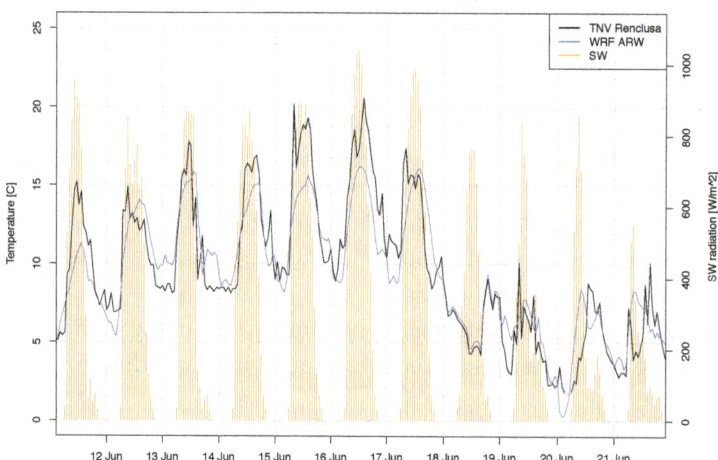

Figure 3. Comparison temperature simulated by the WRF model and measured at Renclusa "telenivometer" (TN1). Orange bars is shortwave radiation simulated by WRF at the same location.

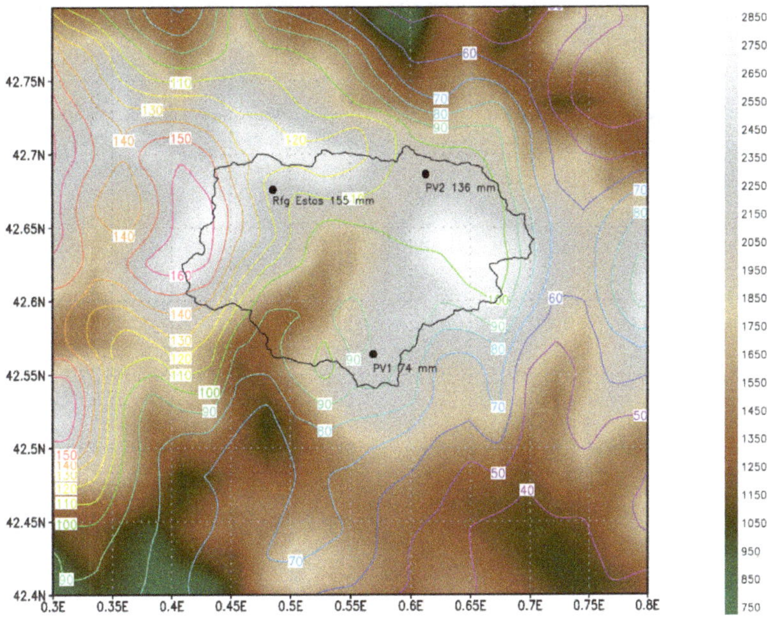

Figure 4. Simulated precipitation [mm] contour map by the WRF, in black is the Ésera catchment. Background colour is model surface elevation as indicated by the legend on the right side. The points indicate accumulated precipitation measurements at two tipping bucket pluviometers and in a totaliser pluviometer at the Refugio de Estós (Estós mountain Hut). The totaliser data is provided by Serrano Notivoli et al. [68], but they have not been verified.

Figure 5 shows the simulated depletion of snow height in three grid cells containing the cosmic ray snow gauge at La Renclusa, Salenques and Eriste (TN1, TN2 and TN3; at 2180, 2350 and 2600 m a.s.l. respectively). There are no data from radiometers or albedometers in the catchment, and therefore no measurements to check the validity of the model radiation inputs. This could cause large errors in the snow model simulation, nonetheless, the depletion curves of snow height are properly simulated during the previous days to the flood, as shown in Figure 5. During the ROS event, melting is slightly underestimated, except for Salenques, the highest site, where an apparent peak due snow accumulation is missed by the model. The Nash-Stuclife efficiency model coefficient [69] gives values ranging between 0.96 and 0.99, confirming the good agreement between measurements and simulations (Table 3). Figure 5 also shows that except in the highest TN (Salenques at 2600 m a.s.l.) snow cover completely disappeared by the end of the period. Figure 6 shows the snow extent retrieved from a Landsat image in 26th June, 2013; and the simulated snow cover at the same date. It confirms the ability of the model to simulate the spatial distribution of snow cover depletion during the previous days and during the ROS event.

During the day when peak flow occurred, the snow energy balance was dominated by a positive influx of energy from turbulent interchange with the atmosphere. Both sensible heat and latent heat transfer were positive, indicating condensation of atmospheric moisture on the snow (Figure 7). Shortwave radiation was second in importance, followed by advective heat transfer from the rain. Net long wave was positive, contrary to most days when there is an effective radiative cooling of the snow pack. Simulated subsurface fluxes were negligible during the ROS event, except for a short period early on the 19th of June when there was a positive flux toward the surface snow pack. For the calculation of heat transfer due to rain (Equation (A28)) it is necessary to know the temperature of the rain. Byers et al. [70] measured rain temperatures and found that after an intense precipitation event rain temperature was always close to air temperature (about 1 °C difference). For this study we assumed that the rain temperature was 1 °C colder than the air temperature. During the time of peak precipitation, simulated lapse rates above the surface were in the range between −0.003 and

-0.0045 °C m^{-1}, thus one degree colder means that the rain had the temperature of a layer between 200 and 300 m above the surface.

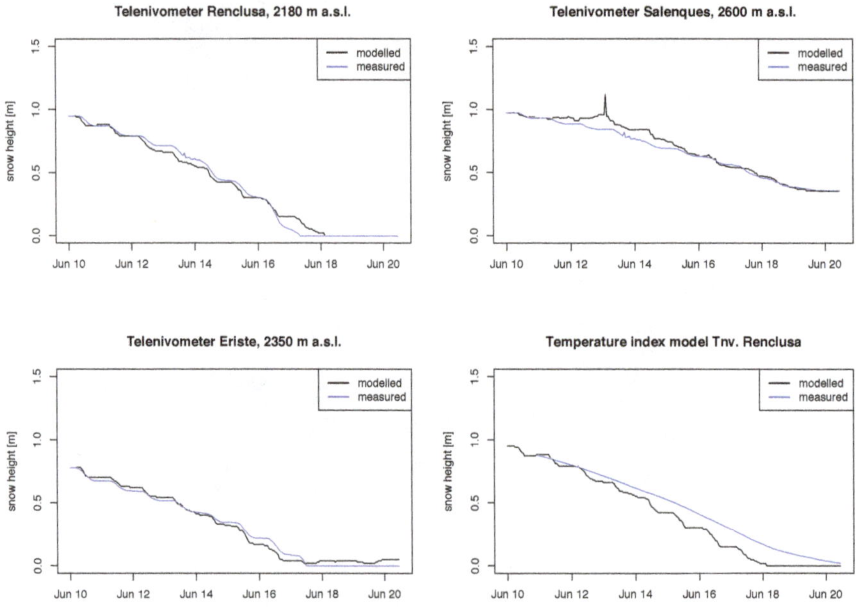

Figure 5. Snow height modelled and measured in three snow gauges (TN1, TN2 and TN3 in map in Figure 1) at the Ésera catchment. The lower panel is the result of a simpler temperature index model applied to the Renclusa site. The melt factor was computed as the best linear fit to previous 24 h accumulated positive degree days and corresponding melt. The calibration period was the week before the ROS event in this case the Nash-Sutcliffe efficiency value degrades to 0.89.

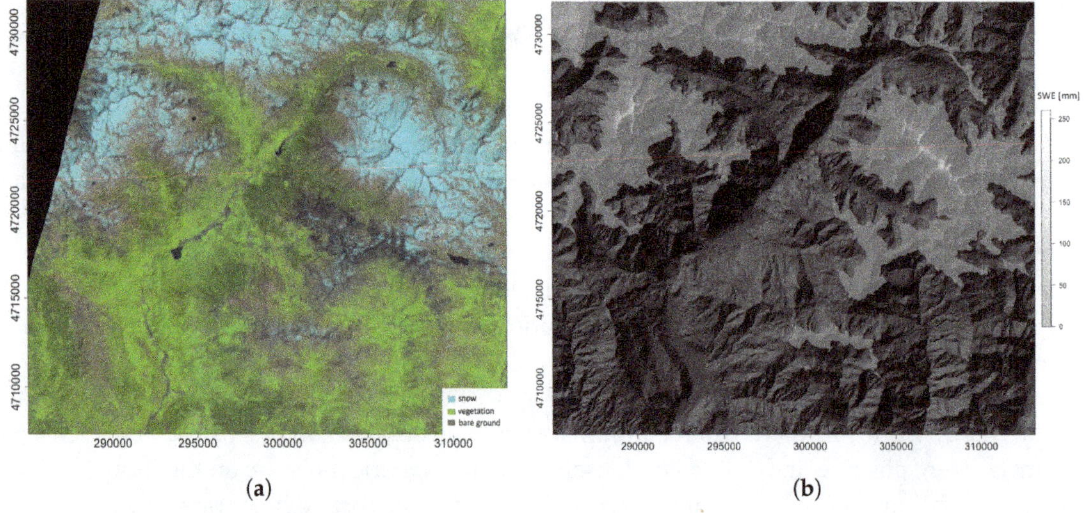

Figure 6. Observed and simulated snow cover after the event. (**a**) Landsat image 26 June 2013; (**b**) Simulated snow water equivalent (SWE) on the same day.

Table 3. Nash-Sutcliffe efficiency values for the measured and simulated snow ablation at three snow gauges.

Renclusa Snow Ablation	Eriste Snow Ablation	Salenque Snow Ablation	Total Discharge
0.9854	0.9835	0.9596	0.7561

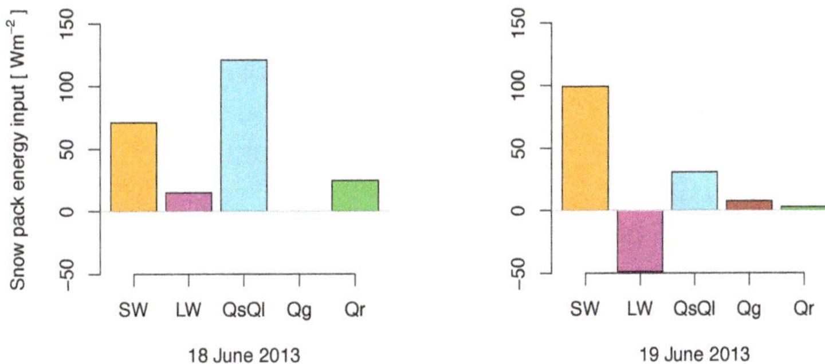

Figure 7. Snow energy balance flux partition. SW is net shortwave radiation, LW is net long wave radiation, QsQl is turbulent heat interchange with the atmosphere, Qg is heat transfer from the subsurface and Qr is advection of heat due to rain. Following convention, positive indicates flux towards the snow and negative flux is energy leaving the snowpack.

Figure 8 shows the observed and simulated discharge at Eriste gauging station, the grey uncertainty shade is ± 30 m^{3}s^{-1} corresponding to the intake of the hydroelectric plants. The peak discharge is well represented in the simulation (Nash-Sutcliffe coefficient of 0.76), but the river flow seems to decrease very slowly, with a secondary peak in precipitation that is larger than the observed variation. Rainfall is the main contribution to peak flow. Snow melt contribution is very well simulated before the ROS event, which adds confidence to the snow model results. Snowmelt was the main contributor to total water discharge during the previous days. Discharge was larger than the average for the season, probably causing saturation of soils and a very fast hydrological response to the rainfall event. During the flood event of the 18–19th of June estimated snowmelt represented 33% of total runoff, on the day of maximum precipitation (the 18th) snowmelt was calculated as 28% of total discharge and 23% at the time of peak flow.

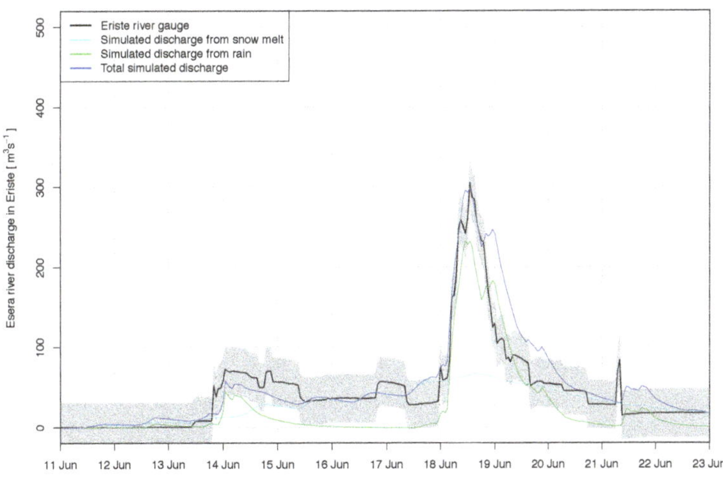

Figure 8. Observed and simulated discharge (total and originated from snowmelt and rain) at Eriste river gauge, error bands in measured discharge are ± 30 m^{3}s^{-1} which is the pipe discharge of the hydroelectric plant.

4. Discussion

We believe that the combination of high resolution numerical weather prediction models, together with energy balance snow models are excellent tools to predict snowmelt and mountain river discharge, both during extreme events such as intense ROS, or during average conditions. We also think that is useful to stress uncertainty and limitations, so that we can improve this kind of simulation in the future. Simulated precipitation from the WRF model approximates the timing and intensity of the ROS event of 18–19th June, 2013 in the Ésera river. There are large discrepancies with local measurements ranging from 8 to 38% difference in accumulated precipitation. The sensor network in the area is too sparse to detect where the spatial distribution is correct or there is a displacement bias, as has been noted in other WRF simulations in the Pyrenees [58], as well as in other regions and for other variables [71]. Pennelly et al. [57] has shown the impact of cumulus parameterizations on the location and magnitude of precipitation events at lower resolution. It is unclear if this effect is propagated to the higher resolution nested domains, even when they can allow for resolving rather than parameterizing convection. The simulation of the precipitation peak is good enough for regional scale simulations, but it would be a great advantage if location accuracy could be improved. A thorough investigation into the best parameterization and best resolution for accurate numerical precipitation forecast in these mountain ranges is strongly recommended.

There is a strong north to south precipitation gradient, both in the measurements and the simulations. There is probably a vertical gradient in precipitation, too. This has been observed in many other mountain regions. For example Liston and Elder [72] use a seasonally variable linear gradient, where precipitation increases with altitude, according to work in the western US by Thornton et al. [73]. Closer to the Pyrenees, in the Alps Lehning et al. [74] have noted the problems of deriving a reliable precipitation gradient, either because of lack of measurements at high altitude [75,76], or because the difficulty of measuring solid precipitation at high elevation and with strong winds [77]. In fact, extensive work in the Swiss Alps reveals precipitation gradients ranging from −69 to +15 mm/km in summer or from −27 to +25 mm/km in winter [78]. We do not yet have enough information to derive a reliable elevation gradient of precipitation in the Pyrenees, and therefore it was not applied to the spatial interpolation of precipitation over the model domain.

The simulation of snow melt at a single point is successful, with a mean Nash-Sutcliffe efficiency value of 0.98 (Table 3) and suggests that the energy balance approach to simulate snow melt is a reliable one. This approach works well both for clear days, when solar radiation is the main energy input, and for overcast and rainy days, when the main energy input are turbulent fluxes with the atmosphere and there is an additional influx of long-wave thermal radiation. There is some discrepancy at the highest snow gauge where the model underestimates the amount of fresh snow on the 13th of June. This is probably due to a simplistic estimation of snow-rain limit which does not take into account the precipitation rate. WRF modelled temperature is very similar to measured temperature, but unfortunately we do not have measured data on short-wave radiation, long-wave radiation nor albedo. A full meteorological station measuring these variables would be of great help to test further the reliability of the WRF model. It is interesting to note that the mean slope of ablation rate changes very little after the 17th of June on the highest snow gauge, even when there is a drop of temperature of almost 10 degrees. The reason may be the increase in downwelling long-wave radiation from a cloudier atmosphere, which compensates for the reduced energy input from lower air temperature and reduced solar radiation. This highlights the remarks of Pomeroy et al. [79] on the unsuitability of simplistic temperature index models to simulate snow ablation and rain on snow episodes. These type of models can perform very satisfactorily when there are hardly any data available (e.g., [80,81]), but are more limited for application ROS events, as shown in lower right panel of Figure 5.

The distributed snowmelt model performs very well when comparing the total output of meltwater to discharge on days when there is no precipitation (Figure 7). This is reassuring and indicates that this modelling tools are also useful for standard operational river discharge forecasts in snow-covered regions. This is useful to optimise planning of hydropower production and to minimise conflicts in water usage. On days when there is heavy precipitation, the sum of rainfall and snow melt agrees well with measured discharge (Nash-Sutcliffe efficiency value of 0.76 overall), but rainfall decays too slowly on the WRF model. During heavy precipitation, rain creates more efficient channels and discharge is faster through snow [2]. An additional reason for this behaviour of the model could be an overestimation of the snow line altitude. Minder et al. [82] have reported observations where the snow line on the mountainside is lower than the estimated snow line in the free atmosphere and provide a theoretical explanation supported by model simulations. The main reasons for the depression of the snow line on the mountain slope are (a) latent cooling from melting precipitation; (b) microphysical melting distance (a 10 mm snow flake would melt 100 m lower than a 5 mm snowflake); and (c) adiabatic cooling of orographically forced air masses. This situation has an important impact on runoff, for example in the Sierra Nevada of California a rise in the snow line of 610 m (1000 ft) would triple the runoff from three mountain catchments [83].

Ideally the storage coefficient to estimate discharge should change with time to reflect snow ripening and snow pack reduction, we have set it to vary linearly with time. Further calibration efforts are hindered by the artificial fluctuation of the river discharge in a catchment with hydropower use. This approach uses an empirical calibration, which is not ideal to simulate extreme events. For future work it would be advisable to get enough information on land cover and soil properties to implement different runoff models that are less sensitive to calibration (for example [79]).

At peak flow during the flood event, simulated discharge due to snow melt is 23% of total discharge, and 28% during the day of maximum precipitation, with rainfall being the main contribution to the flood. Problems caused in the catchment due to excess precipitation were exacerbated by a very high freezing level, which caused most of the precipitation to fall as rain, with only some snowfall towards the end of the event on the highest peaks. We did some tests with the GFS and the WRF models (not shown in this paper) to evaluate the lead times for a reliable warning in future cases. Both models identify well the height of the 0 °C isotherm up to 2.5 days before the event. GFS detects unusual precipitation but lacks resolution to identify the valleys most likely to be affected. WRF underestimates precipitation three days ahead and forecasts it approximately correct 24 h before the event.

5. Conclusions

We have shown that mesoscale numerical weather prediction models as the WRF are useful tools to predict the timing and location of rain on snow and flood events in mountain catchments of complex topography. Estimating the exact amount of precipitation is still the weakest point, although it can be improved by a systematic evaluation of current models spatial resolution and parameterization options [54]. An energy balance of the snow pack coupled with precise treatment of terrain parameters is able to simulate the snow depletion in the catchment and therefore the snowmelt input to discharge. Agreement between simulated and measured discharge is excellent in days without precipitation, which highlights the usefulness of these models for operational flow forecasting. The partition of energy fluxes show that the main input during this ROS event was turbulent fluxes of sensible and latent heat. Both fluxes were positive, implying strong condensation of atmospheric moisture. A simplified linear reservoir model to calculate the timing of the discharge gives satisfactory results, but can be improved by applying models that are less prone to calibration. This, however, requires extensive data on land cover, soil types and depth and snow depth. An additional source of uncertainty is the actual transition between snow and rain, we have used a simplified approach based on the local wet bulb temperature. This gives a satisfactory result but an estimation that includes the precipitation rate could be an improvement. Finding the correct solution is not easy as the identification of the correct snow line is difficult during heavy precipitation and on mountain ranges with sparse instrumentation. We have also found that both global and mesoscale models can give lead times of at least 2.5 days on severe ROS and flood events. The WRF model coupled with an energy balance model of the snow can give satisfactory results 24 h in advance, which can be very helpful to to implement early warning systems and to avoid and mitigate potential flood damages.

Acknowledgments: We are grateful to the Automatic Hydrologic Information System of the Ebro river basin (SAIH, http://195.55.247.237/saihebro/index.php for making meteorological and discharge data easily available online). We thank Lorna Raso for help with GIS processing, map production and the English editing of the final version, and Jose Luís Santiago Somolinos, a geologist and local resident, for first hand information on the flood event, geomorphological evidence on debris and sediment distribution and snow cover at the time. We are also grateful to two anonymous reviewers whose suggestions greatly improved the final manuscript. Part of the research of this paper is framed in the objectives of the project IBERNIEVE (CGL201452599-P) from MINECO, I+D+i Spanish National Plan. meteoexploration.com provided computer servers and HPC time.

Author Contributions: Juan Ignacio López-Moreno conceived the original research idea and contributed to data collection and data analysis. Javier G. Corripio did the atmospheric, snow and discharge modelling and also contributed to data collection and data analysis.

Conflicts of Interest: The authors declare no conflict of interest. There was no funding for this study, therefore the founding sponsors had no role in the design of the study; in the collection, analyses, or interpretation of data; in the writing of the manuscript, and in the decision to publish the results.

Abbreviations

The following abbreviations are used in this manuscript:

ARW	Advanced research WRF core
DEM	Digital Elevation Model
GFS	Global Forecast System
HPC	High Performance Computing
ROS	Rain On Snow
SAIH	Automatic Hydrologic Information System of the Ebro river basin
SLR	Single Linear Reservoir
SWE	Snow Water Equivalent
UTM	Universal Transverse Mercator
WRF	Weather Research and Forecasting Model

Appendix A

Here we describe the detailed components of the energy balance of the snow pack. The incoming global solar radiation on every grid cell of the DEM is calculated as:

$$I_G = I_n \cos\theta f_{sh}(1-\alpha) + I_d f_{skv} + I_{dr}(1-f_{skv}) f_{rsh} \tag{A1}$$

where I_n is direct normal radiation $\cos\theta$ is the cosine of the angle of incidence of the sun on the terrain slope, f_{sh} is a shading factor (0 if in shade, 1 if in sun), α is the snow albedo, I_d is diffuse solar radiation, I_{dr} is diffuse solar radiation reflected from the surrounding slopes, f_{skv} is the sky view factor or fraction of upper hemisphere visible from the DEM cell, and f_{rsh} is an estimation of the ratio of surrounding cells in the shadow.

The ratio of direct to diffuse solar radiation is given by some compilations of the WRF or can be derived from a parametric solar radiation model, we used that of [84,85] and additional work by [86–88] for the estimation on cloudy days. The cosine of incidence of sun on the slope is the dot product

$$\vec{s} \cdot \vec{n}, \tag{A2}$$

where $\vec{s}$ is a unit vector in the direction of the sun:

$$\vec{s} = \begin{pmatrix} -\sin\omega\cos\delta \\ \sin\varphi\cos\omega\cos\delta - \cos\varphi\sin\delta \\ \cos\varphi\cos\omega\cos\delta + \sin\varphi\sin\delta \end{pmatrix} \tag{A3}$$

with φ as latitude, δ is declination, ω is the hour angle and $\vec{n}$ is the unit vector normal to the surface. For a regular DEM of cell size ℓ and elevation values z at row and column i, j respectively, $\vec{n}$ would be:

$$\vec{n} = \begin{pmatrix} 1/2\,\ell\left(z_{i,j} - z_{i+1,j} + z_{i,j+1} - z_{i+1,j+1}\right) \\ 1/2\,\ell\left(z_{i,j} + z_{i+1,j} - z_{i,j+1} - z_{i+1,j+1}\right) \\ \ell^2 \end{pmatrix} \tag{A4}$$

For a detailed description of the sun terrain geometry and the calculation of sky view factor see Corripio [35].

Long wave downwelling radiation $L_{\downarrow}$ is derived from the WRF model and compared to clear sky long-wave flux to determine the degree of cloud cover.

$$L_{\downarrow} = \epsilon_a \sigma T_s^4 f_{skv} + \epsilon_s \sigma T_s^4 (1 - f_{skv}) \tag{A5}$$

where ϵ is emissivity, T is temperature (K) with subscripts a and s for atmosphere and snow respectively and σ is the Stefan-Boltzman constant (5.6704×10^{-8} Wm^{-2}K^{-4}).

From this, atmospheric emissivity is calculated according to Prata [89] as

$$\epsilon_a = 1 - (1 + w_p)\, e^{-(1.2 + 3w_p)^{0.5}} \tag{A6}$$

where precipitable water (w_p) is calculated following an empirical equation given also by Prata [89]:

$$w_p = 46.5 \frac{e_0}{T_a} \tag{A7}$$

where T_a is screen level air temperature in K. The actual vapour pressure is $e_0 = e^* RH$, where e^* is saturated vapour pressure computed following [90] polynomials (Equation (A8)), and RH is relative humidity from 0.0 to 1.0.

$$\begin{aligned}
a_0 &= 6984.505294 \\
a_1 &= -188.9039310 \\
a_2 &= 2.133357675 \\
a_3 &= -1.288580973e-2 \\
a_4 &= 4.393587233e-5 \\
a_5 &= -8.023923082e-8 \\
a_6 &= 6.136820929e-11
\end{aligned}$$

$$e^* = a_0 + T(a_1 + T(a_2 + T(a_3 + T(a_4 + T(a_5 + Ta_6))))) \tag{A8}$$

For the saturation vapour pressure over ice, the polynomials are:

$$\begin{aligned}
T &= T - 273.15 \\
a_0 &= 6.109177956 \\
a_1 &= 5.03469897e-1 \\
a_2 &= 1.886013408e-2 \\
a_3 &= 4.176223716e-4 \\
a_4 &= 5.824720280e-6 \\
a_5 &= 4.838803174e-8 \\
a_6 &= 1.838826904e-10
\end{aligned}$$

The Obukhov's stability length L is defined as:

$$L = \frac{-u_*^3 \rho}{kg\left[\left(\frac{H}{T_a C_p}\right) + 0.61\, E\right]} \tag{A9}$$

where g is the acceleration of gravity [41,42].

The profiles for wind speed, water vapour and heat, considering measurements at one level and at the surface, as:

$$u_* = \frac{\bar{u}k}{\ln\left(\frac{z-d_0}{z_{0m}}\right) - \Psi(\zeta)} \tag{A10}$$

$$E = \frac{(q_a - q_s)a_v\, k\, u_*\, \rho}{\ln\left(\frac{z_q-d_0}{z_{0v}}\right) - \Psi_{sv}(\zeta)} \tag{A11}$$

$$H = \frac{(\theta_a - \theta_s)a_h\, k\, u_*\, \rho\, C_p}{\ln\left(\frac{z_t-d_0}{z_{0h}}\right) - \Psi_{sh}(\zeta)} \tag{A12}$$

Subscripts m, v, h refer to momentum, water vapour and heat respectively.

The Ψ_s functions are stability corrections, which depend on the Obukhov's stability length. There is some disagreement on their values, which have been obtained empirically [91]. The values used here are similar to those used by Marks and Dozier [92], also reported by Brutsaert [91]. Thus, for unstable conditions ($\zeta = (z_u - d_0)/L \leq 0$):

$$\Psi_{sv}(\zeta) = 2\,ln\left(\frac{1+\chi^2}{2}\right) \tag{A13}$$

$$\Psi_{sm}(\zeta) = 2\,ln\left(\frac{1+\chi}{2}\right) + ln\left(\frac{1+\chi^2}{2}\right) - 2\arctan\chi + \frac{\pi}{2} \tag{A14}$$

$$\Psi_{sh}(\zeta) = 2\,ln\left(\frac{1+\chi^2}{2}\right) \tag{A15}$$

with $\chi = (1 - 16\zeta)^{1/4}$.

For stable conditions ($\zeta = (z - d_0)/L > 0$):

$$\Psi_s(\zeta) = -\beta\zeta \quad \text{and} \quad \beta = 5 \tag{A16}$$

where

$$\zeta = \frac{z - d_0}{L} \tag{A17}$$

The transfer mechanisms of momentum and of other scalar admixtures are different at the surface, and consequently the roughness lengths have different values for momentum, water vapour and heat, which were calculated following Andreas [43]. He developed a fitting polynomial of the scalar roughness lengths over snow and sea ice (the coefficients for this are given in Table A1):

$$\ln(z_s/z_0) = b_0 + b_1 \ln R_e + b_2(\ln R_e)^2 \tag{A18}$$

where z_s refers to water vapour or heat depending on the coefficients used. The Reynolds number R_e is a ratio of inertial to viscous forces [93] defined as:

$$R_e = \frac{u_*}{z_0 \nu} \tag{A19}$$

and $\nu = \mu/\rho$ is the kinematic viscosity of air, with μ the coefficient of dynamic viscosity for air.

Table A1. Values of the coefficients in the polynomials (Equation (A18)) that predict z_0/z for temperature and water vapour [43] for smooth, transition and rough surfaces.

	$R_e \leq 0.135$	$0.135 < R_e < 2.5$	$2.5 \leq R_e \leq 1000$
Temperature			
b_0	1.250	0.149	0.317
b_1	–	−0.550	−0.565
b_2	–	–	−0.183
Water vapour			
b_0	1.610	0.351	0.396
b_1	–	−0.628	−0.512
b_2	–	–	−0.180

The air density ρ for a given pressure is computed as the sum of densities of dry air ρ_d and water vapour ρ_v [91,94]:

$$\rho_d = p_d/(R_d * T_a) \tag{A20}$$

$$\rho_v = 0.622 * e_0/(R_d * T_a) \tag{A21}$$

$$\rho = \rho_d + \rho_v \tag{A22}$$

where the partial pressure of water vapour is $p_d = p_z - e_0$ and p_z is atmospheric pressure, R_d is the gas constant for dry air and $0.622 = 18.016/28.966$ is the ratio of the molecular weights of water and dry air. The specific humidity at screen level q_a and on the snow surface q_s is calculated as:

$$q = \rho_v/\rho \tag{A23}$$

For a one dimensional system atmosphere—snow surface layer—subsurface snow pack the thermodynamic equation describing the conservation of energy, neglecting melting and refreezing, can be described as [95,96]:

$$\rho_s c_s \frac{\partial T}{\partial t} = \frac{\partial}{\partial z}\left(k_s \frac{\partial T}{\partial z}\right) - \frac{\partial Q_s}{\partial z} \tag{A24}$$

This equation is applied in a very simplified form following Oke [47] and implemented using finite differences:

$$\frac{\Delta Q_s}{\Delta z} = C_s \frac{\Delta T}{\Delta t} \tag{A25}$$

where ΔQ_s is the flux density energy used to heat the snow at the surface, and C_s is the heat capacity of the snow $C_s = \rho_s c_s$, ρ_s is snow density and c_s is the specific heat of snow (2.09×10^3 Jkg^{-1}K^{-1}). As the main energy input comes from the surface, while the subsurface remains at a relatively stable temperature, we need to consider the thermal conductivity of the snow and the energy losses to subsurface layers:

$$Q_g = -\kappa_{Hs} C_s \frac{\partial T}{\partial z} \simeq -k_s \frac{\Delta T}{\Delta z} \tag{A26}$$

where κ_{Hs} is the thermal diffusivity of the snow and k_s its thermal conductivity.

The volume of snow considered is that affected by the daily temperature changes at the surface. This temperature oscillation decreases exponentially with depth, and its amplitude at any given distance from the surface can be estimated according to Oke [47] by

$$\Delta T_z = \Delta = T_0 e^{-z(\pi \kappa_{Hs} P)^{1/2}} \tag{A27}$$

where P is the wave period. Brutsaert [91] has shown that, roughly, 95% of the wave is damped at a depth of $3(2\kappa_{Hs}/\omega)^{1/2}$, where $\omega = 2\pi/P$.

Heat transfer due to precipitation Q_r is estimated following Brun et al. [46] as:

$$Q_r = \rho C_p P_r (T_r - T_s) \tag{A28}$$

where ρ is rain water density, C_p is specific heat of water, P_r is precipitation rate and T_s is temperature of the snow and T_r temperature of the rain.

References

1. Añel, J.A.; López-Moreno, J.I.; Otto, F.E.L.; Vicente-Serrano, S.; Schaller, N.; Massey, N.; Buisán, S.T.; Allen, M.R. The extreme snow accumulation in the Western Spanish pyrennes during winter and spring 2013. In *Explaining Extreme Events of 2013 from a Climate Perspective*; Herring, S.C., Hoerling, M.P., Peterson, T.C., Stott, P.A., Eds.; Bulletin of the American Meteorological Society: Providence, RI, USA, 2014; Volume 95, Chapter 21, pp. S73–S76, doi: 10.1175/1520-0477-95.9.S1.1.
2. Singh, P.; Spitzbart, G.; Hübl, H.; Weinmeister, H. Hydrological response of snowpack under rain-on-snow events: A field study. *J. Hydrol.* **1997**, *202*, 1–20.
3. Morán-Tejeda, E.; López-Moreno, J.; Stoffel, M.; Beniston, M. Rain-on-snow events in Switzerland: Recent observations and projections for the 21st century. *Clim. Res.* **2016**, *71*, 111–125.
4. Kattelmann, R. Flooding from Rain-on-Snow Events in the Sierra Nevada. In Proceedings of the North American Water and Environment Congress & Destructive Water (ASCE), Anaheim, CA, USA, 22–28 June 1996; Bathala, C., Ed.; pp. 1145–1146.
5. Brunengo, M.J. A method of modeling the frequency characteristics of daily snow amount, for stochastic simulation of rain-on-snowmelt events. In Proceedings of the 58th Annual Western Snow Conference, Sacramento, CA, USA, 17–19 April 1990; Volume 58, pp. 110–121.
6. McCabe, G.J.; Hay, L.E.; Clark, M.P.; McCabe, G.J.; Hay, L.E.; Clark, M.P. Rain-on-Snow Events in the Western United States. *Bull. Am. Meteorol. Soc.* **2007**, *88*, 319–328.
7. Harr, R. Some characteristics and consequences of snowmelt during rainfall in western Oregon. *J. Hydrol.* **1981**, *53*, 277–304.
8. Harr, R.D. Effects of Clearcutting on Rain-on-Snow Runoff in Western Oregon: A New Look at Old Studies. *Water Resour. Res.* **1986**, *22*, 1095–1100.
9. Christner, J.; Harr, R.D. Peak streamflows from the transient snow zone, western Cascades, Oregon. In Proceedings of the 50th Annual Western Snow Conference, Reno, NC, USA, 19–23 April 1982.
10. Bergman, J.A. Rain-on-snow and soil mass failure in the Sierra Nevada of California. In *Landslide Activity in the Sierra Nevada during 1982 and 1983*; DeGraff, J.V., Ed.; USDA Forest Service: San Francisco, CA, USA, 1987; pp. 15–26.
11. Jones, J.A.; Perkins, R.M. Extreme flood sensitivity to snow and forest harvest, western Cascades, Oregon, United States. *Water Resour. Res.* **2010**, *46*, W12512.
12. Sandersen, F.; Bakkehøi, S.; Hestnes, E.; Lied, K. The influence of meteorological factors on the initiation of debris flows, rockfalls, rockslides and rockmass stability. *Publ. Nor. Geotek. Inst.* **1997**, *201*, 97–114.
13. Marks, D.; Kimball, J.; Tingey, D.; Link, T. The sensitivity of snowmelt processes to climate conditions and forest cover during rain-on-snow: A case study of the 1996 Pacific Northwest flood. *Hydrol. Process.* **1998**, *12*, 1569–1587.
14. Van Heeswijk, M.; Kimball, J.; Marks, D. *Simulation of Water Available for Runoff in Clearcut Forest Openings During Rain-on-Snow Events in the Western Cascade Range of Oregon and Washington*; Technical Report; U.S. Geological Survey: Tacoma, WA, USA, 1996.
15. Mazurkiewicz, A.B.; Callery, D.G.; McDonnell, J.J. Assessing the controls of the snow energy balance and water available for runoff in a rain-on-snow environment. *J. Hydrol.* **2008**, *354*, 1–14.
16. Pradhanang, S.M.; Frei, A.; Zion, M.; Schneiderman, E.M.; Steenhuis, T.S.; Pierson, D. Rain-on-snow runoff events in New York. *Hydrol. Process.* **2013**, *27*, 3035–3049.
17. Shanley, J.; Chalmers, A. The effect of frozen soil on snowmelt runoff at Sleepers River, Vermont. *Hydrol. Process.* **1999**, *13*, 1843–1857.
18. Niu, G.Y.; Yang, Z.L. Effects of vegetation canopy processes on snow surface energy and mass balances. *J. Geophys. Res. Atmos.* **2004**, *109*, doi:10.1029/2004JD004884.
19. Westrick, K.J.; Mass, C.F.; Westrick, K.J.; Mass, C.F. An Evaluation of a High-Resolution Hydrometeorological Modeling System for Prediction of a Cool-Season Flood Event in a Coastal Mountainous Watershed. *J. Hydrometeorol.* **2001**, *2*, 161–180.

20. Fiori, E.; Comellas, A.; Molini, L.; Rebora, N.; Siccardi, F.; Gochis, D.; Tanelli, S.; Parodi, A. Analysis and hindcast simulations of an extreme rainfall event in the Mediterranean area: The Genoa 2011 case. *Atmos. Res.* **2014**, *138*, 13–29.
21. Würzer, S.; Jonas, T.; Wever, N.; Lehning, M. Influence of Initial Snowpack Properties on Runoff Formation during Rain-on-Snow Events. *J. Hydrometeorol.* **2016**, *17*, 1801–1815.
22. Wever, N.; Jonas, T.; Fierz, C.; Lehning, M. Model simulations of the modulating effect of the snow cover in a rain-on-snow event. *Hydrol. Earth Syst. Sci.* **2014**, *18*, 4657–4669.
23. Förster, K.; Meon, G.; Marke, T.; Strasser, U. Effect of meteorological forcing and snow model complexity on hydrological simulations in the Sieber catchment (Harz Mountains, Germany). *Hydrol. Earth Syst. Sci.* **2014**, *18*, 4703–4720.
24. Skamarock, W.C.; Klemp, J.B.; Dudhia, J.; Gill, D.O.; Barker, D.M.; Duda, M.G.; Huang, X.Y.; Wang, W.; Powers, J.G. *A Description of the Advanced Research WRF Version 3. NCAR/TN-475+STR*; Technical Report; NCAR: Boulder, CO, USA, 2008.
25. García-Ruiz. J.M.; Puigdefábregas, J.; Creus, J. *Los Recursos Hídricos Superficiales del Alto Aragón (Surface Water Resources of the High Aragón)*; Technical Report; Instituto de Estudios Altoaragoneses: Huesca, Spain, 1985.
26. Martínez de Pisón, E. *Morfoestructuras del Valle de Benasque (Pirineo Aragonés)*; Anales de Geografía de la Universidad Complutense: Madrid, Spain, 1990; pp. 121–148.
27. López-Moreno, J.I.; Beguería, S.; García-Ruiz, J.M. Influence of the Yesa reservoir on floods of the Aragón River, central Spanish Pyrenees. *Hydrol. Earth Syst. Sci.* **2002**, *6*, 753–762.
28. Chueca Cía, J.; Andrés, A.J.; Saz Sánchez, M.; Novau, J.C.; López Moreno, J. Responses to climatic changes since the Little Ice Age on Maladeta Glacier (Central Pyrenees). *Geomorphology* **2005**, *68*, 167–182.
29. Rijckborst, H. Hydrology of the Upper-Garonne basin (Valle de Arán, Spain). *Leidse Geol. Meded.* **1967**, *40*, 1–74.
30. Lampre, F. *Estudio Geomorfológico de Ballibierna, Macizo de la Maladeta, Pirineo Aragonés. Modelado Glaciar y Periglaciar (Geomorphologic Study of Ballibierna, Maladeta Sector, Aragonés Pyrenees. Glacial and Periglacial Forms)*; Technical Report; Consejo de Protección de la Naturaleza de Aragón: Zaragoza, Spain, 1998.
31. Del Barrio, G.; Creus, J.; Puigdefabregas, J. Thermal Seasonality of the High Mountain Belts of the Pyrenees. *Mt. Res. Dev.* **1990**, *10*, 227.
32. López-Moreno, J.I.; Nogués-Bravo, D. Interpolating local snow depth data: An evaluation of methods. *Hydrol. Process.* **2006**, *20*, 2217–2232.
33. Hall, D.; Riggs, G. Normalized-Difference Snow Index (NDSI). In *Encyclopedia of Snow, Ice and Glaciers SE-376*; Singh, V., Singh, P., Haritashya, U., Eds.; Encyclopedia of Earth Sciences Series; Springer: Dordrecht, The Netherlands, 2011; pp. 779–780.
34. Wang, Z.; Erb, A.M.; Schaaf, C.B.; Sun, Q.; Liu, Y.; Yang, Y.; Shuai, Y.; Casey, K.A.; Román, M.O. Early spring post-fire snow albedo dynamics in high latitude boreal forests using Landsat-8 OLI data. *Remote Sens. Environ.* **2016**, *185*, 71–83.
35. Corripio, J.G. Vectorial algebra algorithms for calculating terrain parameters from DEMs and the position of the sun for solar radiation modelling in mountainous terrain. *Int. J. Geogr. Inf. Sci.* **2003**, *17*, 1–23.
36. Corripio, J.G. Insol: Solar Radiation. R Package. 2013. Available online: https://cran.r-project.org/package=insol (accessed on 28 March 2017).
37. Corripio, J.G. Modelling the Energy Balance of High Altitude Glacierised Basins in the Central Andes. Ph.D. Thesis, The University of Edinburgh, Edinburgh, UK, 2003.
38. Corripio, J.G.; Purves, R.S. Surface Energy Balance of High Altitude Glaciers in the Central Andes: The Effect of Snow Penitentes. In *Climate and Hydrology in Mountain Areas*; de Jong, C., Collins, D., Ranzi, R., Eds.; Wiley: Hoboken, NJ, USA, 2005; pp. 15–27.
39. Dadic, R.; Corripio, J.G.; Burlando, P. Mass-balance estimates for Haut Glacier d'Arolla, Switzerland, from 2000 to 2006 using DEMs and distributed mass-balance modeling. *Ann. Glaciol.* **2008**, *49*, 22–26.
40. Iziomon, M.G.; Mayer, H.; Matzarakis, A. Downward atmospheric longwave irradiance under clear and cloudy skies: Measurement and parameterization. *J. Atmos. Sol. Terr. Phys.* **2003**, *65*, 1107–1116.
41. Businger, J.A.; Yaglom, A.M. Introduction to Obukhov's paper on 'Turbulence in an atmosphere with non-uniform temperature'. *Bound. Layer Meteorol.* **1971**, *2*, 3–6.
42. Obukhov, A.M. Turbulence in an Atmosphere with Non-Uniform Temperature. *Bound. Layer Meteorol.* **1971**, *2*, 7–29.

43. Andreas, E.L. A theory for the scalar roughness and the scalar transfer coefficients over snow and sea ice. *Bound. Layer Meteorol.* **1987**, *38*, 159–184.
44. Andreas, E.L. Parameterizing Scalar Transfer over Snow and Ice: A Review. *J. Hydrometeorol.* **2002**, *3*, 417–432.
45. Sexstone, G.A.; Clow, D.W.; Stannard, D.I.; Fassnacht, S.R. Comparison of methods for quantifying surface sublimation over seasonally snow-covered terrain. *Hydrol. Process.* **2016**, *30*, 3373–3389.
46. Brun, E.; Martin, E.; Simon, V.; Gendre, C.; Coleou, C. An energy and mass model of snow cover suitable for operational avalanche forecasting. *J. Glaciol.* **1989**, *35*, 333–342.
47. Oke, T.J. *Boundary Layer Climates*; Methuen: London, UK, 1987.
48. Fukami, H.; Kojima, K.; Aburakawa, H. The extinction and absorption of solar radiation within a snow cover. *Ann. Glaciol.* **1985**, *6*, 118–122.
49. Warren, S.G. Optical properties of snow. *Rev. Geophys. Space Phys.* **1982**, *20*, 67–89.
50. Brock, B.W.; Willis, I.C.; Sharp, M.J. Measurement and parameterisation of albedo variations at Haut Glacier d'Arolla, Switzerland. *J. Glaciol.* **2000**, *46*, 675–688.
51. Oerlemans, J.; Knapp, W.H. A 1 year record of global radiation and albedo in the ablation zone of Morteratschgletscher, Switzerland. *J. Glaciol.* **1998**, *44*, 231–238.
52. Frick, C. The Numerical Modeling of Wet Snowfall Events. Ph.D. Thesis, ETH Zürich, Zürich, Switzerland, 2012.
53. Tobin, C.; Rinaldo, A.; Schaefli, B. Snowfall Limit Forecasts and Hydrological Modeling. *J. Hydrometeorol.* **2012**, *13*, 1507–1519.
54. Bukovsky, M.S.; Karoly, D.J. Precipitation simulations using WRF as a nested regional climate model. *J. Appl. Meteorol. Climatol.* **2009**, *48*, 2152–2159.
55. Cassola, F.; Ferrari, F.; Mazzino, A. Numerical simulations of Mediterranean heavy precipitation events with the WRF model: A verification exercise using different approaches. *Atmos. Res.* **2015**, *164*, 210–225.
56. Fernández-González, S.; Valero, F.; Sánchez, J.L.; Gascón, E.; López, L.; García-Ortega, E.; Merino, A. Numerical simulations of snowfall events: Sensitivity analysis of physical parameterizations. *J. Geophys. Res. Atmos.* **2015**, *120*, 10,130–10,148.
57. Pennelly, C.; Reuter, G.; Flesch, T. Verification of the WRF model for simulating heavy precipitation in Alberta. *Atmos. Res.* **2014**, *135*, 172–192.
58. Trapero, L.; Bech, J.; Lorente, J. Numerical modelling of heavy precipitation events over Eastern Pyrenees: Analysis of orographic effects. *Atmos. Res.* **2013**, *123*, 368–383.
59. Wang, W.; Bruyere, C.; Duda, M.; Dudhia, J.; Gill, D.; Kavulich, M.; Keene, K.; Lin, H.C.; Michalakes, J.; Rizvi, S.; et al. *User's Guides for the Advanced Research WRF (ARW) Modeling System*, version 3; Technical Report; National Center for Atmospheric Research NCAR. 2014. Available online: http://www2.mmm.ucar.edu/wrf/users/docs/user_guide_V3/contents.html (accessed on 28 March 2017).
60. GDAL Development Team. *GDAL—Geospatial Data Abstraction Library*, version 2.1.1; 2016. Available online: http://www.gdal.org/ (accessed on 28 March 2017).
61. Pedersen, J.T.; Peters, J.C.; Helweg, O.J. *Hydrographs by Single Linear Reservoir Model*; Technical Report No. HY5; US Army Corps of Engineers, Hydrologic Engineering Center: Davis, CA, USA, 1980.
62. Chow, V. *Open-Channel Hydraulics*; McGraw-Hill: New York, NY, USA, 1959.
63. Hannah, D.M.; Gurnell, A.M. A conceptual, linear reservoir runoff model to investigate melt season changes in cirque glacier hydrology. *J. Hydrol.* **2001**, *246*, 123–141.
64. Neteler, M.; Bowman, M.H.; Landa, M.; Metz, M. GRASS GIS: A multi-purpose Open Source GIS. *Environ. Model. Softw.* **2012**, *31*, 124–130.
65. Freixes, A.; Monterde, M.; Ramoneda, J. Tracer tests in the Joèu karstic system (Aran Valley, Central Pyrennes, NE Spain). In *Tracer Hydrology 97*; Kranjc, K., Ed.; Balkema: Rotterdam, The Netherlands, 1997; pp. 219–225.
66. Monterde García, M. El sistema kárstico de Uelhs Deth Joèu. *Treb. Soc. Catalana Geogr.* **2001**, *52*, 233–244.
67. Georges, C.; Kaser, G. Ventilated and unventilated air temperature measurements for glacier-climate studies on a tropical high mountain site. *J. Geophys. Res.* **2002**, *107*, ACL 15-1–ACL 15-10.
68. Respuesta Hidrológica al Evento de Precipitación de Junio de 2013 en el Pirineo Central. *Investigaciones Geográficas* **2014**, *62*, 5–21.
69. Nash, J.E.; Sutcliffe, J.V. River flow forecasting through conceptual models part I—A discussion of principles. *J. Hydrol.* **1970**, *10*, 282–290.
70. Byers, H.R.; Moses, H.; Harney, P.J. Measurement of Rain Temperature. *J. Meteorol.* **1949**, *6*, 51–55.

71. Mass, C.F.; Ovens, D.; Westrick, K.; Colle, B.A. Does increasing horizontal resolution produce more skillful forecasts? The results of two years of real-time numerical weather prediction over the Pacific Northwest. *Bull. Am. Meteorol. Soc.* **2002**, *83*, 407–430.
72. Liston, G.E.; Elder, K. A micrometeorological distribution system for high-resolution terrestrial modeling applications (MicroMet). *J. Hydrometeorol.* **2006**, *7*, 217–234.
73. Thornton, P.E.; Running, S.W.; White, M.A. Generating surfaces of daily meteorological variables over large regions of complex terrain. *J. Hydrol.* **1997**, *190*, 214–251.
74. Lehning, M.; Grünewald, T.; Schirmer, M. Mountain Snow Distribution Governed by an Altitudinal Gradient and Terrain Roughness. *Geophys. Res. Lett.* **2011**, doi: 10.1029/2011GL048927.
75. Daly, C.; Halbleib, M.; Smith, J.I.; Gibson, W.P.; Doggett, M.K.; Taylor, G.H.; Curtis, J.; Pasteris, P.P. Physiographically sensitive mapping of climatological temperature and precipitation across the conterminous United States. *Int. J. Climatol.* **2008**, *28*, 2031–2064.
76. Blanchet, J.; Marty, C.; Lehning, M. Extreme value statistics of snowfall in the Swiss Alpine region. *Water Resour. Res.* **2009**, *45*, doi:10.1029/2009WR007916.
77. Sevruk, B. Regional Dependency of Precipitation-Altitude Relationship in the Swiss Alps. *Clim. Chang.* **1997**, *36*, 355–369.
78. Sevruk, B.; Mieglitz, K. The effect of topography, season and weather situation on daily precipitation gradients in 60 Swiss valleys. *Water Sci. Technol.* **2002**, *45*, 41–48.
79. Pomeroy, J.W.; Fang, X.; Marks, D.G. The cold rain-on-snow event of June 2013 in the Canadian Rockies—Characteristics and diagnosis. *Hydrol. Process.* **2016**, *30*, 2899–2914.
80. Boscarello, L.; Ravazzani, G.; Rabuffetti, D.; Mancini, M. Integrating glaciers raster-based modelling in large catchments hydrological balance: The Rhone case study. *Hydrol. Process.* **2014**, *28*, 496–508.
81. Ceppi, A.; Ravazzani, G.; Salandin, A.; Rabuffetti, D.; Montani, A.; Borgonovo, E.; Mancini, M. Effects of temperature on flood forecasting: Analysis of an operative case study in Alpine basins. *Nat. Hazards Earth Syst. Sci.* **2013**, *13*, 1051–1062.
82. Minder, J.R.; Durran, D.R.; Roe, G.H. Mesoscale Controls on the Mountainside Snow Line. *J. Atmos. Sci.* **2011**, *68*, 2107–2127.
83. White, A.B.; Neiman, P.J.; Ralph, F.M.; Kingsmill, D.E.; Persson, P.O.G. Coastal Orographic Rainfall Processes Observed by Radar during the California Land-Falling Jets Experiment. *J. Hydrometeorol.* **2002**, *4*, 264.
84. Bird, R.E.; Hulstrom, R.L. *A Simplified Clear Sky Model for Direct and Diffuse Insolation on Horizontal Surfaces*; Technical Report SERI/TR-642-761; Solar Energy Research Institute: Golden, CO, USA, 1981.
85. Bird, R.E.; Hulstrom, R.L. Review, evaluation and improvements of Direct irradiance models. *Trans. ASME J. Sol. Energy Eng.* **1981**, *103*, 182–192.
86. Liu, B.Y.H.; Jordan, R.C. The interrelationship and characteristics distribution of direct, diffuse and total solar radiation. *Sol. Energy* **1960**, *4*, 1.
87. Erbs, D.G.; Klein, S.A.; Duffie, J.A. Estimation of the diffuse radiation fraction for hourly, daily, and monthly average global radiation. *Sol. Energy* **1982**, *28*, 293–302.
88. Vijayakumar, G.; Kummert, M.; Klein, A.S.; Beckman, W.A. Analysis of short-term solar radiation data. *Sol. Energy* **2005**, *79*, 495–504.
89. Prata, A.J. A new long-wave formula for estimating downward clear-sky radiation at the surface. *Q. J. R. Meteorol. Soc.* **1996**, *122*, 1127–1151.
90. Lowe, P.R. An approximating polynomial for the computation of saturation vapor pressure. *J. Appl. Meteorol.* **1977**, *16*, 100–103.
91. Brutsaert, W. *Evaporation into the Atmosphere: Theory, History, and Applications*; Kluwer Academic: Dordrecht, The Netherlands, 1982; p. 316.
92. Marks, D.; Dozier, J. Climate and energy exchange at the snow surface in the alpine region of the Sierra Nevada. 2. Snow cover energy balance. *Water Resour. Res.* **1992**, *28*, 3043–3054.
93. Stull, R.B. *An Introduction to Boundary Layer Meteorology*; Kluwer Academic: Dordrecht, The Netherlands, 1988; p. 670.
94. Jacobson, M.Z. *Fundamentals of Atmospheric Modeling*; Cambridge University Press: Cambridge, UK, 1999.

95. Koh, G.; Jordan, R. Sub-surface melting in a seasonal snow cover. *J. Glaciol.* **1995**, *41*, 474–482.
96. Greuell, W.; Konzelmann, T. Numerical modelling of the energy balance and the Englacial temperature of the Greenland Ice Sheet. Calculations for the ETH-Camp location (West Greenland, 111 1 a.s.l.). *Glob. Planet. Chang.* **1994**, *9*, 91–114.

Permissions

All chapters in this book were first published in HYDROLOGY, by MDPI; hereby published with permission under the Creative Commons Attribution License or equivalent. Every chapter published in this book has been scrutinized by our experts. Their significance has been extensively debated. The topics covered herein carry significant findings which will fuel the growth of the discipline. They may even be implemented as practical applications or may be referred to as a beginning point for another development.

The contributors of this book come from diverse backgrounds, making this book a truly international effort. This book will bring forth new frontiers with its revolutionizing research information and detailed analysis of the nascent developments around the world.

We would like to thank all the contributing authors for lending their expertise to make the book truly unique. They have played a crucial role in the development of this book. Without their invaluable contributions this book wouldn't have been possible. They have made vital efforts to compile up to date information on the varied aspects of this subject to make this book a valuable addition to the collection of many professionals and students.

This book was conceptualized with the vision of imparting up-to-date information and advanced data in this field. To ensure the same, a matchless editorial board was set up. Every individual on the board went through rigorous rounds of assessment to prove their worth. After which they invested a large part of their time researching and compiling the most relevant data for our readers.

The editorial board has been involved in producing this book since its inception. They have spent rigorous hours researching and exploring the diverse topics which have resulted in the successful publishing of this book. They have passed on their knowledge of decades through this book. To expedite this challenging task, the publisher supported the team at every step. A small team of assistant editors was also appointed to further simplify the editing procedure and attain best results for the readers.

Apart from the editorial board, the designing team has also invested a significant amount of their time in understanding the subject and creating the most relevant covers. They scrutinized every image to scout for the most suitable representation of the subject and create an appropriate cover for the book.

The publishing team has been an ardent support to the editorial, designing and production team. Their endless efforts to recruit the best for this project, has resulted in the accomplishment of this book. They are a veteran in the field of academics and their pool of knowledge is as vast as their experience in printing. Their expertise and guidance has proved useful at every step. Their uncompromising quality standards have made this book an exceptional effort. Their encouragement from time to time has been an inspiration for everyone.

The publisher and the editorial board hope that this book will prove to be a valuable piece of knowledge for researchers, students, practitioners and scholars across the globe.

List of Contributors

Salomon Obahoundje and Komlavi Akpoti
Faculty of Science and Techniques, Master Research Program of the West African Science Service Center on Climate Change and Adapted Land Use (WASCAL), Climate Change and Energy, University abdou Moumouni of Niamey, P.O. Box 10662, Niamey 8000, Niger

Eric Antwi Ofosu and Amos T. Kabo-bah
Department of Energy and Environmental Engineering, University of Energy and Natural Resources, P.O. Box 214, Sunyani, Ghana

Philippe Machetel
Laboratoire Géosciences Montpellier, CNRS/UM2, 34095 Montpellier CEDEX 9, France

David A. Yuen
Minnesota Supercomputing Institute and Department of Earth Sciences, University of Minnesota, 310 Pillsbury Dr. SE, Minneapolis, MN 55455, USA
School of Environmental Studies, China University of Geosciences, 388 Lumo Road,Wuhan 430074, China

Nicholas C. Coops
Faculty of Forestry, University of British Columbia, 2424 Main Mall, Vancouver, BC V6T 1Z4, Canada

Robbie A. Hember
Faculty of Forestry, University of British Columbia, 2424 Main Mall, Vancouver, BC V6T 1Z4, Canada
Pacific Forestry Centre, Canadian Forest Service, Natural Resources Canada, 506West Burnside Road, Victoria, BC V8Z 1M5, Canada

David L. Spittlehouse
Competitiveness and Innovation Branch, Ministry of Forests, Lands and Natural Resources Operations, Victoria, BC V8W 9C2, Canada

Ezéchiel Obada, Josué Zandagba and Amédée Chabi
International Chair inMathematical Physics and Applications (ICMPA), University of Abomey-Calavi (UAC), Cotonou 072 B.P 50, Benin
Laboratory of Applied Hydrology, University of Abomey-Calavi (UAC), Cotonou 01 BP 4521, Benin

Eric Adéchina Alamou
Laboratory of Applied Hydrology, University of Abomey-Calavi (UAC), Cotonou 01 BP 4521, Benin

Abel Afouda
Laboratory of Applied Hydrology, University of Abomey-Calavi (UAC), Cotonou 01 BP 4521, Benin
West African Science Service Center on Climate Change and Adapted Land Use (WASCAL), GRP Climate Change and Water Resources, University of Abomey-Calavi (UAC), Abomey-Calavi BP 2008, Benin

Arpit Chouksey, Vinit Lambey, Bhaskar R. Nikam and Shiv Prasad Aggarwal
Water Resources Department, Indian Institute of Remote Sensing, Indian Space Research Organisation, 4 Kalidas Road, Dehradun-248001, Uttarakhand, India

Subashisa Dutta
Civil Engineering Department, Indian Institute of Technology Guwahati, Guwahati-781039, Assam, India

Gneneyougo Emile Soro, Yao Morton Kouame and Tié Albert Goula Bi
Unit Training and Research in Science and Environment Management, University Nangui Abrogoua, 02 BP 801 Abidjan 02, Abidjan, Ivory Coast

Affoué Berthe Yao
Unit Training and Research in Environment, Université Lorougnon Guédé, BP150 Daloa, Daloa, Ivory Coast

Yohannes Yihdego
Snowy Mountains Engineering Corporation (SMEC), Sydney, New South Wales 2060, Australia
Environmental Geoscience, La Trobe University, Melbourne, Victoria 3086, Australia

John A Webb
Environmental Geoscience, La Trobe University, Melbourne, Victoria 3086, Australia

Babak Vaheddoost
Hydraulic Lab., Istanbul Technical University, Istanbul 34467, Turkey

Hassan A. K. M. Bhuiyan and Jarrett Powers
Science and Technology Branch, Agriculture and Agri-Food Canada, Winnipeg, MB R3C 3G7, Canada

Heather McNairn and Amine Merzouki
Science and Technology Branch, Agriculture and Agri-Food Canada, Ottawa, ON K1A 0C6, Canada

Elias Nkiaka, N. R. Nawaz and Jon C. Lovett
School of Geography, University of Leeds, Leeds LS2 9JT, UK

Torsten Starkloff and Jannes Stolte
Norwegian Institute of Bioeconomy Research (NIBIO), Fredrik A Dahlsvei 20, N-1431 Ås, Norway

Rudi Hessel and Coen Ritsema
Wageningen Environmental Research (Alterra), P.O. Box 47, 6700 AA Wageningen, The Netherlands

Javier G. Corripio
Meteoexploration, Höttinger Gasse 21/17, A-6020 Innsbruck, Austria

Juan Ignacio López-Moreno
Department Procesos Geoambientales y Cambio Global, Instituto Pirenaico de Ecología, C.S.I.C, E-50059 Zaragoza, Spain

Index

www.ingramcontent.com/pod-product-compliance
Lightning Source LLC
Chambersburg PA
CBHW080648200326
41458CB00013B/4774

* 9 7 8 1 6 4 1 1 6 0 5 8 2 *